Organic Materials for Civil Engineering

Organic Materials for Civil Engineering

Editor:

Dr. Tanjina Nur, Ph.D.

ArclerPress

www.arclerpress.com

Organic Materials for Civil Engineering

Editor: Dr. Tanjina Nur, Ph.D.

708 3rd Avenue, 6th Floor
New York
NY 10017
United States of America
www.arclerpress.com

ISBN: 978-1-68094-371-9

Library of Congress Control Number: 2016950935

About the Editor

Dr. Tanjina Nur, Ph.D.

Tanjina finished her PhD in Civil and Environmental Engineering in 2014 from University of Technology Sydney (UTS). Now she is working as Post-Doctoral Researcher in the Centre for Technology in Water and Wastewater (CTWW) and published about eight International journal papers with 80 citations. Her research interest is wastewater treatment technology using adsorption process.

Contents

List of Contributors

Ailian Zhang
School of Civil Engineering, Harbin Institute of Technology, Harbin 150001, China
School of Civil Engineering, Sichuan College of Architectural Technology, Deyang 618000, China

Xiaojian Gao
School of Civil Engineering, Harbin Institute of Technology, Harbin 150001, China

Jiří Witzany
Department of Building Structures, Czech Technical University in Prague, Faculty of Civil Engineering, Thákurova 7, 166 29 Prague, Czech Republic

Radek Zigler
Department of Building Structures, Czech Technical University in Prague, Faculty of Civil Engineering, Thákurova 7, 166 29 Prague, Czech Republic

Jae-Woong Han
Department of Bio-Industry Engineering, Koungju National University, Yesan 340-702, Korea

Ji-Hong Jeon
Department of Environmental Engineering, Andong National University, 1375 Gyeongdong Street, Andong 760-749, Korea

Chan-Gi Park
Department of Rural Construction Engineering, Koungju National University, Yesan 340-702, Korea

Tae-Kyun Kim
Concrete Structural Engineering Laboratory, School of Civil and Environmental Engineering, Yonsei University, Seoul 120-794, Korea

Byung-Yun Kim
Department of Architectural Engineering, Catholic Kwandong University, Gangneung 210-701, Korea

Wei-Jie Fan
Department of Architectural Engineering, Kangwon National University, Chuncheon-Si 200-701, Korea

Xiao-Yong Wang
Department of Architectural Engineering, Kangwon National University, Chuncheon-Si 200-701, Korea

Ki-Bong Park
Department of Architectural Engineering, Kangwon National University, Chuncheon-Si 200-701, Korea

Sung-Wook Yun
Institute of Agriculture & Life Science, Gyeongsang National University, Jinju 660-701, Korea

Phillippe C. Baveye
Laboratory of Soil and Water Engineering, Department of Civil and Environmental Engineering, Rensselaer Polytechnic Institute, 110 8th street, Troy, NY 12180, USA

Chan Yu
Institute of Agriculture & Life Science, Gyeongsang National University, Jinju 660-701, Korea
Department of Agricultural Engineering, Gyeongsang National University, Jinju 660-701, Korea

Bon-Min Koo
Concrete Structural Engineering Laboratory, School of Civil and Environmental Engineering, Yonsei University, A404 Engineering Building #A, 134 Shinchon-dong, Seodaemun-gu, Seoul 120-794, Korea

Jang-Ho Jay Kim
Concrete Structural Engineering Laboratory, School of Civil and Environmental Engineering, Yonsei University, A274 Engineering Building #A, 134 Shinchon-dong, Seodaemun-gu, Seoul 120-794, Korea

Sung-Bae Kim
Concrete Structural Engineering Laboratory, School of Civil and Environmental Engineering, Yonsei University, A404 Engineering Building #A, 134 Shinchon-dong, Seodaemun-gu, Seoul 120-794, Korea
Baytech Korea, Inc., Institute Technology, Joong-il Eines Platz 9F, 513-22, Sangdaewon-dong, Jungwon-gu, Seongnam-si, Gyunggi-do 462-120, Korea

Sungho Mun
Department of Civil Engineering, Seoul National University of Science and Technology, Engineering Building, 232 Gongneung-ro, Nowon-gu, Seoul 139-743, Korea

Panagiotis Frantzis
MIMechE, I. Polyla 13, 542 48 Thessaloniki, Greece

Augusto Cannone Falchetto
Pavement Engineering Centre (ISBS), Technical University of Braunschweig, Beethovenstraße 51b, Braunschweig 38106, Germany

Ki Hoon Moon
Material R&D Division, Samsung C&T Corporation, 5th Floor Daeryung Gangnam Tower, 826-20, Yeoksam 1-Dong, Gangnam-Gu, Seoul 135-935, Korea

Michael P. Wistuba
Pavement Engineering Centre (ISBS), Technical University of Braunschweig, Beethovenstraße 51b, Braunschweig 38106, Germany

Ben Liu, Junan Shen
Road Engineering Research Center, Suzhou University of Science and Technology, Suzhou 215011, China

Xuyan Song
Road Engineering Research Center, Suzhou University of Science and Technology, Suzhou 215011, China

Tao Ma
School of Transportation, Southeast University, Nanjing 210096, China
Department of Civil and Environmental Engineering, Rutgers, The State University of New Jersey, Piscataway, NJ 08854, USA

Hao Wang
Department of Civil and Environmental Engineering, Rutgers, The State University of New Jersey, Piscataway, NJ 08854, USA

Yongli Zhao
School of Transportation, Southeast University, Nanjing 210096, China

Xiaoming Huang
School of Transportation, Southeast University, Nanjing 210096, China

Yuhui Pi
School of Transportation, Southeast University, Nanjing 210096, China

Peiqiang Cui
State Key Laboratory of Silicate Materials for Architectures, Wuhan University of Technology, Wuhan 430070, China

Shaopeng Wu
State Key Laboratory of Silicate Materials for Architectures, Wuhan University of Technology, Wuhan 430070, China

Fuzhou Li
State Key Laboratory of Silicate Materials for Architectures, Wuhan University of Technology, Wuhan 430070, China

Yue Xiao
State Key Laboratory of Silicate Materials for Architectures, Wuhan University of Technology, Wuhan 430070, China

Honghua Zhang
State Key Laboratory of Silicate Materials for Architectures, Wuhan University of Technology, Wuhan 430070, China

Xu Jiang
Department of Bridge Engineering, College of Civil Engineering, Tongji University, Shanghai 200092, China

Xuhong Qiang
Department of Structural Engineering, College of Civil Engineering, Tongji University, Shanghai 200092, China

Henk Kolstein
Faculty of Civil Engineering and Geosciences, Delft University of Technology, Delft 2628CN, The Netherlands

Frans Bijlaard
Faculty of Civil Engineering and Geosciences, Delft University of Technology, Delft 2628CN, The Netherlands

Asim Hassan Ali
Center for Transportation Research, Faculty of Engineering, University of Malaya, 50603 Kuala Lumpur, Malaysia

Nuha S. Mashaan
Center for Transportation Research, Faculty of Engineering, University of Malaya, 50603 Kuala Lumpur, Malaysia

Mohamed Rehan Karim
Center for Transportation Research, Faculty of Engineering, University of Malaya, 50603 Kuala Lumpur, Malaysia

Sivakumar Anandan
VIT University, Vellore, Tamil Nadu 632 014, India

Southararajan Vallarasu Manoharan
VIT University, Vellore, Tamil Nadu 632 014, India

Thirumurugan Sengottian
VIT University, Vellore, Tamil Nadu 632 014, India

Shuhua Liu
State Key Laboratory of Water Resources and Hydropower Engineering Science, Wuhan University, Wuhan 430072, China

Lihua Li
School of Civil Engineering and Architecture, Hubei University of Technology, Wuhan 430068, China

Zhigang Wang
State Key Laboratory of Water Resources and Hydropower Engineering Science, Wuhan University, Wuhan 430072, China

Jiafeng Wang
State Key Laboratory of Water Resources and Hydropower Engineering Science, Wuhan University, Wuhan 430072, China

Meijuan Rao
Department of Civil and Earth Resources Engineering, Kyoto University, Kyoto 615-8540, Japan

Pan Zhang
Department of Civil Engineering, Tsinghua University, Beijing 100084, China

Xinmiao Meng
Department of Civil Engineering, Tsinghua University, Beijing 100084, China

Lieping Ye
Department of Civil Engineering, Tsinghua University, Beijing 100084, China

Chao Wu
Department of Civil Engineering, Monash University, Clayton, VIC 3800, Australia

Peng Feng
The Key Laboratory of Civil Engineering Safety and Durability of China Education Ministry, Department of Civil Engineering, Tsinghua University, Beijing 100084, China

Yu Bai
Department of Civil Engineering, Monash University, Clayton, VIC 3800, Australia

Ye Lu
Department of Civil Engineering, Monash University, Clayton, VIC 3800, Australia

Preface

This book presents wide descriptions of organic materials used in all civil engineering projects, in terms of their scientific and technical background, including the regulations that cover their use and their predicted useful life. Organic materials are essentially perceived as bonding additives in the field of construction. These types of materials include adhesive for concrete and mortar, bitumens and asphalt for road surfacing, polymers for formulating products for repair and structural gluing or admixtures for a more compact concrete.

Firstly, we discuss the feasibility of a heat resistant adhesive for strengthening concrete with carbon sheets and stress state analysis and failure mechanisms of masonry columns reinforced with fibre reinforced polymer (FRP). Then mechanical and permeability characteristics of Latex-modified pre-packed pavement repair concrete as a function of the rapid-set binder content and material performance and animal clinical studies on performance-optimized Hwangtoh mixed mortar and concrete are presented in details. We also try to find out evaluation of the chemical and mechanical properties of hardening high-calcium fly ash blended concrete and effect of industrial by-products on unconfined compressive strength of solidified organic marine clayey soils. In the middle part of the book, we focus on material and structural performance evaluations of Hwangtoh admixtures and recycled PET fiber-added eco-friendly concrete and fabrication of cement, polymer, and metal-matrix fibre composites by gravity-assisted liquid-binder Infiltration. Microstructural analysis and rheological modelling of asphalt mixtures and changes in nanoscaled mechanical and rheological properties of asphalt binders caused by aging are also discussed. Then we present the strength mechanism and influence factors for cold recycled asphalt mixture and investigation on using SBS and Active Carbon Filler to reduce the VOC emission from bituminous materials. At the end, we try to describe analysis on adhesively-bonded joints of FRP-steel composite bridge under combined loading and investigations of physical and rheological properties of aged rubberised bitumen. Also corrosion effects on the strength properties of steel fibre reinforced concrete containing slag and corrosion inhibitor and study on strength and microstructure of cement pastes containing limestone powder under flowing acid solution condition are also discussed. Then mechanical analysis of stress distribution in a carbon fiber-reinforced polymer rod bonding anchor and epoxy enhanced by recycled milled carbon fibres in adhesively-bonded CFRP for structural strengthening are presented briefly.

Editor
Dr. Tanjina Nur, Ph.D.

INTRODUCTION

Organic materials are substances in the soil that were once alive. Dead plants, animals, bacteria and fungi are all examples of organic material. Organic material is distinguished from organic matter by the amount of decomposition that has taken place. Organic matter has decomposed as much as it can. Approaching civil engineering materials from a physico-chemical point of view is probably the simplest way to bring together products that are extremely different in their design as well as in their use. However, it is not just the method but other economic, technological and scientific realities that brings them closer. To do this, we need to add the real environmental and health impacts that they could actually cause or likely to cause in the collective unconscious. We then come straight to the point of sustainable development, the concept which drives development and research in the beginning of the 21st century.

Organic Polymer

Organic polymers belong to a family of materials whose industrial development is very recent. It is believed that the first synthetic "plastic material" was developed in 1862 by the English chemist Parkes by mixing sulfuric acid and nitric acid with cotton wool. The nitrocellulose thus obtained was stabilized with castor oil and camphor, and dyes were added and various objects were produced using this mixture. However, for production on an industrial scale, this formula had to be modified slightly and "Parkesine" was forgotten, to be replaced by "celluloid", developed by the American Hyatt brothers. It is believed that they developed this product in 1869 for a competition organized by New York City to discover a substitute material for ivory in the manufacture of billiard balls.

Definitions

A polymer is a substance composed of macromolecules whose structure is characterized by a large number of repetitions of groups of atoms, called structural units, repeat units, monomeric units or constitutional units. These macromolecules are molecules with very high molecular weight. The macromolecular chains are composed of an array of a very large number of constitutional units linked together by covalent (and therefore very strong) bonds. They are bound by secondary bonds of lower energy but whose overall importance accounts for the originality of these materials. Macromolecular compounds generally have heterogeneous molecular weight. This property results from the random nature of most synthesis reactions. The average molecular weight and the distribution of molecular weights have a significant influence on the technical characteristics (elasticity modulus, breaking strength, impact strength, etc.) and the forming conditions of polymers. It is therefore important to specify this distribution. One way to do so is to consider the average molecular weights in number and in weight, the comparison of which helps determine the degree of heterogeneity in the molecular weight of the polymer.

We can thus define:

– the numerical average molecular weight Mn by the expression:

$$\overline{M_n} = \frac{\sum_i n_i M_i}{\sum_i n_i}$$

where n_i is the number of molecules with degree of polymerization i and M_i is the molecular weight of the molecule with degree of polymerization i;

– the weight-average molecular weight $\overline{M_P}$ by the expression:

$$\overline{M_p}=\frac{\sum_i n_i M_i^2}{\sum_i n_i M_i}$$

– the polymolecularity or polydispersity index I by the expression:

$$I=\frac{\overline{M_p}}{\overline{M_n}}$$

This index, which is equal to 1 for a strictly isomolecular polymer, is always greater for real compounds and can vary up to 30 or 50.

The most common methods to determine $\overline{M_n}$ are osmometry, cryometry and ebulliometry. The best technique to determine $\overline{M_p}$ is by light diffusion.

The technological properties of polymers depend highly on the distribution of molecular weights but also on the macromolecular structure, which we will discuss now.

To use an organic polymer-based material correctly, the user needs some data on the macromolecular structure as well as on the synthesis of polymers.

Types and Properties of Bituminous Materials

Importance and background of Bituminous Material:

Bitumen is an essential component of any pavement and is used widely throughout the world. It can be termed as the building block of the pavements without which all the pavement materials would behave independently and thus will be deemed useless. Almost ninety percent of bitumen is used in road construction. It is usually available in dark colors ranging from brown to black. The main purpose of bitumen in flexible pavements is to strongly bind and hold the other pavement components together and provide a smooth and leveled surface for the moving vehicles. Bitumen is a naturally occurring material and is found in large quantities in the solid or semi-solid forms of petroleum. It is also manufactured artificially in vast amounts globally.

Bitumen mixed with some other materials has always been used as a sealant and adhesive material over the ages. It was also widely used in the water proofing of boats and ships as it is insoluble in water. Highest applications of bitumen are found in the construction filed for the construction of roads, airports etc., in the hydraulic field for the construction of water tanks, dams, bridges etc., is also used in battery making, tyre making and for the thermal and acoustic insulation purposes.

Properties of Bituminous Materials:

The high demand of bituminous materials is owing to the number of their first-rate advantageous properties listed below:

- Adhesive: Binds together all the components without bringing about any positive or negative changes in their properties.
- Water proof: Bitumen is insoluble in water and can serve as an effective sealant
- Strong: Though the coarse aggregates are the main load bearing component in a pavement, bitumen or asphalt also play a vital role in distributing the traffic loads to the layers beneath.
- Durable: Bitumen lives up to twenty years if maintained properly throughout the pavement life.
- Versatile: Bitumen is a relatively easy to use material because of its thermoplastic property. It can be spread easily along the underlying pavement layers as it liquefies when heated making the job easier and hardens in a solid mass when cooled.
- Economical: It is available in cheaper rates almost all over the world

Types of Bituminous Materials:

Depending upon the temperature and other factors various types of bitumen are found and used throughout the world.

- *Cutback Bitumen:* Cut-back bitumens are those which are prepared with the addition of a volatile to reduce the thickness of the binder.
- *Fluxed Bitumen:* Fluxed bitumens are those bitumen which are prepared by the addition of relatively nonvolatile oils to reduce the viscosity of the binder.
- *Modified Bitumen:* Modified bituminous binder are those whose properties such as cohesive strength, adhesive property, elasticity or viscosity have been modified by the use of one or combined chemical agents.

- *Asphalt:* Asphalt is a mixture of aggregates both fine (sand and filler) and coarse (stone) and a bituminous binder. It typically contains approximately 4-7% of bitumen. Asphalt is primarily used in road construction and its properties depend upon the type, size and amount of aggregate used in the mixture, all of which can be adjusted to provide the required properties for the desired application.

Owing to their high thermoplastic properties, bitumen is always recommended to be used wearing all protective gadgets. Even a small drop of highly heated bitumen can cause severe burns to exposed body parts. Therefore high care is needed in the handling of any type of bitumen to avoid any mishaps during site work.

Bituminous road constructions steps:

1. Preparation of the existing base course layer

The existing surface is prepared by removing the pot holes or rust if any. The irregularities are filled in with premix chippings at least a week before laying surface course. If the existing pavement is extremely way, a bituminous leveling course of adequate thickness is provided to lay a bituminous concrete surface course on a binder course instead of directly laying it on a WBM.

2. Application of Tuck Coat

It is desirable to lay AC layer over a bituminous base or binder course. A tack coat of bitumen is applied at 6.0 to 7.5 kg per 10 sq.m area, this quantity may be increased to 7.5 to 10 kg for non-bituminous base.

3. Preparation and placing of Premix

The premix is prepared in a hot mix plant of a required capacity with the desired quality control. The bitumen may be heated upto 150 – 177 deg C and the aggregate temperature should not differ by over 14 deg C from the binder temperature. The hot mixed material is collected from the mixture by the transporters, carried to the location is spread by a mechanical paver at a temperature of 121 to 163 deg C. the camber and the thickness of the layer are accurately verified. The control of the temperatures during the mixing and the compaction are of great significance in the strength of the resulting pavement structure.

4. Rolling

A mix after it is placed on the base course is thoroughly compacted by rolling at a speed not more than 5km per hour. The initial or break down rolling is done by 8 to 12 tonnes roller and the intermediate rolling is done with a fixed wheel pneumatic roller of 15 to 30 tonnes having a tyre pressure of 7kg per sq.cm. the wheels of the roller are kept damp with water.

The number of passes required depends on the thickness of the layer. In warm weather rolling on the next day, helps to increase the density if the initial rolling was not adequate. The final rolling or finishing is done by 8 to 10 tonne tandem roller.

5. Quality control of bituminous concrete construction

The routine checks are carried out at site to ensure the quality of the resulting pavement mixture and the pavement surface.

Periodical checks are made for

a) Aggregate grading

b) Grade of bitumen

c) Temperature of aggregate

d) Temperature of paving mix during mixing and compaction.

At least one sample for every 100 tonnes of the mix discharged by the hot mix plant is collected and tested for above requirements. Marshall tests are also conducted. For every 100 sq.m of the compacted surface, one test of the field density is conducted to check whether it is atleast 95% of the density obtained in the laboratory. The variation in the thickness allowed is 6mm per 4.5m length of construction.

6. Finished surface

The AC surface should be checked by a 3.0 m straight edge. The longitudinal undulations should not exceed 8.0 mm and the number of undulations higher than 6.0 mm should not exceed 10 in a length of 300 m. The cross-traffic profile should not have undulations exceeding 4.0mm.

4 Different Forms of Bitumen Used In Road Construction

1. **Cutback Bitumen:** Normal practice is to heat bitumen to reduce its viscosity. In some situations preference is given to use liquid binders such as cutback bitumen. In cutback bitumen suitable solvent is used to lower the viscosity of the bitumen. From the environmental point of view also cutback bitumen is preferred. The solvent from the bituminous material will evaporate and the bitumen will bind the aggregate. Cutback bitumen is used for cold weather bituminous road construction and maintenance. The distillates used for preparation of cutback bitumen are naphtha, kerosene, diesel oil, and furnace oil. There are different types of cutback bitumen like rapid curing (RC), medium curing (MC), and slow curing (SC). RC is recommended for surface dressing and patchwork. MC is recommended for premix with less quantity of fine aggregates. SC is used for premix with appreciable quantity of fine aggregates.
2. **Bitumen Emulsion:** Bitumen emulsion is a liquid product in which bitumen is suspended in a finely divided condition in an aqueous medium and stabilized by suitable material. Normally cationic type emulsions are used in India. The bitumen content in the emulsion is around 60% and the remaining is water. When the emulsion is applied on the road it breaks down resulting in release of water and the mix starts to set. The time of setting depends upon the grade of bitumen. The viscosity of bituminous emulsions can be measured as per IS: 8887-1995. Three types of bituminous emulsions are available, which are Rapid setting (RS), Medium setting (MS), and Slow setting (SC). Bitumen emulsions are ideal binders for hill road construction. Where heating of bitumen or aggregates are difficult. Rapid setting emulsions are used for surface dressing work. Medium setting emulsions are preferred for premix jobs and patch repairs work. Slow setting emulsions are preferred in rainy season.
3. **Bituminous Primers:** In bituminous primer the distillate is absorbed by the road surface on which it is spread. The absorption therefore depends on the porosity of the surface. Bitumen primers are useful on the stabilized surfaces and water bound macadam base courses. Bituminous primers are generally prepared on road sites by mixing penetration bitumen with petroleum distillate.
4. **Modified Bitumen:** Certain additives or blend of additives called as bitumen modifiers can improve properties of Bitumen and bituminous mixes. Bitumen treated with these modifiers is known as modified bitumen. Polymer modified bitumen (PMB)/ crumb rubber modified bitumen (CRMB) should be used only in wearing course depending upon the requirements of extreme climatic variations. The detailed specifications for modified bitumen have been issued by IRC: SP: 53-1999. It must be noted that the performance of PMB and CRMB is dependent on strict control on temperature during construction. The advantages of using modified bitumen are as follows
   - Lower susceptibility to daily and seasonal temperature variations
   - Higher resistance to deformation at high pavement temperature
   - Better age resistance properties
   - Higher fatigue life for mixes
   - Better adhesion between aggregates and binder
   - Prevention of cracking and reflective cracking

Requirements of Bitumen

The desirable properties of bitumen depend on the mix type and construction. In general, Bitumen should posses following desirable properties.

- The bitumen should not be highly temperature susceptible: during the hottest weather the mix should not become too soft or unstable, and during cold weather the mix should not become too brittle causing cracks.
- The viscosity of the bitumen at the time of mixing and compaction should be adequate. This can be achieved by use of cutbacks or emulsions of suitable grades or by heating the bitumen and aggregates prior to mixing.
- There should be adequate affinity and adhesion between the bitumen and aggregates used in the mix.

Differences between Bitumen and Tar Used in Road Construction

The following are the differences between bitumens and tars.

1. The weathering properties of bitumens are superior to those of tars. Generally greater deterioration is produced in tar than in bitumen when exposed to equal weather conditions. Bitumens have a better durability and resistance to weathering than tars.

2. Tars are more susceptible to temperature changes than bitumens. Hot weather will soften a tar surface more than a surface made with bitumen of the same viscosity, and it will become more brittle at low temperatures than bitumen. In other words tar becomes brittle in cold weather and the surface treated with tar is apt to bleed in hot weather if a little extra quantity has been used. Tar is therefore considered unsuitable for locations with wide temperature changes. Gritting or surface dressing can be delayed a little where bitumen has been used but not with tar.

3. Surface dressings with bitumen are more prone to failure by water displacement than those made with tar. Tars generally adhere better than bitumens on wet aggregate.

4. Tars can be brought to a spraying condition at lower temperatures than those needed for bitumens; stones need not be heated to high temperatures.

5. Setting time for tars is more than that of bitumens and this property is useful in the production of pre-coated aggregate which can be transported to large distances or kept for some time before spreading.

6. Tars harden much quicker than bitumens.

7. Tars have higher specific gravity than bitumens and lower viscosity and these properties give them greater penetrating power and which are more marked during summers. Higher viscosities can generally be used with tars than with bitumens.

8. Tars produce a less slippery surface than bitumens.

9. Bitumens have a tendency to stay at or just near the surface resulting in a rich and fat surface.

10. Roads built of bitumen need constant traffic to be maintained in good order; otherwise the surface will crack and reduce the life of the road.

11. Tars make harder surfaces (but such surfaces are brittle) than bitumens and should be preferred for roads in areas where bullock carts or other hard tyred traffic predominates. Bitumens make more elastic surfaces and are better suited for pneumatic traffic. Hardening of bitumens is very gradual.

12. Tar is more suitable for dense fine grained surfaces and bitumen for more open surfaces.

13. Volume of tar required is about 10 percent less than that of bitumen for the same type of road work.

14. Tar is cheaper than bitumen.

15. A primer is not generally needed with tar.

16. Road tars do not dissolve in a petroleum distillate such as petrol, kerosene, diesel oil. As such tar carpets have proved to be good material for parking sites as it remains unaffected by spillage of oil and petrol from automobiles

Requirements of Bitumen

The desirable properties of bitumen depend on the mix type and construction. In general, Bitumen should posses following desirable properties.

- The bitumen should not be highly temperature susceptible: during the hottest weather the mix should not become too soft or unstable, and during cold weather the mix should not become too brittle causing cracks.

- The viscosity of the bitumen at the time of mixing and compaction should be adequate. This can be achieved by use of cutbacks or emulsions of suitable grades or by heating the bitumen and aggregates prior to mixing.
- There should be adequate affinity and adhesion between the bitumen and aggregates used in the mix.

Tests on bitumen

There are a number of tests to assess the properties of bituminous materials. The following tests are usually conducted to evaluate different properties of bituminous materials.

1. Penetration test
2. Ductility test
3. Softening point test
4. Specific gravity test
5. Viscosity test
6. Flash and Fire point test
7. Float test
8. Water content test
9. Loss on heating test

Penetration test

It measures the hardness or softness of bitumen by measuring the depth in tenths of a millimeter to which a standard loaded needle will penetrate vertically in 5 seconds. BIS had standardised the equipment and test procedure. The penetrometer consists of a needle assembly with a total weight of 100g and a device for releasing and locking in any position. The bitumen is softened to a pouring consistency, stirred thoroughly and poured into containers at a depth at least 15 mm in excess of the expected penetration. The test should be conducted at a specified temperature of 25 ^{o} C. It may be noted that penetration value is largely influenced by any inaccuracy with regards to pouring temperature, size of the needle, weight placed on the needle and the test temperature. A grade of 40/50 bitumen means the penetration value is in the range 40 to 50 at standard test conditions. In hot climates, a lower penetration grade is preferred. The Figure 1 shows a schematic Penetration Test setup.

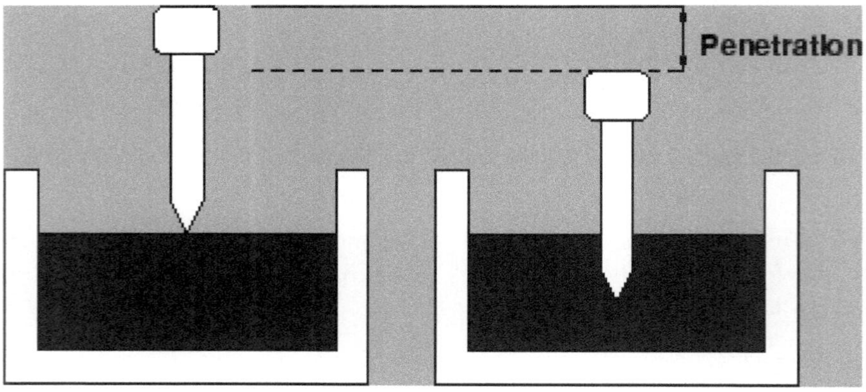

Figure 1: Penetration Test Setup

Ductility test

Ductility is the property of bitumen that permits it to undergo great deformation or elongation. Ductility is defined as the distance in cm, to which a standard sample or briquette of the material will be elongated without breaking. Dimension of the briquette thus formed is exactly 1 cm square. The bitumen sample is heated and poured in the mould assembly placed on a plate. These samples with moulds are cooled in the air and then in water bath at 27 ^{o} C temperature. The excess bitumen is cut and the surface is leveled using a hot knife. Then the mould with assembly containing sample is kept in water bath of the ductility machine for about 90 minutes. The sides of the moulds are removed, the clips are hooked on the machine and the machine is operated. The distance up to the point of breaking of thread is the ductility value which is reported in cm. The ductility value gets affected by factors such as pouring temperature, test temperature, rate of pulling etc. A minimum ductility value of 75 cm has been specified by the BIS. Figure. 2 shows ductility moulds to be filled with bitumen.

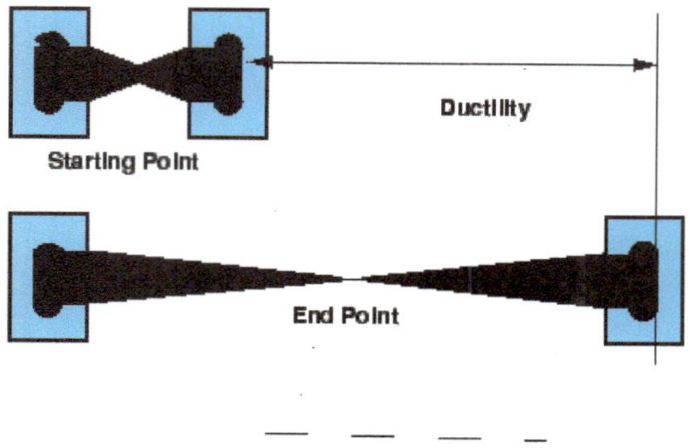

Figure 2: Ductility Test

Softening point test

Softening point denotes the temperature at which the bitumen attains a particular degree of softening under the specifications of test. The test is conducted by using Ring and Ball apparatus. A brass ring containing test sample of bitumen is suspended in liquid like water or glycerin at a given temperature. A steel ball is placed upon the bitumen sample and the liquid medium is heated at a rate of 5 ^{o} C per minute. Temperature is noted when the softened bitumen touches the metal plate which is at a specified distance below. Generally, higher softening point indicates lower temperature susceptibility and is preferred in hot climates. Figure. 3 shows Softening Point test setup.

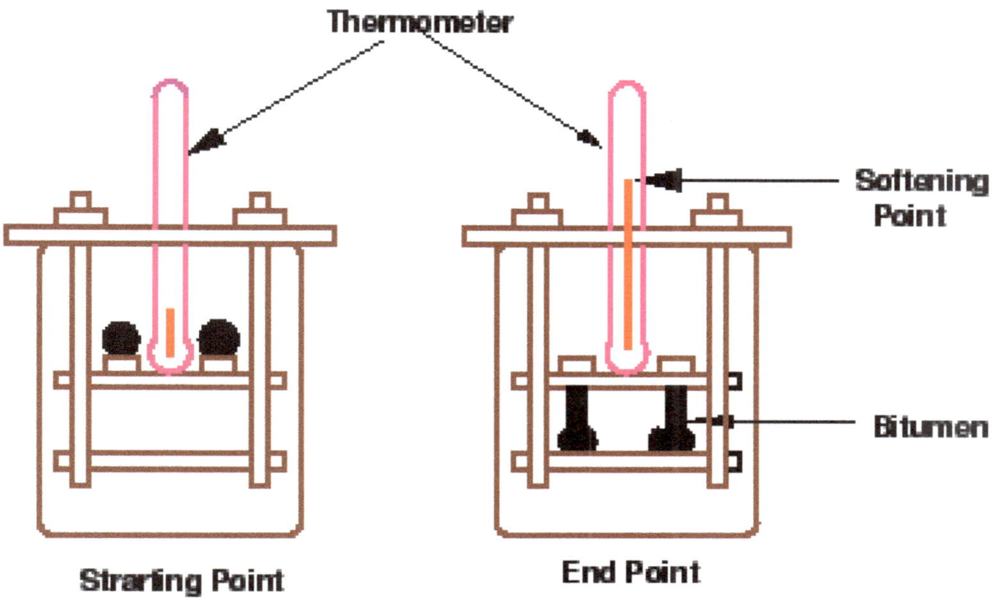

Figure 3: Softening Point Test Setup

Specific gravity test

In paving jobs, to classify a binder, density property is of great use. In most cases bitumen is weighed, but when used with aggregates, the bitumen is converted to volume using density values. The density of bitumen is greatly influenced by its chemical composition. Increase in aromatic type mineral impurities cause an increase in specific gravity.

The specific gravity of bitumen is defined as the ratio of mass of given volume of bitumen of known content to the mass of equal volume of water at 27 ^{o} C. The specific gravity can be measured using either pycnometer or preparing a cube specimen of bitumen in semi-solid or solid state. The specific gravity of bitumen varies from 0.97 to 1.02.

Viscosity test

Viscosity denotes the fluid property of bituminous material and it is a measure of resistance to flow. At the application temperature, this characteristic greatly influences the strength of resulting paving mixes. Low or high viscosity during compaction or mixing has been observed to result in lower stability values. At high viscosity, it resist the compactive effort and thereby resulting mix is heterogeneous, hence low stability values. And at low viscosity instead of providing a uniform film over aggregates, it will lubricate the aggregate particles. Orifice type viscometers are used to indirectly find the viscosity of liquid binders like cutbacks and emulsions. The viscosity expressed in seconds is the time taken by the 50 ml bitumen material to pass through the orifice of a cup, under standard test conditions and specified temperature. Viscosity of a cutback can be measured with either 4.0 mm orifice at 25 ^{o} C or 10 mm orifice at 25 or 40 ^{o} C.

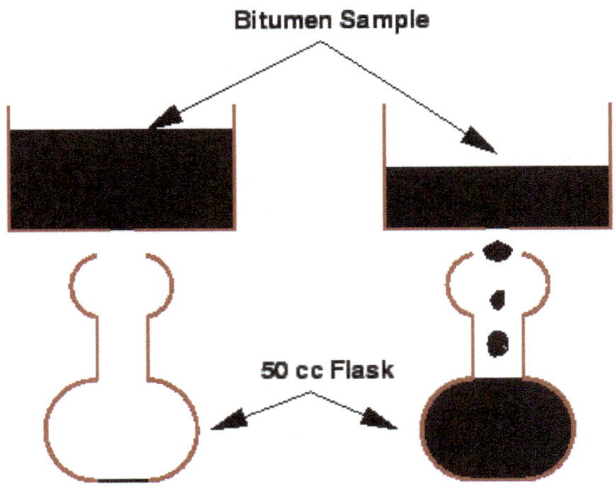

Figure 4: Viscosity Test

Flash and fire point test

At high temperatures depending upon the grades of bitumen materials leave out volatiles. And these volatiles catches fire which is very hazardous and therefore it is essential to qualify this temperature for each bitumen grade. BIS defined the flash point as the temperature at which the vapor of bitumen momentarily catches fire in the form of flash under specified test conditions. The fire point is defined as the lowest temperature under specified test conditions at which the bituminous material gets ignited and burns.

Float test

Normally the consistency of bituminous material can be measured either by penetration test or viscosity test. But for certain range of consistencies, these tests are not applicable and Float test is used. The apparatus consists of an aluminum float and a brass collar filled with bitumen to be tested. The specimen in the mould is cooled to a temperature of 5 °C and screwed in to float. The total test assembly is floated in the water bath at 50 °C and the time required for water to pass its way through the specimen plug is noted in seconds and is expressed as the float value.

Water content test

It is desirable that the bitumen contains minimum water content to prevent foaming of the bitumen when it is heated above the boiling point of water. The water in a bitumen is determined by mixing known weight of specimen in a pure petroleum distillate free from water, heating and distilling of the water. The weight of the water condensed and collected is expressed as percentage by weight of the original sample. The allowable maximum water content should not be more than 0.2% by weight.

Loss on heating test

When the bitumen is heated it loses the volatility and gets hardened. About 50gm of the sample is weighed and heated to a temperature of 163 °C for 5hours in a specified oven designed for this test. The sample specimen is weighed again after the heating period and loss in weight is expressed as percentage by weight of the original sample. Bitumen used in pavement mixes should not indicate more than 1% loss in weight, but for bitumen having penetration values 150-200 up to 2% loss in weight is allowed.

Table 1: Tests for Bitumen with IS codes

Type of test	Test Method
Penetration Test	IS: 1203-1978
Ductility test	IS: 1208-1978
Softening Point test	IS: 1205-1978
Specific gravity test	IS: 1202-1978
Viscosity test	IS: 1206-1978
Flash and Fire Point test	IS: 1209-1978
Float Test	IS: 1210-1978
Determination of water content	IS: 1211-1978
Determination of Loss on heating	IS:1212-1978

CHAPTER 1

THE FEASIBILITY OF MODIFIED MAGNESIA-PHOSPHATE CEMENT AS A HEAT RESISTANT ADHESIVE FOR STRENGTHENING CONCRETE WITH CARBON SHEETS

Ailian Zhang and Xiaojian Gao

Keywords: Strengthening concrete, carbon fiber sheets, epoxy resin, modified magnesia-phosphate cement (MPC), resistance to elevated temperatures

INTRODUCTION

Carbon fiber reinforced polymer (CFRP) has become in recent years one of the state-of-the-art materials in repairing and/or strengthening structural concrete elements since it has numerous advantages such as extremely high strength to weight ratio, versatility, and resistance to electrochemical corrosion [1]. However, CFRP materials are too sensitive to elevated temperatures [2]. Deterioration in mechanical and/or bond properties can be expected at temperatures approaching the glass transition temperature (T_g) of the organic epoxy adhesive [2], which is typically lower than 100 °C [3,4,5]. The replacement of organic epoxy adhesive with some inorganic adhesives is a promising solution. The inorganic adhesives such as magnesium oxychloride cement (MOC) and alkali activated cementitious materials [6] were used for organic epoxy adhesives replacement due to the favorable binding properties, non-toxic, low cost and excellent heat-resistance.

Several studies have been performed on the use of alkali activated cementitious materials as a sole binding material to expand the applied range of CFRP [7,8,9]. Kurtz et al. [7] reported that the alkali activated cementitious material increases the strength and stiffness of reinforced concrete beams as effectively as organic epoxy, with a minor reduction in ductility. Toutanji et al. [8] presented that the increase of fatigue strength of concrete beam due to the use of alkali activated cementitious materials pasting carbon fiber sheets reaches as high as 55% when being compared with unstrengthened beams. Katakalos et al. [9] reported that reinforced concrete beams strengthened by alkali activated cementitious materials have good fatigue performance. On the other hand, alkali activated cementitious material shows a significant shrinkage, as many studies reported [10], which could easily generate cracking between carbon fiber sheets and concrete specimen. MOC cement has many properties superior to the conventional Portland cements, including higher compressive and tensile strength, better fire resistance, lower shrinkage and creep and better durability, but the application has been limited by its poor water resistance, warping deformation, efflorescence, cracking and reinforced corrode by chloride ion [11,12].

Magnesia-phosphate cement (MPC) is based on the chemical reaction of acidic liquid or phosphate with a solid basic dead-burned magnesia oxide powder [13]. This cement has several advantages such as high early strength, rapid setting and hardening, excellent bonding with old Portland cement concrete [14,15,16,17]. The use of this cement as a rapid patch repair material in construction field has been well studied in

Citation: Zhang and Xiaojian Gao, The Feasibility of Modified Magnesia-Phosphate Cement as a Heat Resistant Adhesive for Strengthening Concrete with Carbon Sheets, doi:10.3390/app6060178.

the past decades [14,18,19]. However, this material has seldom been used as an adhesive for concrete strengthening, especially in high temperature environments.

This paper seeks to investigate the bond properties of the modified MPC between externally bonded carbon fiber sheets and concrete at elevated temperatures when being compared with epoxy resin. The compressive strength of the modified MPC does not decrease at the temperature of 600 °C [20]. The specimens bonded by different adhesives (epoxy resin and the modified MPC) were heated to different temperature levels of 105 °C, 200 °C, 300 °C, 400 °C and 500 °C and maintained for 3 h. The bond strength and failure types of specimens after different high temperature exposures were measured.

EXPERIMENTAL

Raw Materials

The modified MPC was prepared by blending dead-burned magnesia powder (MgO), ammonium dihydrogen orthophosphate (ADP, $NH_4H_2PO_4$), borax (B), wollastonite powder ($CaSiO_3$) and tap water together. The first three ingredients were industrial-grade of purity 95%–99%. The wollastonite consists of 45.6% of CaO and 48% of SiO_2, mainly in amorphous form and has an average specific surface area of 842.7 m^2/kg. Based on the former study (unpublished) [20], three binder proportions were designed as presented in Table 1. The molar ratio of Wollastonite and MgO to ADP was kept at 9:1. MA1 is the control mix for MPC. Based on the control mix, wollastonite was added to replace 5% and 10% (in molar percent) of MgO dosage. The mass ratio of water to powder was kept at 0.15. The borax was added by 5% of MgO weight to control setting time. The organic adhesive used in this paper was a two-part epoxy resin. The epoxy resin consists of parts A and B that are blended at a weight ratio of 8:2, and stirred until a homogeneous mixture is obtained. Unidirectional carbon fiber sheets were used. The mechanical properties of the carbon fiber sheets and epoxy resin provided by the manufacture are shown in Table 2.

Table 1. Details of the designed binder proportions.

Mix	Wollastonite (Molar Percent)	MgO (Molar Percent)	Weight Dosage		
			Wollastonite	MgO	ADP
MA1 (control)	0	100	0	100	31.9
MA2 (5%)	5	95	15.3	100	33.6
MA3 (10%)	10	90	32.2	100	35.5

Table 2. Mechanical properties of carbon fiber sheets and epoxy resin.

Property	Carbon Fiber Sheets	Epoxy Resin
Elastic Modulus (GPa)	244	19
Tensile strength (MPa)	4125	39.12
Areal Density (g/m^2)	300	-
Nominal Thickness (mm)	0.167	-

For interfacial bond testing, concrete with dimensions of 100 mm × 100 mm × 100 mm was prepared in the laboratory, respectively, with the following three mixing proportions: cement (319, 370 and 425 kg/m^3), water (179, 170 and 170 kg/m^3), sand (755, 784 and 801 kg/m^3), coarse aggregate (1042, 1043 and 1051 kg/m^3) and superplasticizer (5.42, 4.25 and 8.50 kg/m^3). The concrete specimens were cured for 28 days in a fog room before the bond strength test. To protect the carbon fiber sheets from fire damage during the high temperature treatment, a tunnel fire coating was used which is produced by the Beijing Mercutio Fire Source Factory (China). This tunnel fire coating has a density of 600kg/m^3, a specific heat 1000 J/(kg·K) and a heat conductivity coefficient 0.12W/(m·K).

Testing Methods

Bond Strength

The bond property between carbon fiber sheets and concrete is a key factor controlling the behaviour of the strengthened concrete elements. There are many different experimental methods used for determining the bond strength. Yao *et al.* [21] classified the existing measurement approaches into the following three types (see Figure 1): (a) single-shear test; (b) double-shear test; and a (c) beam test. Due to the simplicity and reliability [22], the double-shear test was adopted in this study for measuring the bond strength between the carbon fiber sheets and concrete specimen.

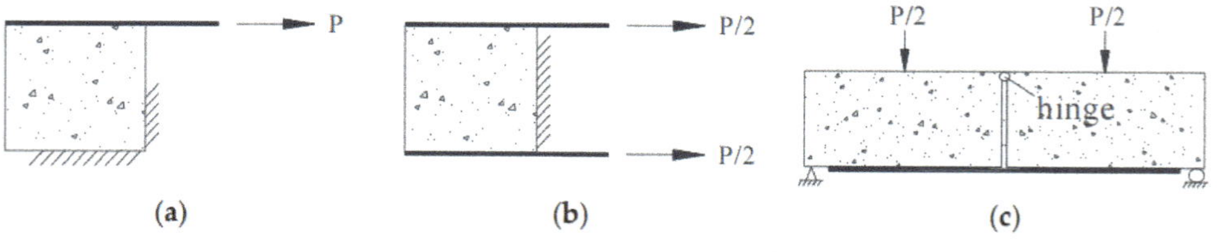

Figure 1. Sketch of test methods for interfacial bond properties: (**a**) single-shear test; (**b**) double-shear test; (**c**) beam test.

During the double-shear test, the loading was imposed on the prepared specimen by using a computer controlled electronic universal testing machine until failure, under a displacement control at a speed of 0.2 mm/min. The critical load at failure was recorded and the bond strength can be calculated according to the following equation:

$$\tau = \frac{P_u}{2b_f L_f}. \quad (1)$$

In the above equation, τ = bond strength (MPa); P_u = critical load imposed on the specimen at failure (kN); b_f = the bonding width of carbon fiber sheets (mm); and L_f = the bonding length of carbon fiber sheets (mm).

Preparation of Specimens

For the double-shear test, the carbon fiber sheets were pasted on a pair of opposite surfaces for each concrete specimen and the procedure (as shown in Figure 2) was described as below.

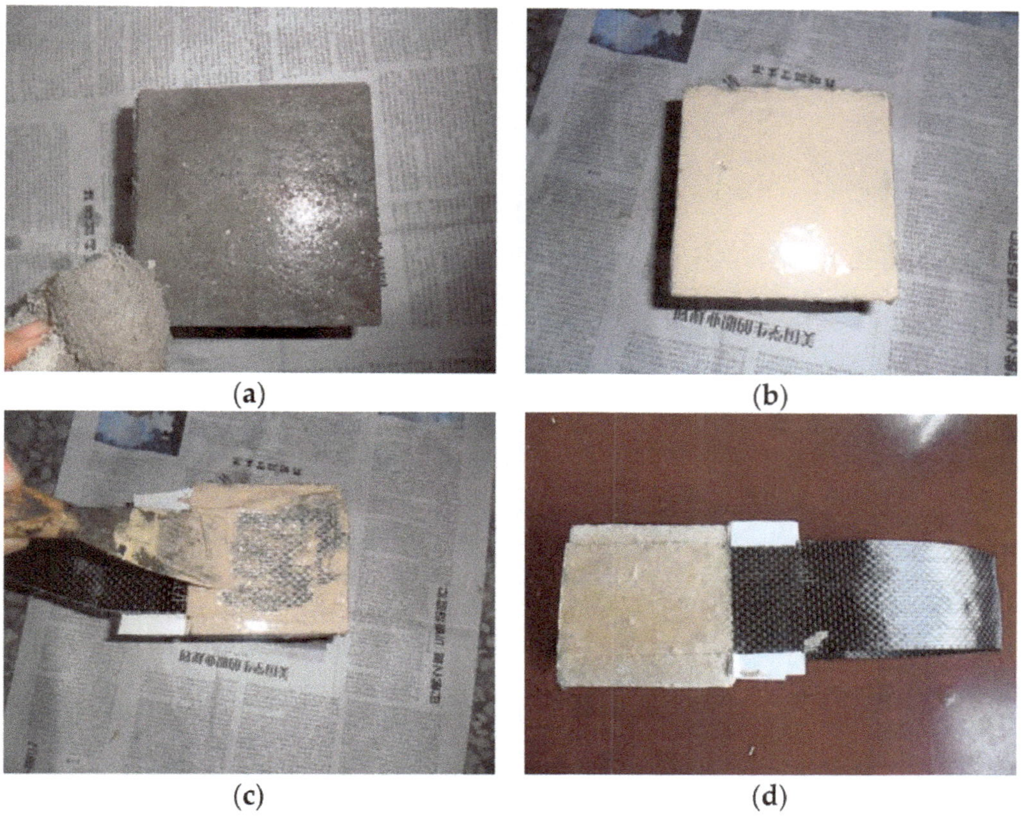

Figure 2. Process of pasting the carbon fiber sheets: (**a**) cleaning up the concrete surface; (**b**) trowelling the primer layer of magnesia-phosphate cement (MPC); (**c**) pasting the carbon fiber sheets; and (**d**) trowelling the surface layer of MPC.

(a) The concrete surface was cleaned using a brush to remove dust particles and then was wet by some water.

(b) A one-millimetre-thick primer layer of MPC or epoxy resin was trowelled to the concrete surface uniformly by using a spatula. Then, the specimen was kept to cure until MPC or epoxy resin started to lose flowability or become thicker.

(c) The carbon fiber sheets were pressed onto the above-preprocessed surfaces by using a ribbed roller ensuring the required bond width and length of 70 mm and 100 mm respectively. At the same time, air bubbles entrapped in the MPC or epoxy resin layer were rolled out by this treatment.

(d) To provide the same protection for carbon fiber sheets from fire damage during high temperature treatment, a 2-mm thick layer of MPC was coated on the outer surface of the carbon fiber sheets for all the specimens.

The prepared specimen (Figure 3a) was kept curing for different days at room temperature and then the double-shear testing was conducted. The fixture for the double-shear test is shown in Figure 3b.

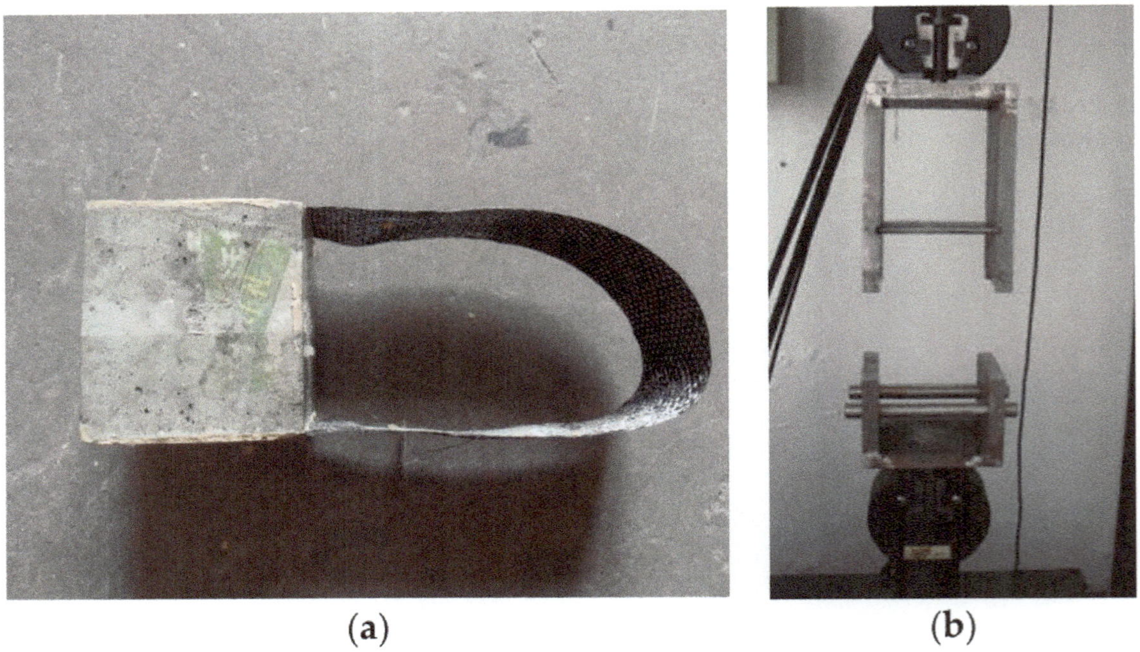

Figure 3. Specimen and fixture of the double-shear test: (a) specimen; (b) fixture.

Elevated Temperature Treatment

After 28 days of room temperature curing, the upper unanchored zone of carbon fiber sheets on the strengthened concrete specimens, which was not coated by MPC mixture, was wrapped with the tunnel fire coating. After the tunnel fire coating became completely dry, the specimens were firstly exposed to 105 °C in an oven for 24 h. The purpose of this treatment is to avoid explosive spalling of concrete caused by the evaporation of free water under high temperatures. Then, the dried specimens were placed in a high temperature furnace and heated up respectively to 200 °C, 300 °C, 400 °C and 500 °C for 3 h with the temperature increasing rate of 3 °C/min and then cooled to room temperature. After the high temperature treatment, the tunnel fire coating was cleaned up and then the bond strength measurement was carried out. Three specimens were tested for each mixture after every temperature exposure and the average value was evaluated as the bond strength.

Microscopic Characterization

Scanning Electron Microscope (SEM, FEI CO., Quanta200F, Hillsboro, TX, USA) was employed to examine the morphology and microstructure of the interfacial connection between the carbon fiber sheets and adhesives. Images were produced using the Environmental SEM in a high-vacuum mode with spot size 4.7, aperture 4 and an accelerating voltage of 20 kV. In order to observe the overall topography and microstructure of the attachment, fracture sections of all samples at room temperature were used for this analysis.

RESULTS AND DISCUSSION

Failure Modes

During double-shear tests, special attention was paid to failure modes. Based on the experimental results of bond property between carbon fiber sheets and concrete surface after room temperature curing or elevated temperature exposure, several failure modes were found in this study as follows (as shown in Figure 4): (a) delamination of the carbon fiber sheets, which is the dominate failure mode under both room temperature and elevated temperature exposure lower than 200 °C in this study; (b) complete debonding failure between the adhesive and the concrete substrate, which was only occurred for several MPC specimens when the concrete has a low strength grade or the epoxy resin specimens [23]; (c) partial debonding failure between the adhesive and the concrete substrate (this failure can be attributed to minor eccentricity of loading or nonhomogeneity of adhesive matrix, or the poorer quality of concrete surface preparation) [24]; (d) failure in the adhesive layer, due to the adhesive cracking at high temperature; (e) failure at the end anchorage zones, due to minor eccentricity of loading; and (f) peeling off of the carbon fiber sheets, due to the decrease of adhesive bond strength at high temperature. The failure modes (d) and (e) have been reported in literature [25]. For the specimens failed through failure mode (a), the real bond strength cannot be obtained; therefore, the following reported bond strength for such failure specimens is lower than the real bond strength.

Figure 4. Typical failure modes observed from double-shear tests: (a) delamination of the carbon fiber sheets; (b) complete debonding; (c) partial debonding; (d) failure in the adhesive layer; (e) failure at the end anchorage zones; and (f) peeling off of the carbon fiber sheets.

Bond Properties at Room Temperature

For interfacial bond properties affected by types of adhesives, the used concrete specimen has a 28-day compressive strength of 30.95 MPa. The bond strength for MPC and epoxy resin specimens cured for 28 days at room temperature is shown in Figure 5. The bond strength of MPC at 28 days is much higher than that of the epoxy resin. The influence of the type of matrix has also been analysed referring to the load versus displacement, as reported in Figure 6. From Figure 6, it can be noted that the slope of MPC specimens is a little larger than that of epoxy resin specimens.

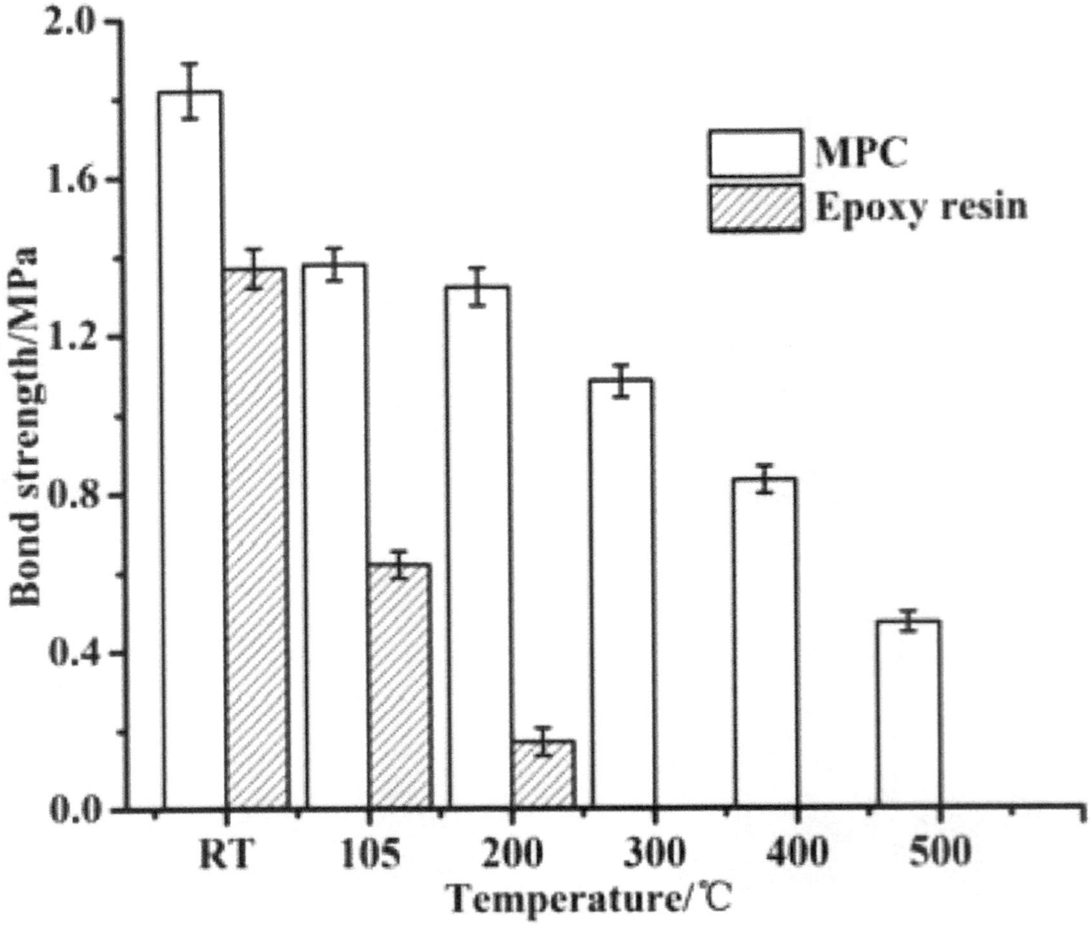

Figure 5. Bond strength of modified MPC and epoxy resin after exposure to different temperature.

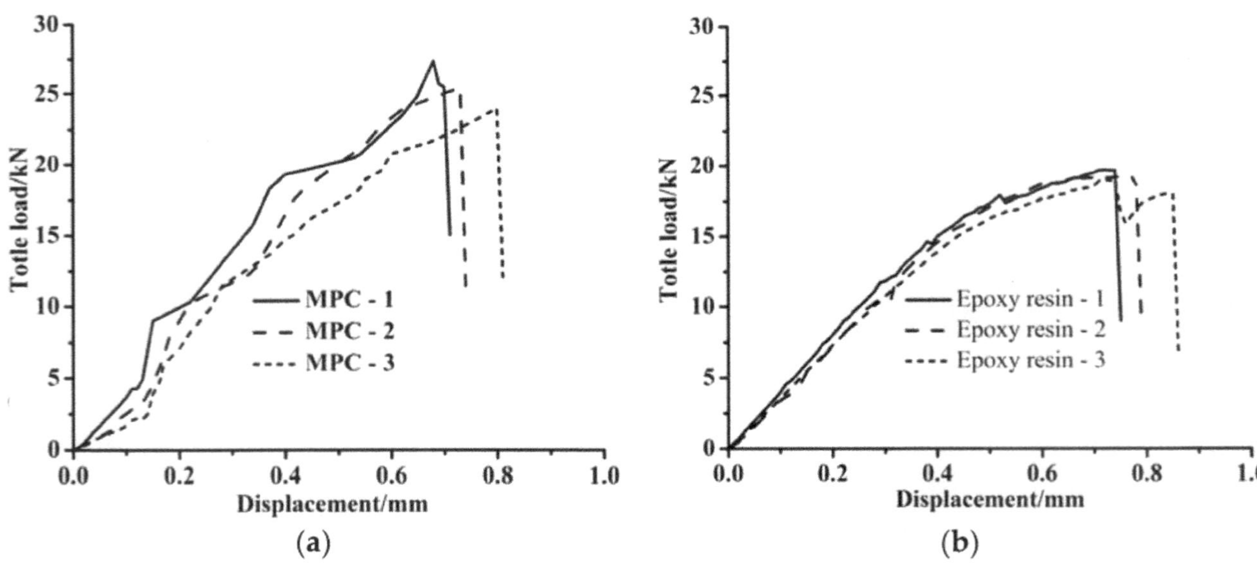

Figure 6. Load-displacement curves at room temperature: (**a**) MPC specimens; (**b**) epoxy resin specimens.

For the MPC specimens at room temperature, the failure mode is failure mode (a), perhaps due to the high viscosity of MPC mixture that failed to fully penetrate into the carbon fiber sheets. Therefore, the real bond strength of MPC should be not less than this tested value. The failure mode is failure mode (b) for the epoxy resin specimens, perhaps due to the good adhesive performance of the epoxy resin. However, the MPC specimens failed through interlaminar slip of fiber strips instead of complete debonding. MPC specimens exhibit higher bond strength than epoxy resin at ambient temperature. In addition, the bonding is very tight between the CFRP and adhesives for both the samples of the MPC and the epoxy resin cured for 28 days (see Figure 7). It may be feasible to strengthen concrete members with carbon fiber sheets bonded with the MPC mixture.

(a) (b)

Figure 7. Interfacial connection between carbon fiber sheets and adhesives: (**a**) MPC; (**b**) epoxy resin.

For interfacial bond properties affected by concrete strength, three different concretes were carried out with compressive strength of 30.95 MPa, 42.67 MPa and 49.89 MPa, respectively. MPC was used as the adhesive, and the curing age under room temperature was also considered. The results of bond strength are indicated in Table 3. With the increasing compressive strength of concrete substrate, the bond strength is increased slightly. This may be attributed to the fact that the higher compressive strength leads to the higher shear strength of concrete surface. The average bond strength at three days and seven days is about 82% and 93% of that at 28 days, respectively. It could be concluded that MPC has advantages of rapid hardening and high early strength.

Table 3. Influence of concrete strength and curing age on bond strength of magnesia-phosphate cement (MPC).

Compressive Strength of Concrete $f_{cu,m}$ (MPa)	Curing Age (Days)	Bond Strength (MPa)
30.95	3	1.50
	7	1.71
	28	1.82
42.67	3	1.61
	7	1.75
	28	1.88
49.89	3	1.68
	7	1.87
	28	2.01

Failure Modes after Exposure to High Temperatures

Experimental Phenomena

The carbon fiber sheets have no obvious change when the temperature is less than 105 °C, but oxidation occurs when heating up to higher than 105 °C. The higher the temperature is, the deeper the oxidation degree is. For MPC specimens, a strong pungent smell and white smoke was emitted from the exhaust port of the furnace when the temperature was higher than 105 °C. The primary reason for this is considered to be the decomposition of hydration products of MPC. Explosive spalling of concrete specimens occurred at 500 °C for one group of MPC specimens. The epoxy resin exhibits a color of pale yellow at ambient temperature, and it gradually changes to dark yellow and ultimately black with the temperature increase. There is a burning smell when heating up to more than 105 °C, which is in accordance with the seriously degraded bond performance of epoxy resin specimens at higher temperature.

Failure Modes

The failure appearance of MPC and epoxy resin specimens after exposure to high temperatures is shown in Figure 8 and Figure 9, respectively. For the MPC specimens at 105 °C, they failed through the falling-off of small lumps of concrete, perhaps due to minor eccentricities of loading or nonhomogeneity of adhesive matrices. The MPC specimens failed in the adhesive layer for all of the treatments under high temperatures ranging from 200 °C to 400 °C, perhaps due to the adhesive layer cracking. The MPC specimens failed through rupture of fiber strips, perhaps mainly due to the strength degradation of MPC or the adhesive cracking at high temperature. The complete debonding failure between the adhesive and the concrete substrate happened for the epoxy resin specimens under temperatures from room temperature to 200 °C, perhaps due to the good adhesive performance of the epoxy resin. However, peeling off of the carbon fiber sheets happened for the epoxy resin specimens at 300 °C due to the serious deterioration of epoxy resin.

Figure 8. Failure modes of MPC specimens after exposure to high temperatures: (**a**) 105 °C; (**b**) 200 °C; (**c**) 300 °C; (**d**) 400 °C; (**e**) 500 °C.

Figure 9. Failure modes of epoxy resin specimens after exposure to high temperatures: (**a**) 105 °C; (**b**) 200 °C; (**c**) 300 °C.

Bond Strength after Exposure to High Temperatures

Influence of Wollastonite Addition

Effect of wollastonite content on the bond strength of MPC exposed to different temperature is shown in Figure 10. For all the mixtures without and with the addition of wollastonite, the bond strength of MPC reduces significantly when the temperature increases to 105 °C from room temperature, but then almost remains constant from 105 °C to 200 °C. This drop in room temperature (RT)-105 °C can be mainly attributed to the loss of crystallization water of hydration products [20]. With the increasing temperature from 200 °C to 500 °C, the bond strength decreases gradually. When being compared with the room temperature samples, the exposure at a temperature of 500 °C leads to about 75% reduction in bond strength. At every temperature level, the 10% wollastonite specimen showed the highest strength. Therefore, the addition of wollastonite could help to improve the bond strength of MPC at elevated temperatures.

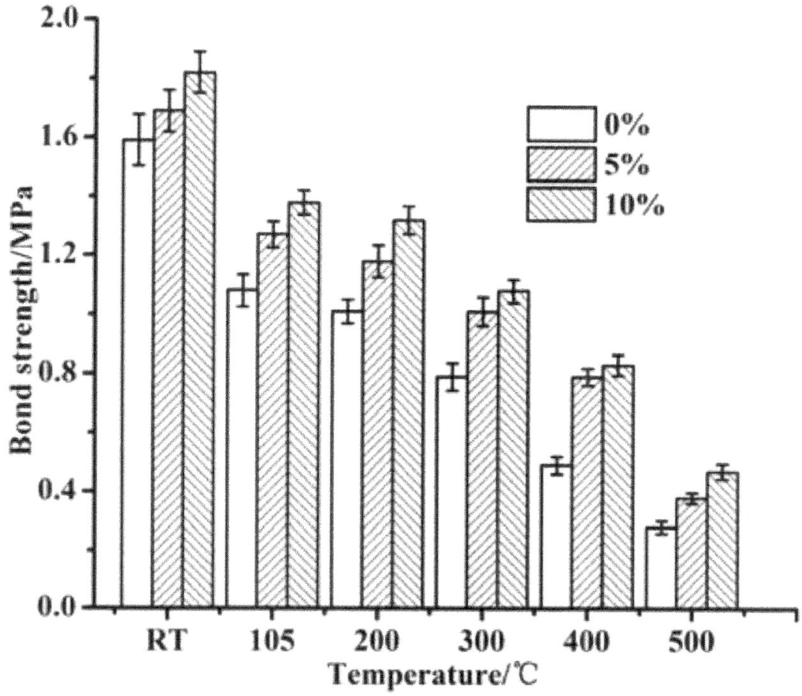

Figure 10. Effect of wollastonite content on the bond strength of MPC after exposure to different temperatures.

Comparison between Modified MPC and Epoxy

Double-shear tests were carried out on nine groups of specimens after exposure to temperatures from room temperature to 500 °C for evaluating the effect of temperature on bond strength of different types of adhesives (MPC and epoxy resin). The specified compressive strength of the modified MPC was 32 MPa at 28 days (the addition of 10% wollastonite). Three specimens were tested at each temperature and bond strength was evaluated as the average of bond strength obtained from these three tests. The bond strength of modified MPC and epoxy resin variation with temperature were shown in Figure 5. It can be clearly found that epoxy resin exhibits lower bond strength than MPC at ambient temperature, and retains much lower bond strength after exposures of temperatures ranging from 105 °C to 500 °C.

For the epoxy resin specimens, more than a half of bond strength was lost when the temperature increased from room temperature to 105 °C. When the temperature was increased to 200 °C, the bond strength of epoxy resin specimens decreased to 0.17 MPa, which was accompanied with serious carbonation. Full carbonation with almost 0 MPa was detected for the epoxy resin specimen after exposure to temperature of 300 °C, while only 40% reduction of bond strength was found for the MPC specimen.

The load versus displacement curves of specimens at selected temperatures are given in Figure 11. The behavior was basically linear until an abrupt load drop took place. The effect of elevated temperatures was substantial on the load-displacement curves of the epoxy resin specimens, as shown in Figure 11c,d for 105 °C and 200 °C. The thermal exposure also affected the load-displacement curves of the MPC specimens (Figure 11a,b), while the degree of influence was relatively minor [26].

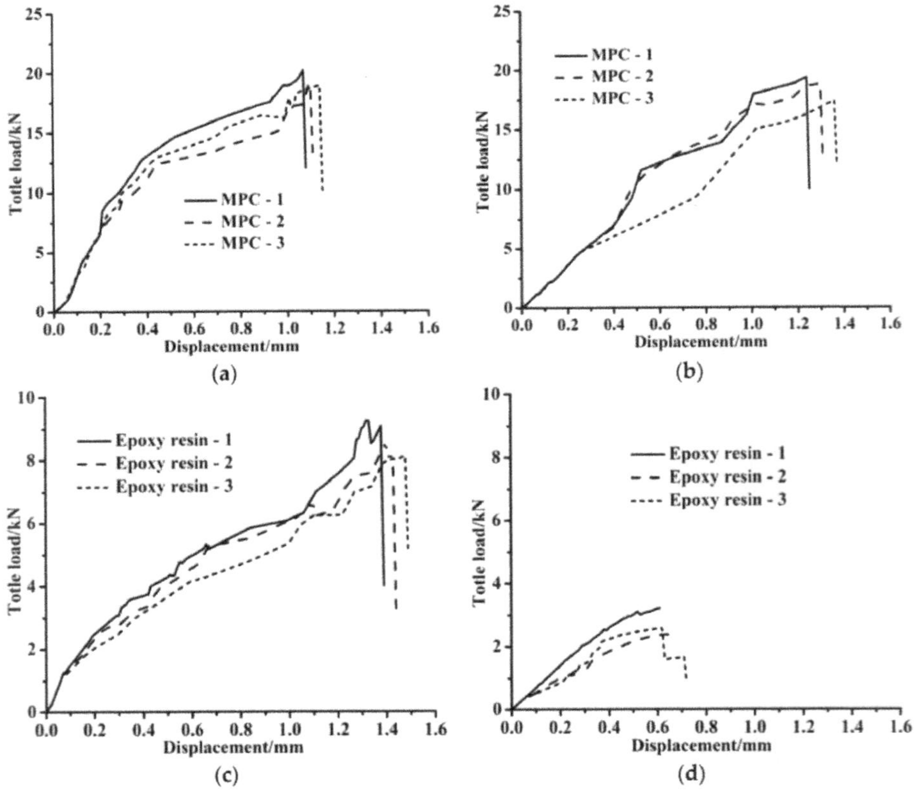

Figure 11. Load-displacement curves at high temperatures: (**a**) MPC at 105 °C; (**b**) MPC at 200 °C; (**c**) epoxy resin at 105 °C; (**d**) epoxy resin at 200 °C.

CONCLUSIONS

- In the double-shear test, six key failure modes occurred for concrete specimens at ambient and elevated temperatures. The delamination of carbon fiber sheet is the dominant one for the MPC specimens under temperatures below 200 °C. The MPC specimens failed in the adhesive layer at higher temperatures. The complete debonding failure between the adhesive and the concrete substrate happened for the epoxy resin specimens under temperatures lower than 200 °C. Peeling off of the carbon fiber sheet happened for the epoxy resin specimens after exposure to 300 °C.
- With the increasing compressive strength of concrete specimen, the higher bond strength between carbon fiber sheets and concrete specimen was measured. The bond strength for the MPC specimens is slightly higher than that for the epoxy resin specimen at ambient temperature, and the former reveals a much higher residual bond strength than the latter after exposure to temperatures from 105 °C to 500 °C.
- Although the MPC specimens failed through interlaminar slip of fiber strips instead of complete debonding, the improved bond strength under ambient temperature and higher temperatures revealed that the modified magnesia-phosphate cement could be a good substitute for epoxy resin in repairing and/or strengthening structural elements. Furthermore, the use of the modified MPC as the binder between the carbon fiber sheets and concrete can be less expensive and an ecologically friendly alternative.

ACKNOWLEDGMENTS

The authors would like to acknowledge the financial support of the Program for New Century Excellent Talents in University (NCET-12-0157) (Beijing, China).

AUTHOR CONTRIBUTIONS

Xiaojian Gao conceived and designed the experiments, and Ailian Zhang performed the experiments. Both of the authors analyzed the data and contributed to the writing of the paper.

Conflicts of Interest

The authors declare no conflict of interest.

REFERENCES

1. Gamage, J.C.P.H.; Al-Mahaidi, R.; Wong, M.B. Bond Characteristics of CFRP Plated Concrete Members under Elevated Temperatures. *Compos. Struct.* 2006, *75*, 199–205.
2. Tadeu, A.; Branco, F. Shear tests of steel plates epoxy-bonded to concrete under temperature. *J. Mater. Civil. Eng.* 2000,*12*, 74–80.
3. Ahmed, A.; Kodur, V.K.R. Effect of Bond Degradation on Fire Resistance of FRP-strengthened Reinforced Concrete Beams. *Compos. Part B Eng.* 2011, *42*, 226–237.
4. Barnes, R.; Fidel, J. Performance in fire of small-scale CFRP strengthened concrete beams. *J. Compos. Constr.* 2006, *10*, 503–508.
5. Williams, B.; Bisby, L.A.; Kodur, V.K.R.; Green,

M.F.; Chowdhury, E. Fire insulation schemes for FRP-strengthened concrete slabs. *Compos. Part A Appl. Sci. Manuf.* 2006, *37*, 1151–1160.

6. Cheng, T.W.; Chiu, J.P. Fire-resistant Geopolymer Produced by Granulated Blast Furnace Slag. *Miner. Eng.* 2003, *16*, 205–210.
7. Kurtz, S.; Balaguru, P. Comparison of Inorganic and Organic Matrices for Strengthening of RC Beams with Carbon Sheets. *J. Struct. Eng. ASCE* 2001, *127*, 35–42.
8. Toutanji, H.; Deng, Y. Comparison between Organic and Inorganic Matrices for RC Beams Strengthened with Carbon Fiber Sheets. *J. Compos. Constr.* 2007, *11*, 507–513.
9. Katakalos, K.; Papakonstantinou, C.G. Fatigue of Reinforced Concrete Beams Strengthened with Steel-Reinforced Inorganic Polymers. *J. Compos. Constr.* 2009, *13*, 103–112.
10. Melo, A.A.; Cincotto, M.A.; Repette, W. Drying and autogenous shrinkage of pastes and mortars with activated slag cement. *Cem. Concr. Res.* 2008, *38*, 565–574.
11. Li, Y.; Yu, H.; Zheng, L.; Wen, J.; Wu, C.; Tan, Y. Compressive strength of fly ash magnesium oxychloride cement containing granite wastes. *Constr. Build. Mater.* 2013, *38*, 1–7.
12. Tan, Y.; Liu, Y.; Grover, L. Effect of phosphoric acid on the properties of magnesium oxychloride cement as a biomaterial. *Cem. Concr. Res.* 2014, *56*, 69–74.
13. Abdelrazig, B.; Sharp, J.; El-Jazairi, B. The chemical composition of mortars made from magnesia-phosphate cement.*Cem. Concr. Res.* 1988, *18*, 415–425.
14. Seehra, S.; Gupta, S.; Kumar, S. Rapid setting magnesium phosphate cement for quick repair of concrete pavements-characterization and durability aspects. *Cem. Concr. Res.* 1993, *23*, 254–266.
15. Yang, Q.; Wu, X. Factors influencing properties of phosphate cement-based binder for rapid repair of concrete. *Cem. Concr. Res.* 1999, *29*, 389–396.
16. Yang, Q.; Zhu, B.; Wu, X. Characteristics and durability test of magnesium phosphate cement-based material for rapid repair of concrete. *Mater. Struct.* 2000, *33*, 229–234.
17. Shi, C.; Yang, J.; Yang, N.; Chang, Y. Effect of waterglass on water stability of potassium magnesium phosphate cement paste. *Cem. Concr. Compos.* 2014, *53*, 83–87.
18. Hall, D.A.; Stevens, R.; El-Jazairi, B. The effect of retarders on the microstructure and mechanical properties of magnesia-phosphate cement mortar. *Cem. Concr. Res.* 2001, *31*, 455–465.
19. Qiao, F.; Chau, C.K.; Li, Z. Property evaluation of magnesium phosphate cement mortar as patch repair material.*Constr. Build. Mater.* 2010, *24*, 695–700.
20. Gao, X.J.; Zhang, A.L.; Li, S.X.; Sun, B.C.; Zhang, L.C. Effects of wollastonite addition on resistance to high temperatures of magnesia phosphate cement paste. *Mater. Struct.* in press.
21. Yao, J.; Teng, J.G.; Chen, J.F. Experimental study on FRP-to-concrete bonded joints. *Compos. Part B Eng.* 2004, *36*, 99–113.
22. Yuan, H.; Teng, J.G.; Seracino, R.; Wu, Z.S.; Yao, J. Full-range behavior of FRP-to-concrete bonded joints. *Eng. Struct.*2004, *26*, 553–565.
23. Pham, H.B.; Al-Mahaidi, R. Modelling of CFRP-concrete shear lap tests. *Constr. Build. Mater.* 2007, *21*, 727–735.
24. Van Gemert, D. Force transfer in epoxy bonded steel concrete joints. *Int. J. Adhes. Adhes.* 1980, *1*, 67–72.
25. Kendall, K. Crack propagation in lap shear joints. *J. Phys. D Appl. Phys.* 1975, *8*, 512–522.
26. Kim, Y.; Siriwardanage, T.; Hmidan, A.; Seo, J. Material characteristics and residual bond properties of organic and inorganic resins for CFRP composites in thermal exposure. *Constr. Build. Mater.* 2014, *50*, 631–641.

CHAPTER 2

STRESS STATE ANALYSIS AND FAILURE MECHANISMS OF MASONRY COLUMNS REINFORCED WITH FRP UNDER CONCENTRIC COMPRESSIVE LOAD

Jiří Witzany and Radek Zigler

Keywords: Masonry, brick, stone, CFRP, CFRCM, external reinforcement, near surface reinforcement, experimental testing

INTRODUCTION

In the last two decades, fiber reinforced polymer (FRP) materials have found their place among traditional reinforcement and stabilization methods for masonry structures [1]. Strengthening with high-strength carbon fibers belongs to novel, partially non-invasive and reversible progressive methods based on limiting the appearance and development of tensile cracks resulting from horizontal tensile stresses caused by the contraction and mutual interaction of masonry components with different displacement properties. These are, above all, applications for strengthening structures against bending, tension, and shear and, to a lesser extent, in increasing the load-bearing capacity by wrapping the load-bearing members (mainly columns and pillars [2,3]). FRP materials also play an important role in strengthening structures in terms of seismic safety [4,5,6,7].

FRP materials offer numerous advantages for strengthening and stabilizing historic, mainly masonry structures, by virtue of their low weight, high effectiveness, and potential reversibility [8]. To date, the application of FRP composites in the restoration of historic and heritage buildings has been primarily focused on stabilizing vertical load-bearing and vaulted structures to resist the effects of horizontal loads due to induced and natural seismicity [9,10].

Research into the reinforcement of masonry structures with FRP materials has only in recent years been aimed at increasing the load-bearing capacity of vertical structures (columns and pillars) by wrapping [11,12]. The bulk of research addresses brick masonry [11,13], with only a few research studies taking interest in stone masonry as well [14,15,16,17,18].

The bond strength between the reinforcing composite layer and the strengthened masonry is the key parameter significantly influencing the resultant effectiveness of the strengthening of masonry structures with FRP materials [19,20]. This issue has been the subject of extensive experimental [21,22] and theoretical research [23,24] during the last few years. Some research studies have been focused on the effect of the binder (mortar) on the resultant bond strength between the FRP material and the masonry substrate [25,26,27], the design of new and suitable theoretical models describing the behavior of FRP materials in terms of their adhesion to the substrate [28], or on the experimental and theoretical investigation of using new, flexible binders for gluing externally bonded FRP reinforcement [29].

Citation: Jiří Witzany and Radek Zigler, Stress State Analysis and Failure Mechanisms of Masonry Columns Reinforced with FRP under Concentric Compressive Load, doi:10.3390/polym8050176.

NUMERICAL ANALYSIS AND EXPERIMENTAL VERIFICATION OF FAILURE MECHANISMS IN COMPRESSED MASONRY COLUMNS

The experimental research was performed on test pieces of brick masonry with dimensions (width × thickness × height) of approximately 286–290 mm × 286–290 mm × 940–1020 mm and stone masonry with dimensions of approximately 530–550 mm × 530–560 mm × 1730–1770 mm with cement mortar as a binder (Figure 1). The test pieces included reference masonry columns without any FRP application, fabricated as "unreinforced", *i.e.*, without passive wrapping of an FRP fabric-based composite, masonry columns "reinforced" with non-pre-stressed FRP-based composite strips applied to their surface with an adhesive bridge of either epoxy resin or a special polymer-modified cement mixture, or reinforced by overall wrapping, and masonry columns reinforced with FRP-based composite strips inserted into horizontal grooves cut into the masonry (Figure 2). An overview of test pieces used is presented in Table 1 and Table 2.

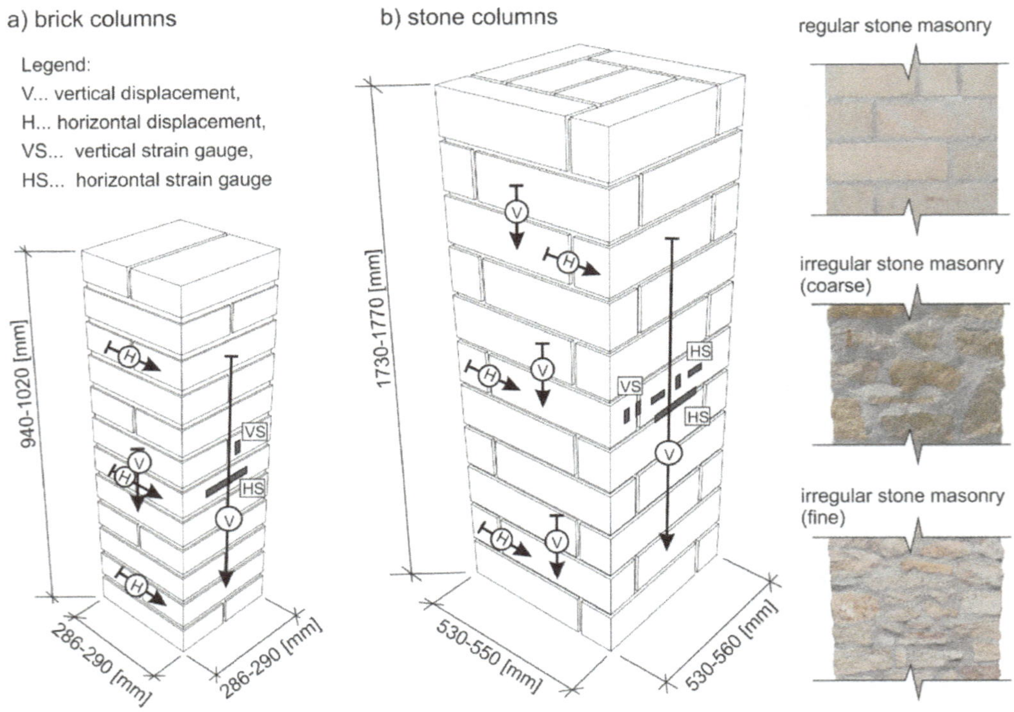

Figure 1. Scheme of test pieces (dimensions, measurement gauge positions. Arrows indicate the direction and length of measurement gauges). (**a**) Brick columns, (**b**) Stone columns.

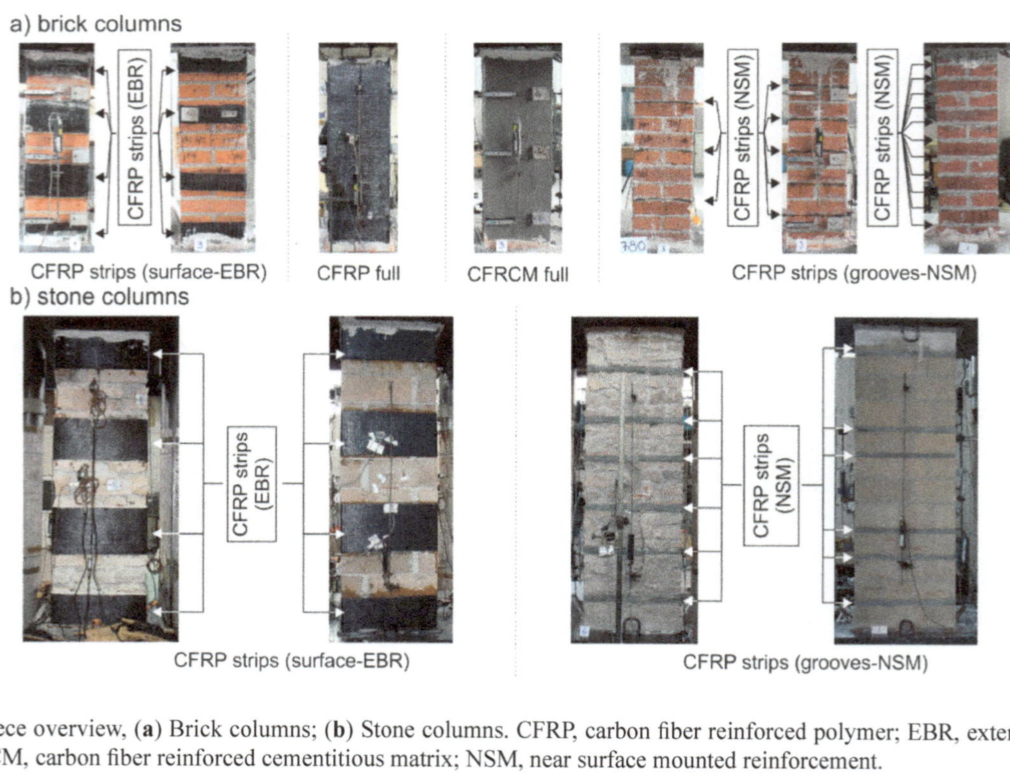

Figure 2. Test piece overview, (**a**) Brick columns; (**b**) Stone columns. CFRP, carbon fiber reinforced polymer; EBR, externally bonded reinforcement; CFRCM, carbon fiber reinforced cementitious matrix; NSM, near surface mounted reinforcement.

Table 1. List of test pieces (clay brick columns), dimensions, and component strength.

Label	Dimensions [mm]	f_b [1] [MPa]	f_m [2] [MPa]	f_k [3] [MPa]	Reinforcement
CB_URM(02)	290 × 290 × 1020	15.0 [4]	2.0 [4]	4.51	Unreinforced (URM)
CB_CFRP_S(11)					CFRP strips on surface (EBR)
CB_URM(51a)					Unreinforced (URM), loaded up to first crack
CB_CFRP_W(51b)					CFRP overall wrapping with initial cracks
CB_URM(54a)	290 × 290 × 920				Unreinforced (URM), loaded up to first crack
CB_CFRP_S(54b)					CFRP strips on surface (EBR) with cracks
CB_CFRP_L(56)					CFRP strips in grooves (NSM), every bed joint
CB_URM(57)		17.28	1.5 [4]	4.57	Unreinforced (URM)
CB_CFRP_W(58)		17.03	1.96	4.9	CFRP overall wrapping
CB_CFRP_S_R20(61)	286 × 286 × 870	11.60	1.88	3.70	CFRP strips on surface (EBR), column corner radius 20 mm
CB_CFRP_S_R35(62)			1.79	3.64	CFRP strips on surface (EBR), column corner radius 35 mm
CB_CFRP_S_R50(63)			1.84	3.67	CFRP strips on surface (EBR), column corner radius 50 mm
CB_CFRP_W_R50(64)			1.84	3.67	CFRP overall wrapping, column corner radius 50 mm
CB_CFRP_S_R85(65)			1.59	3.51	CFRP strips on surface (EBR), column corner radius 85 mm
CB_CFRP_L_J3(66)	290 × 290 × 890	11.60	1.23	3.25	CFRP strips in grooves (NSM), every 3rd bed joint
CB_CFRP_L_J2(67)				3.25	CFRP strips in grooves (NSM), every 2nd bed joint
CB_CFRCM_W(68)			1.5 [4]	3.45	CFRCM overall wrapping
CB_URM(69)			2.24	3.89	Unreinforced (URM)
CB_CFRP_S_H75_1/3(70)	288 × 288 × 875		1.75	3.62	CFRP strips on surface (EBR), 75 mm at 1/3
CB_CFRP_S_H150_1/2(71)			1.68	3.57	CFRP strips on surface (EBR), 150 mm at 1/2
CB_CFRP_S_H75_1/2(72)			1.53	3.47	CFRP strips on surface (EBR), 75 mm at 1/2
CB_CFRCM_S(73)			1.5 [4]	3.46	CFRCM strips on surface (EBR)
CB_CFRP_S_R65(A02)	286 × 286 × 890		2.08	3.81	CFRP strips on surface (EBR), column corner radius 65 mm
CB_CFRP_S_R85(A03)				3.81	CFRP strips on surface (EBR), column corner radius 85 mm

[1] Experimentally obtained compressive strength of masonry units pursuant to ČSN EN 1996-1-1 [30]; [2] Experimentally obtained compressive strength of mortar pursuant to ČSN EN 1996-1-1 [30]; [3] Characteristic compressive strength of masonry pursuant to ČSN EN 1996-1-1 [30]; [4] Compressive strength value according to the manufacturer's specification. Label explanation: CB—clay brick columns, URM—unreinforced, CFRP—reinforced with carbon fabric bonded by epoxy resin, CFRCM—reinforced with carbon mesh bonded by stabilized cement matrix, S—reinforcing strips, W—reinforced by overall wrapping, L—reinforced by lamellae inserted in horizontal grooves (bed joints), J—every Xth bed joint reinforced, H—height of reinforcing strip (mm), (X)—original number (in parenthesis) of the test specimen.

Table 2. List of test pieces (stone columns), dimensions, and component strength.

Label	Dimensions [mm]	f_b [1] [MPa]	f_m [2] [MPa]	f_k [3] [MPa]	Reinforcement
ST_URM_IREG(K01)	530 × 530 × 1750	22.73	2.0 [4]	1.99 [5]	Unreinforced (URM), coursed sandstone masonry (irregular units)
ST_URM_REG(K02)	545 × 545 × 1770	19.93		4.20	Unreinforced (URM), coursed sandstone masonry (regular units)
ST_CFRP_S_IREG(K03)	550 × 560 × 1735	22.73		1.99 [5]	CFRP strips on surface (EBR), coursed sandstone masonry (irregular units)
ST_CFRP_S_REG(K04)	548 × 545 × 1760	19.93		4.20	CFRP strips on surface (EBR), coursed sandstone masonry (regular units)
ST_CFRP_S_IREG(K07a)	550 × 560 × 1735	19.51	3.03	0.94 [5]	CFRP strips on surface (EBR), coursed sandstone masonry (irregular units)
ST_URM_MIX_IREG(K09)	550 × 560 × 1735	20.82	2.14	1.92 [5]	Unreinforced (URM), coursed mixed (sandstone, marlstone) masonry (irregular units)
ST_CFRP_L_IREG(K13)	550 × 560 × 1730	19.51	2.25	0.86 [5]	CFRP strips in grooves (NSM), coursed sandstone masonry (irregular units)
ST_CFRP_L_MIX_IREG(K15)	550 × 550 × 1730	22.84	1.29	1.93 [5]	CFRP strips in grooves (NSM), coursed sandstone masonry (irregular units)

[1] Experimentally obtained compressive strength of masonry units pursuant to ČSN EN 1996-1-1 [30]; [2] Experimentally obtained compressive strength of mortar pursuant to ČSN EN 1996-1-1 [30]; [3] Characteristic compressive strength of masonry pursuant to ČSN EN 1996-1-1 [30]; [4] Compressive strength value according to the manufacturer's specification; [5] Characteristic compressive strength of masonry pursuant to ČSN 731101 (change B) [31]. Label explanation: ST—stone columns, URM—unreinforced, CFRP—reinforced with carbon fabric bonded by epoxy resin, S—reinforcing strips, L—reinforced by lamellae inserted in horizontal grooves, MIX—mixed stone masonry, REG—regular masonry units, IREG—irregular masonry units, (X)—original number (in parenthesis) of the test specimen.

The main objective of the experimental research was to verify the effectiveness (strength performance) of different reinforcing methods according to their parameters (height and distance of reinforcing strips, the number and distance of reinforcing lamellae, influence of used materials or technology *etc.*). Special interest within the experimental research was focused on verifying the effectiveness of additional column reinforcement to tensile crack damage from the initial phase (*i.e.*, vertical tensile cracks passing through 3 to 6 rows of masonry). The reinforcing methods were chosen with the aim of minimizing possible interference into the structure (surface or near-surface application of FRP materials).

The columns were reinforced by wrapping them in non-pre-stressed strips of high-strength carbon fiber fabric (Tyfo®SCH-41 fabric) glued onto the reinforced columns with Tyfo® S thixotropic two-component epoxy resin (FRP system), or by non-pre-stressed strips of mesh made of high-strength carbon fiber Ruredil X Mesh C10 and inorganic stabilized cement matrix M25 (FRCM system). The material characteristics of these carbon fabrics, epoxy adhesive and cement matrix are presented in Table 3 and Table 4. The composite strips were 75 or 150 mm in height and were placed at 3 or 4 levels—at the top and bottom of the column and in the middle (3 levels, axial distance of strips was 450 mm) or at third distances along the column's height (4 levels, axial distance of strips was 300 mm). Another case of reinforcement was overall wrapping of the column in a non- pre-stressed fabric. In this case, a continuous piece of fabric (Tyfo® SCH-41 fabric) was placed over the whole height of the column (glued by Tyfo® S epoxy resin). In the final case of reinforcement, on-site pre-laminated strips (lamellae) were inserted into grooves cut into the masonry (pre-laminated strip width was 40 mm, thickness was 5–7 mm and their surface was covered in fine quartz sand for better adhesion. Groove depth was 50–60 mm and thickness was 15–20 mm) filled with polymer cement mortar (Betosan Superfix TH f).

Table 3. Material characteristics of carbon fabric and epoxy resin (FRP).

Material characteristic	Dry fibers	Epoxy resin	Composite properties
Ultimate Tensile Strength in Primary Fiber Direction (MPa)	3790	72.4	986
Tensile Modulus (GPa)	230	3.18	95.8
Elongation at Break (%)	1.7	5.0	1.0
Ultimate Tensile Strength Normalized by Thickness (MPa/mm)	–	–	986
Tensile Modulus Normalized by Thickness (GPa/mm)	–	–	95.8
Nominal Laminate Thickness (mm)	–	–	1.0
Flexural Strength (MPa)	–	123.4	123.4
Flexural Modulus (GPa)	–	3.12	3.12

Table 4. Material characteristics of carbon mesh and cement matrix (FRCM).

Material characteristic	Dry fibers	Cement matrix
Ultimate Tensile Strength in Primary Fiber Direction (MPa)	4800	8.6 [1]
Ultimate Compressive Strength (MPa)	-	25.56 [1]
Tensile Modulus (GPa)	240	-
Elongation at Break (%)	1.8	-
Density (g/cm^3)	1.78	-
Thickness (mm)	0.047	-

All values according to the manufacturer; [1] Experimentaly obtained values pursuant to ČSN EN 1996-1-1 [30].

Prior to the application of reinforcing fabrics, the surface of the reinforced columns was levelled and cleaned, stripped of non-cohesive parts, and in the case of stonework columns of coursed rubble masonry and quarried sandstone (irregular sandstone blocks), the surface was levelled where the reinforcing fabric was to be applied. In all cases the column corners were rounded (with a fillet radius of 20 mm).

Reinforcement was performed using a fabric with a unidirectional arrangement of fibers applied in a single layer. Thus, the composite consisted of one layer of fabric and an outer and inner layer of epoxy resin. Both ends of the fabric were connected by an overlap (at least 250 mm in the case of clay brick masonry columns and 400 mm in the case of stone masonry columns). The shear strength of the epoxy resin was relevant for determining the anchoring length so that it would match the corresponding tensile strength of the composite strip. Experimental loading mostly resulted in the fabric's failure across its surface area, not at the mutual connection of the ends of the fabric forming the overlap.

The test pieces were gradually loaded with incremental loading steps of 60 kN up to their failure. To verify the permanent displacement component, unloading to the basic loading value of 60 kN (10% of the assumed ultimate load of unreinforced columns of brick or stone masonry) was performed after each third loading step. The ultimate load on the columns was identified as a sudden drop in loading accompanied by a loss in the ability of the loaded member to transfer compressive loads as well as the destruction and disintegration of the column masonry. The testing device required the loading process to be controlled by force, thus the post peak phase of the column's behavior could not be captured.

The masonry displacement and strain values during loading were monitored by means of linear displacement gauges (LVDT) and resistance strain gauges (see Figure 1). To avoid undesirable distortion of stress values obtained from the strain gauges, the minimum thickness of epoxy resin at the location of mounted strain gauges was strictly monitored. The displacement gauges were removed before reaching the ultimate load (in the last loading step) due to the possibility of their destruction. Partial results of the experimental research assessment are presented in Table 5 and Table 6, and graphically displayed in Figure 3, Figure 4, Figure 5, Figure 6, Figure 7 and Figure 8. Examples of characteristic masonry column failures at reaching the ultimate load are shown in Figure 9 and Figure 10.

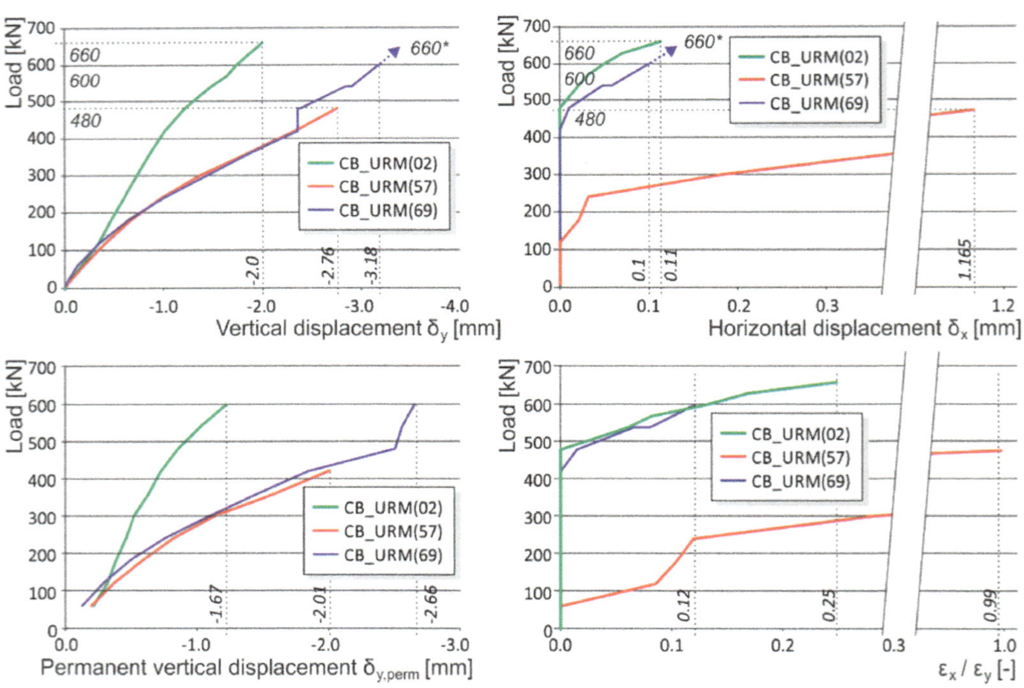

Figure 3. Experimentally obtained load displacement diagrams—unreinforced columns (CB_URM). * are the ultimate loads reached during the experimental testing. Arrows indicate that the corresponding displacement values are not measured due to the removal of measuring gauges before the last loading step (possible destruction).

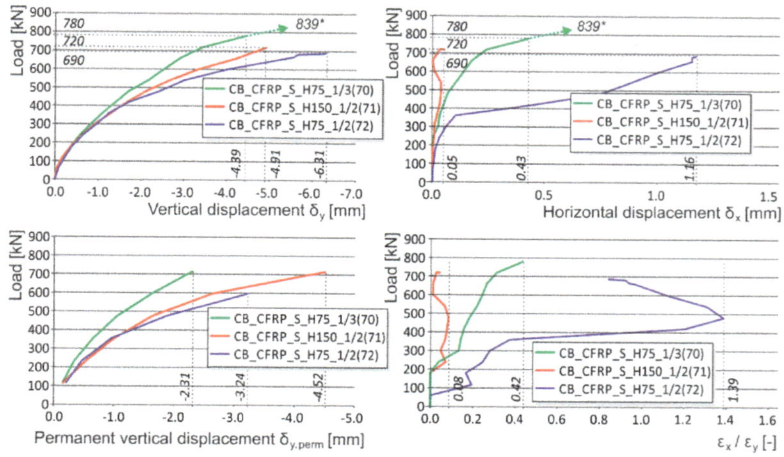

Figure 4. Experimentally obtained load displacement diagrams—strip placement and width influence (CB_CFRP_S_H75_1/3, CB_CFRP_S_H150_1/2, CB_CFRP_S_H75_1/2). * are the ultimate loads reached during the experimental testing. Arrows indicate that the corresponding displacement values are not measured due to the removal of measuring gauges before the last loading step (possible destruction).

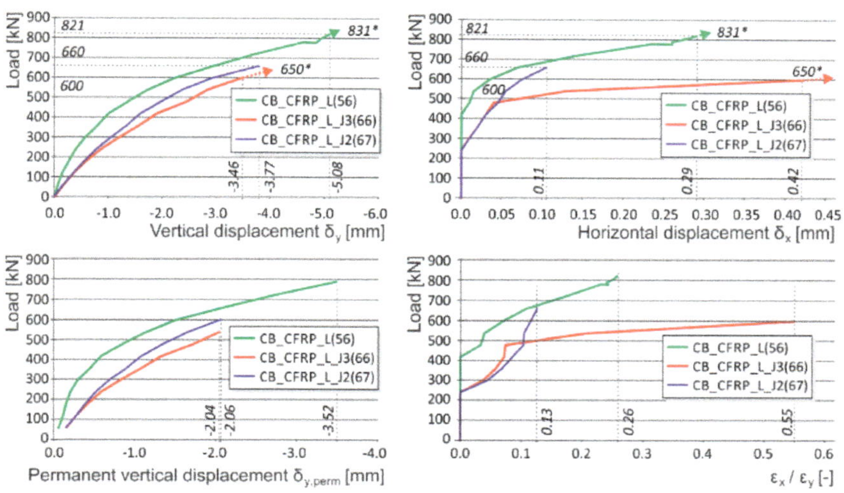

Figure 5. Experimentally obtained load displacement diagrams—NSM reinforcement placement (CB_CFRP_L, CB_CFRP_L_J3, CB_CFRP_L_J2). * are the ultimate loads reached during the experimental testing. Arrows indicate that the corresponding displacement values are not measured due to the removal of measuring gauges before the last loading step (possible destruction).

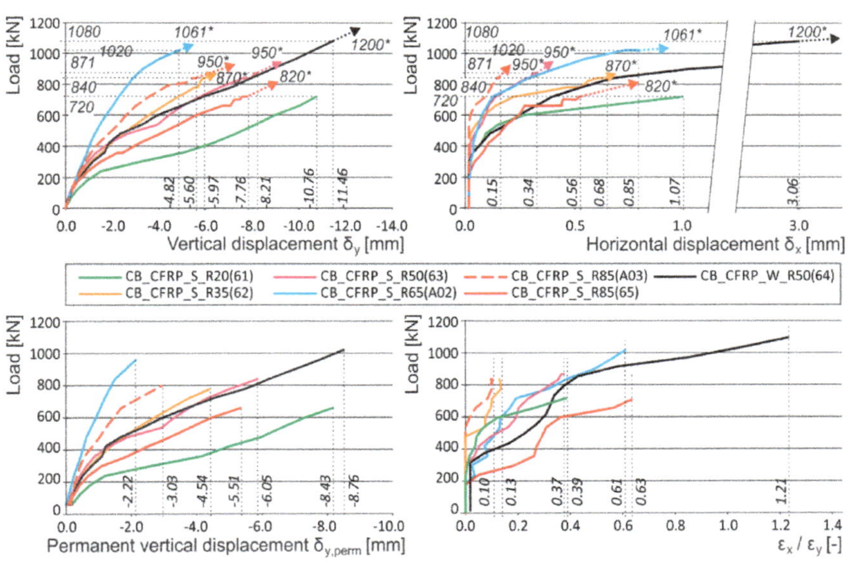

Figure 6. Experimentally obtained load displacement diagrams—corner radius influence (CB_CFRP_S_R20, CB_CFRP_S_R35, CB_CFRP_S_R50 CB_CFRP_S_R65, CB_CFRP_S_R85, CB_CFRP_W_R50). * are the ultimate loads reached during the experimental testing. Arrows indicate that the corresponding displacement values are not measured due to the removal of measuring gauges before the last loading step (possible destruction).

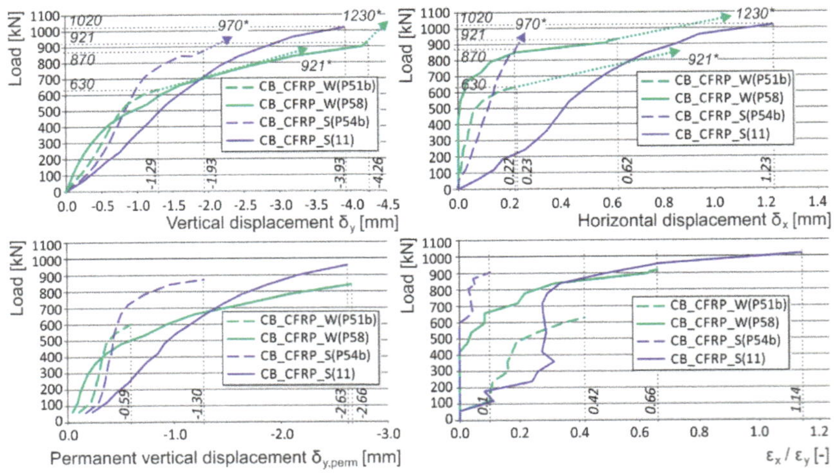

Figure 7. Experimentally obtained load displacement diagrams—initial crack influence (CB_CFRP_W, CB_CFRP_S). * are the ultimate loads reached during the experimental testing. Arrows indicate that the corresponding displacement values are not measured due to the removal of measuring gauges before the last loading step (possible destruction).

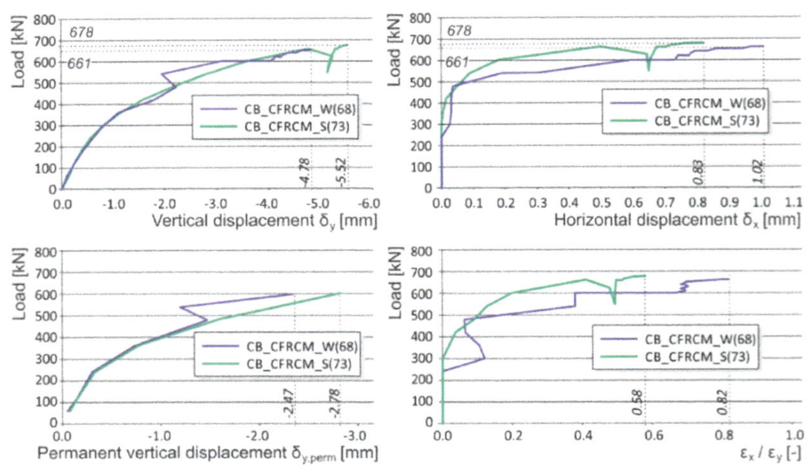

Figure 8. Experimentally obtained load displacement diagrams—FRCM reinforcement (CB_CFRCM_W, CB_CFRCM_S).

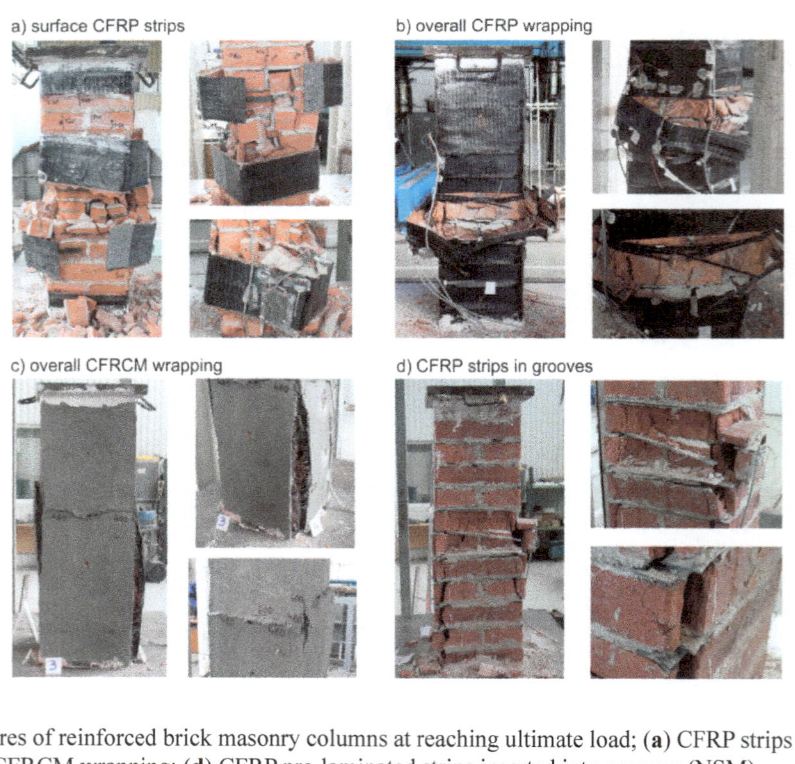

Figure 9. Characteristic failures of reinforced brick masonry columns at reaching ultimate load; (**a**) CFRP strips on surface (EBR); (**b**) overall CFRP wrapping; (**c**) overall CFRCM wrapping; (**d**) CFRP pre-laminated strips inserted into grooves (NSM).

Figure 10. Characteristic failures of reinforced brick masonry columns at reaching ultimate load; (**a**) CFRP strips on surface (EBR); (**b**) CFRP pre-laminated strips inserted into grooves (NSM).

Table 5. Summary of experimental loading results—clay brick columns.

Label	Ultimate vertical displacement [1] [mm]	Ultimate horizontal displacement [1] [mm]	N_{theor} [4] [kN]	N_{exp} [kN]	N_{exp}/N_{theor}
CB_URM(02)	2.0	0.11	341.4	660	1.93
CB_CFRP_S(11)	3.93	1.23	634.6	1,020	1.61
CB_URM(51a) [2]	– [2]	– [2]	– [3]	540 [2]	– [3]
CB_CFRP_W(51b)	1.29	0.22	727.3	921	1.27
CB_URM(54a) [2]	– [2]	– [2]	– [3]	780 [2]	– [3]
CB_CFRP_S(54b)	1.93	0.23	718.5	970	1.35
CB_CFRP_L(56)	5.08	0.42	– [3]	831	– [3]
CB_URM(57)	2.76	1.16	313.0	480	1.54
CB_CFRP_W(58)	4.26	0.62	803.5	1230	1.53
CB_CFRP_S_R20(61)	10.76	1.07	488.3	720	1.48
CB_CFRP_S_R35(62)	5.97	0.68	497.6	870	1.75
CB_CFRP_S_R50(63)	8.21	0.34	509.2	950	1.87
CB_CFRP_W_R50(64)	11.46	3.06	680.6	1200	1.76
CB_CFRP_S_R85(65)	7.76	0.56	489.7	820	1.67
CB_CFRP_L_J3(66)	3.46	0.42	– [3]	650	– [3]
CB_CFRP_L_J2(67)	3.77	0.11	– [3]	660	– [3]
CB_CFRCM_W(68)	4.09	1.02	– [3]	680	– [3]
CB_URM(69)	3.18	0.1	269.9	660	2.45
CB_CFRP_S_H75_1/3(70)	4.39	0.43	367.5	839	2.29
CB_CFRP_S_H150_1/2(71)	4.91	0.05	378.4	720	1.90
CB_CFRP_S_H75_1/2(72)	6.31	1.16	301.2	690	2.29
CB_CFRCM_S(73)	5.52	0.83	– [3]	678	– [3]
CB_CFRP_S_R65(A02)	4.82	0.85	528.2	1061	2.01
CB_CFRP_S_R85(A03)	5.60	0.15	522.9	950	1.82

[1] Last measured values before removal of the measuring gauges (due to their possible destruction); [2] Columns loaded only up to the appearance of first cracks (40%–60% of ultimate load); [3] There are no known formulas for determining the theoretical load-bearing capacity; [4] Theoretical load-bearing capacity pursuant to ČSN EN 1996 1-1 [30], and CNR-DT 200 R1/2013 [32].

Table 6. Summary of experimental loading results—stone columns.

Label	Ultimate vertical displacement [1] [mm]	Ultimate horizontal displacement [1] [mm]	N_{theor} [2] [kN]	N_{exp} [kN]	N_{exp}/N_{theor}
ST_URM_IREG(K01)	3.56	1.83	505.2	600	1.19
ST_URM_REG(K02)	3.42	0.22	1122.8	2280	2.03
ST_CFRP_S_IREG(K03)	5.45	1.14	1073.7	1360	1.27
ST_CFRP_S_REG(K04)	3.59	0.39	1899.0	2641	1.39
ST_CFRP_S_IREG(K07a)	6.17	1.09	600.0	1111	1.85
ST_URM_MIX_IREG(K09)	2.83	1.36	532.7	832	1.56
ST_CFRP_L_IREG(K13)	5.01	1.64	237.7	900	3.78
ST_CFRP_L_MIX_IREG(K15)	5.38	0.78	524.5	900	1.72

[1] Last measured values before removal of the measuring gauges (due to their possible destruction); [2] Theoretical load-bearing capacity pursuant to ČSN EN 1996 1-1 [30], and CNR-DT 200 R1/2013 [32].

The theoretical value of a column's load bearing capacity was, in the case of unreinforced columns, identified pursuant to ČSN EN 1996 1-1 [30]:

$$N_{\text{theor,URM}} = \Phi_{\text{i,m}} \cdot f_{\text{d}} \cdot A_{\text{m}} \text{ [kN]} \tag{1}$$

where $\Phi_{i,m}$ is the reduction factor at the top (bottom) or in the middle of the column reflecting slenderness end eccentricity [-], f_d is the design compressive strength of masonry [kPa], and A_m is the loaded horizontal gross cross-sectional area [m^2].

$$f_{\text{d}} = \frac{f_{\text{k}}}{r_{\text{m}}} \text{ [kPa]} \tag{2}$$

where f_k is the characteristic compressive strength of masonry [kPa] according to the values for compressive strength of masonry units (f_b) and mortar (f_m), γ_m is the partial factor for masonry material [-], including mortar execution, moisture content, uncertainties in geometry, and overall degradation pursuant to ISO 13822:2010 [33].

At the top or bottom of the column

$$\varphi_{\text{i}} = 1 - 2 \cdot \frac{e_{\text{i}}}{t} \text{ [-]} \tag{3}$$

where e_i is the load eccentricity at the top or bottom [m], taking into account also the eccentricity (if any) resulting from horizontal loads (e_{he}) and the initial eccentricity (e_{init}) based on the effective height of the column (h_{ef}/450), and t is the thickness of the column (dimension in the appropriate direction) [m].

In the middle of the column height

$$\varphi_{\text{m}} = A_1 \cdot e^{-\frac{u^2}{2}} \text{ [-]} \tag{4}$$

where

$$A_1 = 1 - 2 \cdot \frac{e_{\text{mk}}}{t} \text{ [-]} \tag{5}$$

$$u = \frac{\lambda - 0.063}{0.73 - 1.17\frac{e_{\text{mk}}}{t}} \text{ [-]} \tag{6}$$

where

$$\lambda = \frac{h_{\text{ef}}}{t_{\text{ef}}}\sqrt{\frac{f_{\text{k}}}{E}} \text{ [-]} \tag{7}$$

$$e_{\text{mk}} = e_{\text{hm}} + e_{\text{init}} + e_{\text{k}} \text{ [m]} \tag{8}$$

where e_{mk} is the eccentricity in the middle height of the column [m], e_{hm} is the eccentricity at mid-height resulting from horizontal loads [m], e_{init} is the initial eccentricity (h_{ef}/450) [m], and e_k is the eccentricity due to creep [m].

$$e_k = 0.002 \cdot \varphi_{¥} \cdot \frac{h_{\text{ef}}}{t_{\text{ef}}}\sqrt{t \cdot e_m} [m] \tag{9}$$

In the case of FRP reinforced masonry columns, the theoretical value of load bearing capacity was identified pursuant to CNR-DT 200 R1/2013 [32]:

$$N_{\text{theor,CRFRP}} = \frac{1}{\gamma_{\text{Rd}}} \cdot A_{\text{m}} \cdot f_{\text{mcd}} \text{ [kN]} \tag{10}$$

where γ_{Rd} is the partial factor [-] equal to 1.10 for a confined column, A_m is loaded horizontal gross cross-sectional area of confined member [m^2], and f_{mcd} is the design compressive strength of the FRP confined member.

$$f_{\text{mcd}} = f_{\text{md}} \left[1 + k' \cdot \left(\frac{f_{\text{1,eff}}}{f_{\text{md}}}\right)^{\alpha_1}\right] \text{ [kP}_{\text{a}}\text{]} \tag{11}$$

where k' is the non-dimensional coefficient [-], f_{md} is the design compressive strength of unconfined masonry [kPa], $f_{1,eff}$ is the effective confining pressure [kPa], and α_1 is the coefficient [-] equal to 0.5.

$$k' = \alpha_2 \cdot \left(\frac{g_{\text{m}}}{1000}\right)^{\alpha_3} \text{ [-]} \tag{12}$$

where g_m is the masonry mass-density [kg/m^3], and α_2 and α_3 are coefficients equal to 1.

$$f_{\text{1,eff}} = k_{\text{H}} \cdot k_{\text{V}} \cdot f_1 \text{ [kPa]} \tag{13}$$

where k_H is the coefficient of horizontal efficiency [-], k_V is the coefficient of vertical efficiency [-], and f_l is the lateral confining pressure [kPa].

$$f_1 = \frac{2 \cdot t_{\text{f}} \cdot b_{\text{f}} \cdot E_{\text{f}}}{\max\{b, h\} \cdot p_{\text{f}}} \cdot \varepsilon_{\text{fd,rid}} \text{ [kPa]} \tag{14}$$

where t_f is the FRP thickness [m], b_f is the FRP strip width [m], p_f is the center-to-center spacing of FRP strips [m], E_f is the Young modulus of FRP sheet elasticity [kPa], $\varepsilon_{fd,rid}$ is the reduced design value of FRP strain measured at a columns collapse [-], and b and h are the column's cross-sectional dimensions [m].

$$k_{\text{H}} = 1 - \frac{b'^2 + h'^2}{3 \cdot A_{\text{m}}} \text{ [-]} \tag{15}$$

where b' and h' are the column's dimensions reduced by the corner radius ($b - 2r_c$).

$$k_{\text{V}} = \left(1 - \frac{p_{\text{f}}}{2 \cdot \min\{b, h\}}\right)^2 \text{ [-]} \tag{16}$$

for continuous confinement k_V is equal to 1

$$\varepsilon_{\text{fd,rid}} = \min\{\eta_s \cdot \varepsilon_{\text{fk}}/\gamma_{\text{f}}; 0.004\} \text{ [-]} \tag{17}$$

where η_s is the environmental conversion factor [-], ranging between 0.85 to 0.95 for carbon fibers bonded with epoxy resin (internal or external exposure, respectively), ε_{fk} is the ultimate strain of FRP sheets [-], γ_f is the partial factor of FRP sheets [-] (ranging between 1 and 1.50, depending on the limit state and on the failure mode—whether debonding occurs or not), and 0.004 is the conventional strain limit.

Note: The following load displacement diagrams are presented without the unloading branches (made during the loading process after each third loading step) for better readability. The vertical displacement plotted is a mean value of two measurements from opposite sides of the columns (long measuring bases were used for capturing the overall masonry behavior). Horizontal displacement was measured at the middle height of the column, where the biggest horizontal displacement of the column was expected.

Failure Mechanism of Brick and Stone Masonry Columns under Concentric Compressive Load

Clarification of brickwork and stonework column failure mechanisms relative to their loading and layout is the basic starting-point for designing masonry structures, their reinforcement and method for stabilization, as well as for deriving accurate formulae for identifying their ultimate loads. Based on analysis of the experimental test results and numerical analyses, the following conclusions about the failure mechanism of compressed masonry columns under concentric load were drawn:

- The masonry failure process under concentric compressive load—its failure mechanism and a gradual loss in its ability to transfer loads—occurs in one or two phases with a different failure mechanism seen in each of these phases.
- In the phase when cracks appear and develop (phase I), failure due to "tension and shear" mostly applies (cracks arise in the direction of the principal compressive stress). In the second phase of progressive development and expansion of cracks (phase II), it is masonry failure due to "compression" that applies. In the case of unreinforced or weakly ("ineffectively") reinforced concentrically compressed masonry structures with different masonry unit and binder properties, the failure mechanism solely or largely includes phase I. The appearance and development of vertical tensile and shear cracks precedes the ultimate load of masonry in compression being reached and is accompanied by the disruption of the integrity (rupture) of the column due to the effect of transverse tensile forces (Figure 11).
- During the initial phase I, when tensile and shear cracks running in the direction of compression trajectories arise and gradually develop throughout the masonry, redistribution and non-uniform distribution of compressive normal stresses in the horizontal cross sections of masonry appear. The transverse (horizontal) tensile (or shear) stresses cause gradual "subdivision" of the column into individual parts defined by vertically running tensile and shear cracks. During this process, the tensile strength of masonry units primarily applies (Figure 12).

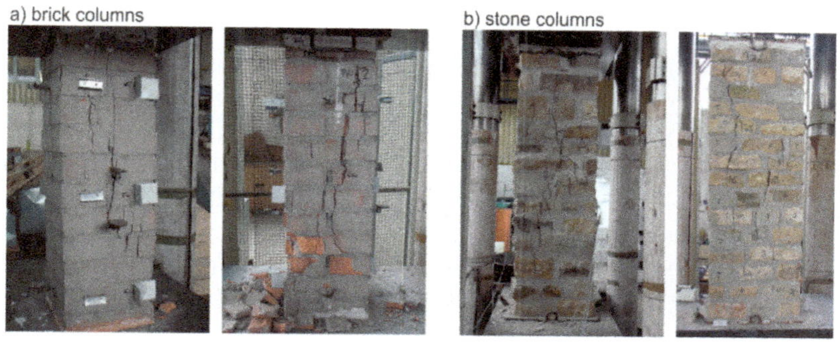

Figure 11. Typical failure of unreinforced compressed masonry columns; (**a**) brick columns; (**b**) stone columns.

Figure 12. Typical failure of unreinforced compressed masonry columns—horizontal section.

Failure Mechanism of FRP Reinforced Columns

Based on analyses of the experimental research results, we may say that in the initial failure phase—the appearance of micro and hairline cracks (10^{-3} to 10^{-1} mm in width)—in agreement with the "masonry–composite" interaction pattern and the relatively low tensile strength of ceramic or natural masonry units and other influences, the effect of a non-pre-stressed wrapping composite is very low (Figure 13). It is only after the appearance of cracks (in the order of 10^{-1} mm or bigger) under higher loading levels that the favorable effect of passive wrapping of column masonry applies more intensely.

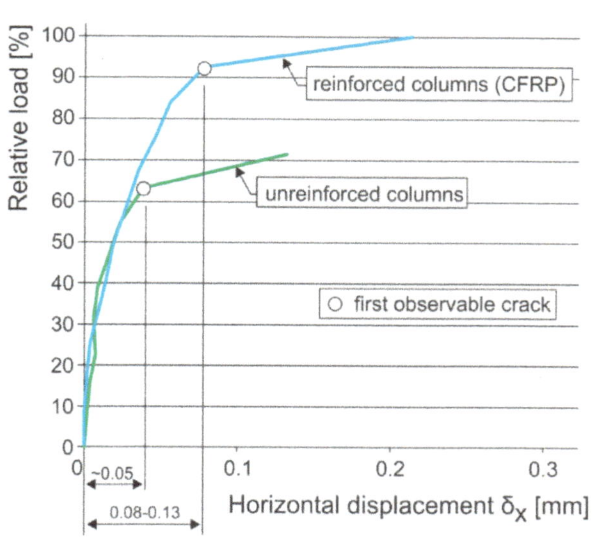

Figure 13. Horizontal displacement comparison of compressed reinforced and unreinforced structures.

- In the case of masonry columns weakly ("insufficiently") reinforced with partial non-pre-stressed fabric-based composite wrapping strips, the gradual propagation and development process of tensile and shear cracks in the column masonry (phase I) is accompanied by a gradual growth in tensile stresses in this composite due to the effect of "expanding" masonry. Horizontal transverse displacement of masonry is caused by transverse stresses due to the effect of masonry contraction and the effect of mutual interaction between masonry units and the bed joint filler. Despite the growing horizontal and vertical displacements δ_x and δ_y, the masonry column is still able to transfer the growing compressive load due to its reinforcement (stabilization) with discrete passive wrapping in a composite that takes up part of the tensile stresses (Figure 14 and Figure 15).

  Figure 14 and Figure 15 represent the results of an Abaqus [34] linear FEM analysis. Masonry was modelled as two separate materials—bricks (E_{bricks} = 2500 MPa, η_{bricks} = 0.2) and mortar (E_{mortar} = 500 MPa, η_{mortar} = 0.15)—with solid three-dimensional hexahedral finite elements (8-node trilinear brick—C3D8R), and reinforcing FRP strips (E_{mortar} = 140000 MPa, η_{mortar} = 0.3) modelled with hexahedral shell elements (SC8R). The column was supported at the bottom by prescribed displacement and rotational constraints. The load was applied in the form of imposed vertical displacement at the top of the column.

- After tensile crack appearance (in the order of 10^{-1} mm and bigger), there is progressive growth in horizontal strains ε_x, the propagation and development of tensile cracks, which become the major agents in the failure of masonry structures under compressive load. The initial value of the $\varepsilon_x/\varepsilon_y$ ratio in unreinforced columns in the phase preceding the appearance of the first (hairline) cracks lies in the interval of 0.05 to 0.1. When ultimate load in concentric compression is reached, the value of the $\varepsilon_x/\varepsilon_y$ ratio in unreinforced columns lies in the interval of 0.35 to 0.8, while in the case of columns reinforced with carbon fiber composites lies in the interval of 1.0 to 1.8 (Figure 16).

- In the case of concentrically compressed masonry columns effectively reinforced (strengthened) with a composite based on high-strength fibers and epoxy resin (which takes up part of the transverse tensile forces in the masonry due to the effect of the forced masonry displacement), the failure mode corresponding to phase II applies during the stage when the ultimate load in compression is approached and reached. In this phase of column masonry failure, the compressive strength of the masonry units applies to a greater extent. Attainment of ultimate load is accompanied by the disintegration and crushing of the masonry units and the bed joint filler and, successively, by the total failure (disintegration) of the masonry.

- Masonry failure due to the exhaustion of masonry unit compressive strength was observed, for example, in coursed masonry stonework columns with irregular masonry units and thin bed joints (*ca.* 12–15 mm), and in brickwork or stonework columns with effective wrapping in carbon fabric composites during the experimental research. Figure 17displays the characteristic failure mode of masonry units and the bed joint filler due to tensile cracks and crushing. It is the consequence of efficient interaction between the carbon fabric wrapping and masonry during the phase of active vertical tensile crack development in the column masonry and the partial failure and disintegration of masonry units (phase I and phase II).

- In phase II, the adhesive bond in the contact joint between the carbon fabric and masonry usually fails, with the carbon fabric's reaction to the column masonry's forced displacement exerted by growing horizontal displacements concentrated in the areas close to the column's edges. First, perimeter parts of the column loaded with biaxial stress states are damaged—separated by tensile cracks. In the central part of the masonry cross section exposed to stress states by triaxial compression/tension, the masonry completely disintegrates (mainly the masonry units). In this failure mechanism, the compressive strength of individual components (mainly masonry units) primarily applies. This failure mechanism mainly applies in cases with effective wrapping of a masonry column—overall wrapping or wrapping in strips optimally distributed along the column's length. With increasing distance between and lower height of carbon fabric wrapping strips, as well as lower efficiency of wrapping strips in which lower ultimate load values in concentric compression are reached, the failure mechanism occurring during the first phase (column failure due to vertical tensile cracks) predominantly applies.

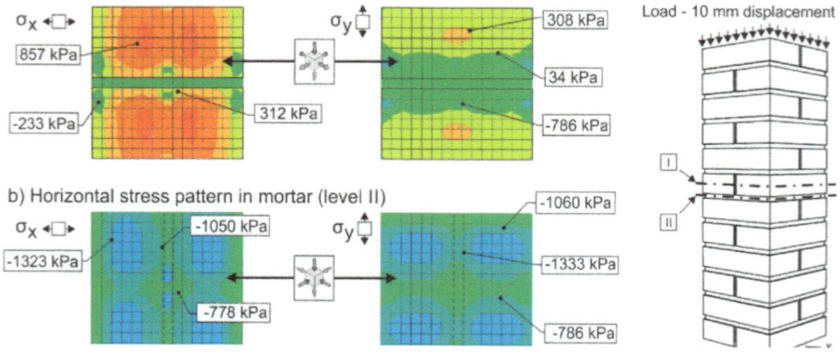

Figure 14. Stress pattern in an unreinforced masonry column. The colors in picture indicate the stress levels in the cross section of the column. Red, orange and yellow colors mean tensile stresses, green and blue colors mean compressive stresses. I and II (levels) are positions of the cross sections.

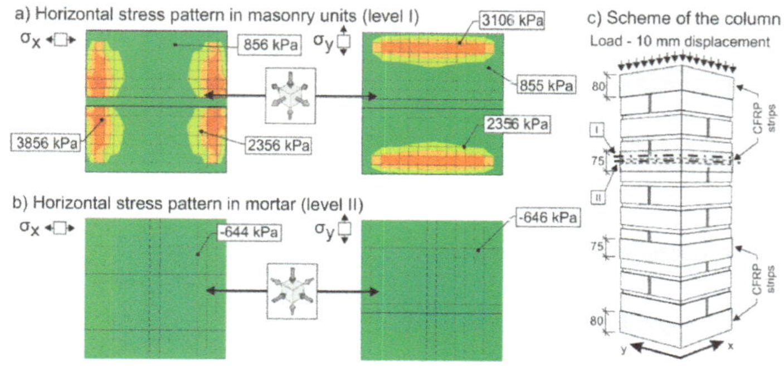

Figure 15. Stress pattern in a reinforced (CFRP strips, EBR) masonry column. The colors in picture indicate the stress levels in the cross section of the column. Red, orange and yellow mean tensile stresses, green and blue mean compressive stresses. I and II (levels) are positions of the cross sections.

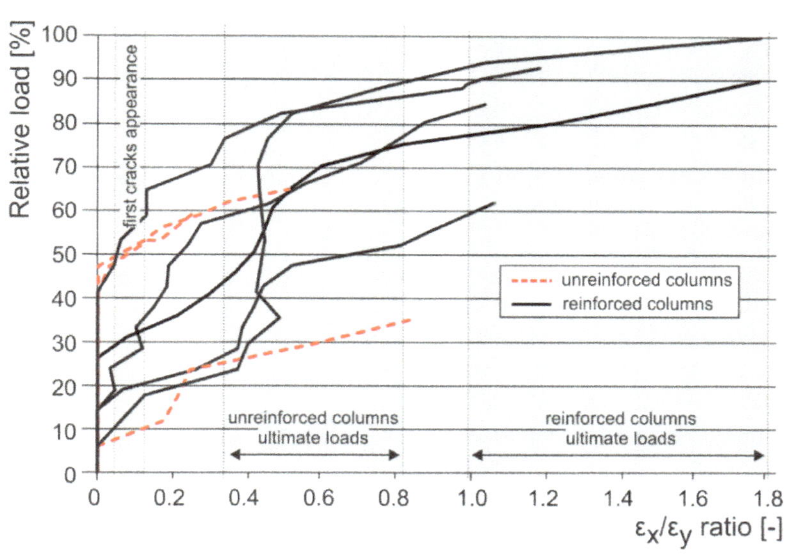

Figure 16. Vertical to horizontal strain ratio comparison of compressed reinforced and unreinforced structures—pseudo-ductility growth.

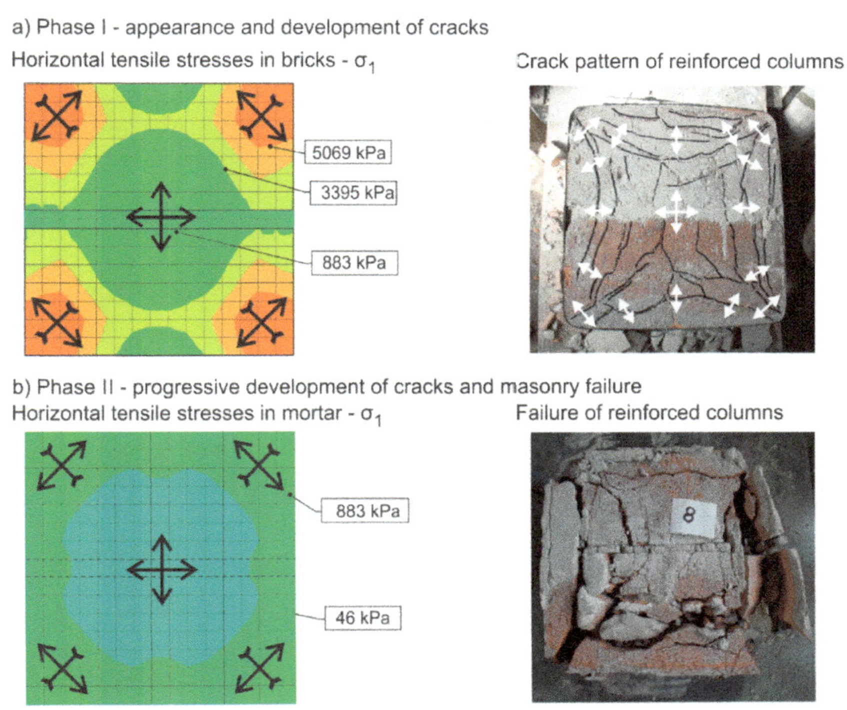

Figure 17. Characteristic failure mode of masonry units and the bed joint filler due to tensile cracks and crushing; (**a**) Phase I—appearance and development of cracks; (**b**) Phase II—progressive development and masonry column collapse (for Abaqus [34] numerical analysis detail see Figure 14). The colors in picture indicate the stress levels in the cross section of the column. Red, orange and yellow colors mean tensile stresses, green and blue colors mean compressive stresses. Arrows indicate the direction of the stresses (tensile and compressive).

Passive FRP Wrapping Efficiency

The experimental research into brick masonry columns pointed out the option of enhancing the displacement and strain characteristics and the ultimate load (see Table 5) of masonry columns under concentric compressive load reinforced with a composite of passive wrapping based on high-strength fabrics:

- In the case of "optimum" reinforcement of a masonry column with composite strips, *i.e.*, with a distance between strips smaller than or equal to no more than 1.5 times the composite strip height, there was an increase in vertical displacements δ_y to 247%, horizontal displacements δ_x to 742%, and ultimate load to 136% compared to the values reached for unreinforced masonry columns.
- In the case of overall wrapping of a masonry column in carbon composites, where the failure mechanism (corresponding to phase II) prior to reaching the ultimate load may be assumed, the growth in vertical displacements δ_y amounted to 161%, horizontal displacements δ_x 400%, and ultimate load 179%, compared to the values for unreinforced masonry columns.
- The experimental research into masonry columns damaged (before reinforcement) by tensile cracks up to (<1.5 mm) in width and extending to *ca.* 30%–40% of the column's height and then additionally reinforced with FRP composites, pointed out the possibility of achieving the comparable mechanical and strain characteristics of identically reinforced undamaged masonry columns (without the initial cracks) through this additional reinforcement (Figure 18). This is particularly evident in phase I of the columns failure. In this phase, the transverse wrapping of a masonry column in composites with high load-bearing capacity is relevant and the effect of overall column reinforcement is especially dramatic (Figure 19). This finding agrees with the "composite–masonry" interaction pattern during which the composite is activated with effect only during the phase when tensile cracks develop (10^{-1} mm and bigger).
- The passive wrapping of column masonry damaged by tensile and shear cracks in non-pre-stressed fabrics made from high-strength fibers represents an efficient stabilization means by preventing further propagation and development of cracks, and securing the structural function of masonry damaged by tensile cracks.
- The effectiveness of column masonry passive wrapping in a carbon fiber composite depends on the extent of damage to individual columns and must be assessed separately for each case of masonry failure.
- The predominantly positive effect on stress states in triaxial compression of mortar in bed joints in brick masonry after passive wrapping cannot be applied to irregular stone or mixed masonry comprising irregular, roughly dressed masonry blocks with fragments and sharp-edged, undressed, quarry stone masonry units. Here, cracking and masonry damage occur in areas where local stress states with extreme stress values (σ_x, σ_y, τ_{xy}) arise, e.g., around the masonry layout of irregular, sharp-edged units with a relatively higher modulus of elasticity than the surrounding relatively more yielding binder (Figure 20). These effects are usually unfavorably manifested, mainly at higher masonry loading values in compression (under loads exceeding 60% of ultimate load). As a consequence, these effects reduce the ultimate strength of stone masonry in compression (Figure 21).
- A special category is represented by compressed masonry columns in which, due to the properties and shape of masonry units, their binder and layout (regular masonry units with identical mechanical properties, thin bed joints with a filler of approximately the same mechanical characteristics as those of the masonry units), no transverse tensile forces that usually compensate the tendency of individual masonry layers towards different transverse displacements under concentric compressive load arise. In these reinforced masonry columns, the transverse displacement of masonry loaded by compressive forces is caused solely by contraction. The effectiveness of passive wrapping of column masonry in these cases is significantly lower compared to coursed masonry with quarry stone blocks (Figure 22).

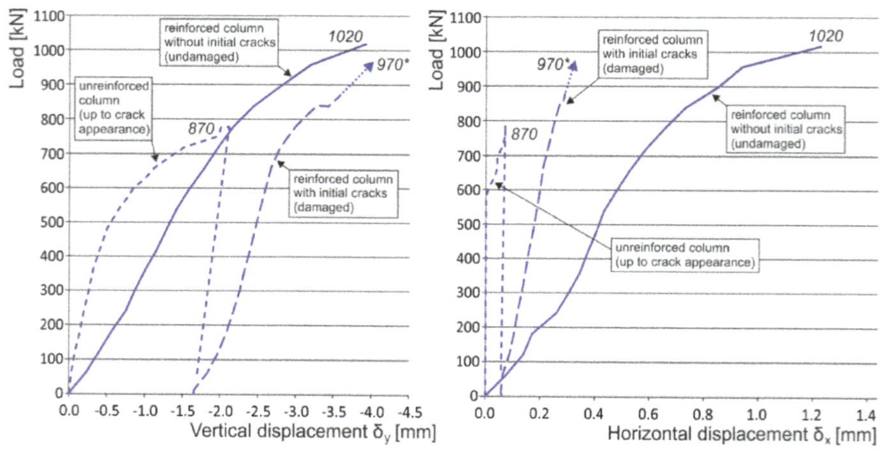

Figure 18. Experimentally obtained load displacement diagrams—masonry columns reinforced with CFRP strips (EBR) with and without the initial cracks. * are the ultimate loads reached during the experimental testing. Arrows indicate that the corresponding displacement values are not measured due to the removal of measuring gauges before the last loading step (possible destruction).

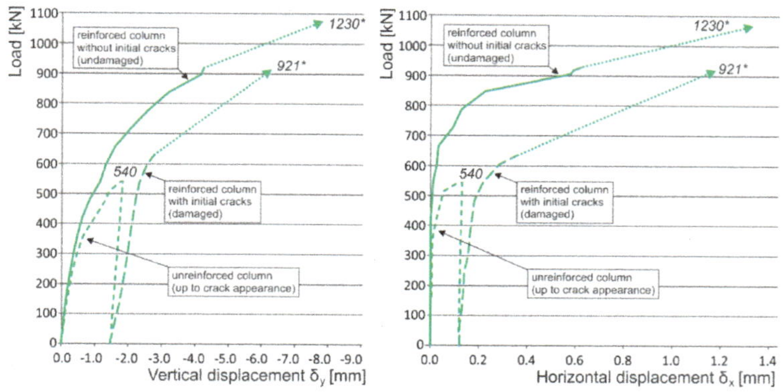

Figure 19. Experimentally obtained load displacement diagrams—masonry columns reinforced with overall CFRP wrapping with and without the initial cracks. * are the ultimate loads reached during the experimental testing. Arrows indicate that the corresponding displacement values are not measured due to the removal of measuring gauges before the last loading step (possible destruction).

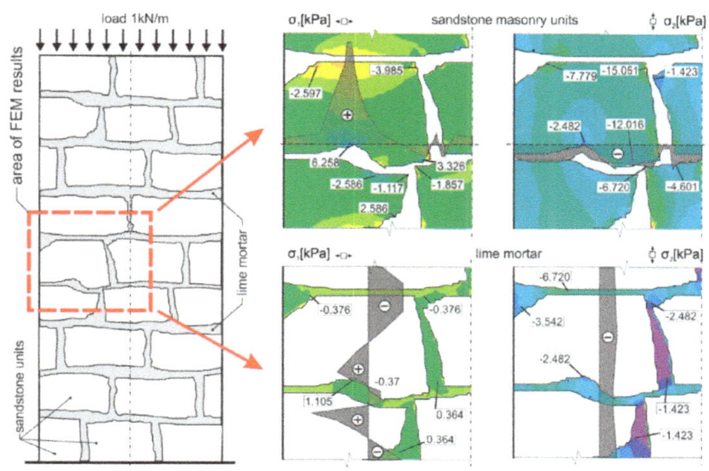

Figure 20. Critical points of vertical and horizontal concentration of stresses in masonry made of irregular masonry units (sandstone blocks and lime mortar; linear finite element (FEM) analysis using FEAT2000 [35] software. Masonry was modelled as two materials—stone blocks ($E_{sandstone}$ = 15000 MPa, $\eta_{sandstone}$ = 0.2) and mortar (E_{mortar} = 1500 MPa, η_{mortar} = 0.15)—with plane triangular elements. The column was supported at the bottom by prescribed displacement and rotational constraints. Load was applied as linear vertical force at the top of the column). The colors in picture indicate the stress levels in the column. For the σ_1 principal stress in horizontal direction, orange, yellow and light green colors mean compressive stresses, dark green color means tensile stresses, for the σ_2 principal stress in vertical direction, orange, yellow and light green colors mean higher values of compressive stresses and light blue, blue, dark blue and purple mean lower values of compressive stresses.

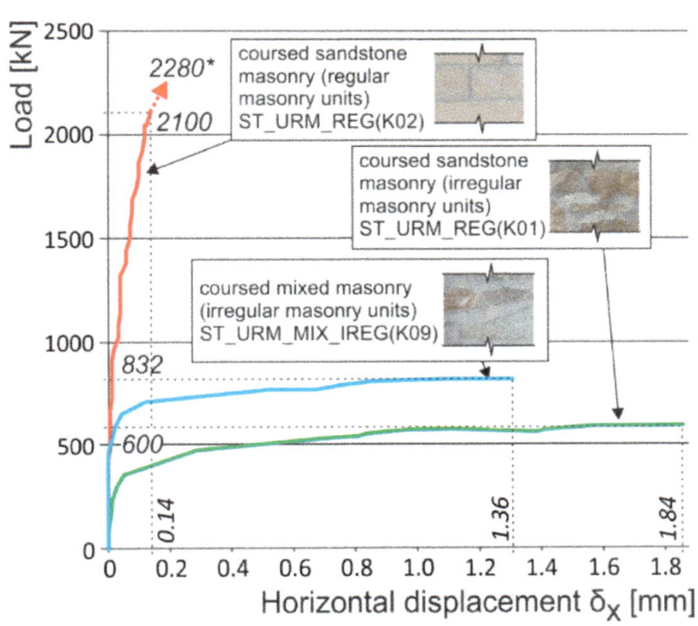

Figure 21. Experimentally obtained load displacement diagrams for different types of masonry units (shape and material).

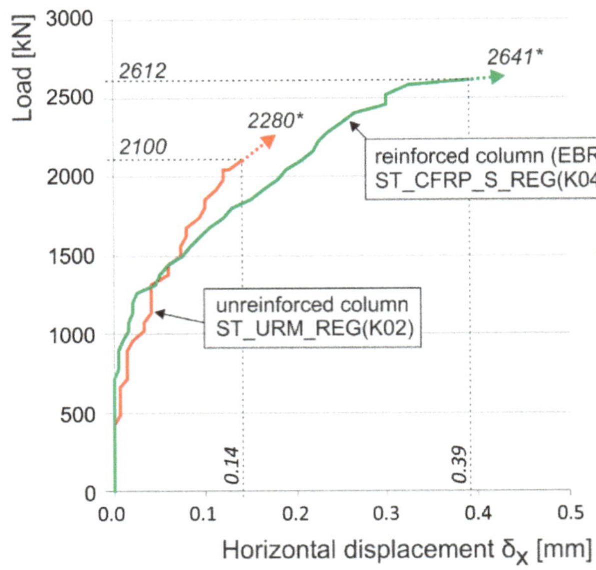

Figure 22. Experimentally obtained load displacement diagrams of unreinforced and reinforced (CFRP, EBR) stone masonry columns—regular masonry units. * are the ultimate loads reached during the experimental testing. Arrows indicate that the corresponding displacement values are not measured due to the removal of measuring gauges before the last loading step (possible destruction).

Ultimate Strength of Masonry Reinforced with Composites under Concentric Compressive Load

The analysis of the experimental research results and the failure mechanism of concentrically compressed masonry units indicated that the design procedure of masonry units practiced to-date based solely on the strength of individual masonry components in compression does not correspond to or encompass the interaction between individual masonry components and the masonry failure mechanism under concentric compressive load. The main drawback to calculation formulae may be the fact that standards (ČSN EN 1996-1-1 [30]) completely ignore or insufficiently consider the tensile strength of masonry components, which is the major agent in the appearance and development of tensile cracks. The full exploitation of tensile strength, masonry, or its individual components may be assumed in cases of coursed masonry with regular masonry units and a bed of low (≤15 mm) joint filler with approximately the same mechanical and displacement properties, and a good bond:

- In the case of columns of regular, coursed rubble masonry with dressed stone and thin bed joints reinforced by composite strips, the experimental research demonstrated that the tensile "capacity" of the composite was not efficiently exploited and, due to this, their ultimate load does not significantly increase. In the case of the above masonry, transverse displacements and forced transverse stresses due to the effect of passive wrapping are relatively very low as a consequence of the "mortar bed–masonry unit" lower intensity interaction.
- The ultimate load of a reinforced stonework column under concentric compressive load hardly differs from the ultimate load of an unreinforced column (see Figure 22). The failure occurs by reaching the ultimate displacement and strain of stone masonry in compression δ_{ym} accompanied by the local failure of masonry in compression—crushing and disintegration of masonry units and the binder. The experimentally measured, relatively low horizontal displacements δ_{xm} also agree with the failure mechanism of stonework columns with regular, dressed blocks and coursed masonry (phase II). This fact must be considered when calculating the compressive strength of stone masonry reinforced with FRP fabrics, in which the above described failure mechanism under compressive loading applies. Also, the relatively slight effect achieved by wrapping a column in fabrics of high-strength carbon or glass fibers may be objectively assumed.
- In the case of coursed or irregular stone masonry (column) with undressed or roughly dressed stone blocks and larger bed joints (>20–25 mm) under concentric compressive load, the application of the failure mechanism characterized by the appearance and development of tensile cracks running approximately in the direction of compression trajectories—phase I—may be assumed. It is usually accompanied by the gradual growth in transverse horizontal displacements δ_x in stone column masonry, together with a significant redistribution of compressive loading over the masonry cross section. In the case of an efficient "masonry–fabric (composite)" interaction, the ultimate load of stone masonry in compression dramatically rises compared to the ultimate load of unreinforced stone masonry (Figure 23) due to the effect of reinforcement by fabrics applied as wrapping bands (EBR) to the masonry or as strips inserted into horizontal joints (NSM).

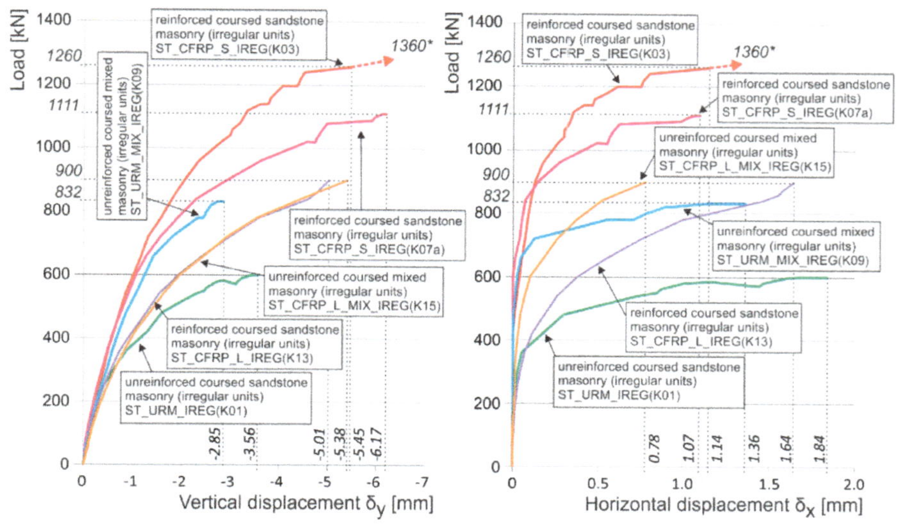

Figure 23. Experimentally obtained load displacement diagrams of unreinforced and reinforced (CFRP, EBR and NSM) stone masonry columns made from irregular masonry blocks. * are the ultimate loads reached during the experimental testing. Arrows indicate that the corresponding displacement values are not measured due to the removal of measuring gauges before the last loading step (possible destruction).

CONCLUSIONS

The failure mechanism of compressed masonry structures is dramatically affected by the size and shape of masonry units, their mechanical characteristics and, last but not least, the masonry bond. For this reason, when calculating a reinforced masonry column's ultimate load, specific cases of brick masonry structures, but mainly stone and historic structures, must be considered separately and the masonry layout (size of masonry units, bond, bed joint thickness, *etc.*) taken into account. Apart from accurately determining the mechanical properties of individual masonry components, a reliable specification of the theoretical value of masonry ultimate load requires respect towards the masonry failure mechanism. The stabilizing and reinforcing effect of fabrics based on inorganic fibers significantly applies to masonry structures (columns, walls) under concentric compressive load whose failure mechanism is characterized by the appearance and development of vertical tensile cracks accompanied by a growth in horizontal displacements of masonry. This effect is particularly pronounced in masonry where the masonry unit–mortar interaction applies (clay brick masonry with lime mortar) and in masonry with a stronger tendency to transverse deformation (irregular stone masonry with thicker mortar joints). To a lesser extent, this effect can be expected in masonry with a thi0n or zero bed joint or in masonry where mortar has the same characteristics as masonry units (cement 00mortar, *etc.*).

ACKNOWLEDGMENTS

The paper was written with the support of the Czech Ministry of Culture's research project NAKI DF12P01OVV037 "Progressive non-invasive methods of the stabilization, conservation and strengthening of historical structures and their parts with fiber- and nanofibre-based composite materials".

AUTHOR CONTRIBUTIONS

Jiří Witzany was the main researcher of the NAKI project who supported all the experimental and theoretical analysis as well as the applications performed. He developed the experimental campaign and theoretical analysis and contributed to the evaluation of results. Radek Zigler participated in the experimental campaign and contributed to the evaluation of experimental and theoretical results. He also made the literature review, prepared the images, and put together the final form of the paper.

Conflicts of Interest

The authors declare no conflict of interest.

REFERENCES

1. Saadatmanesh, H. Extending service life of concrete and masonry structures with fiber composites. *Constr. Build. Mater.* **1997**, *11*, 327–335.
2. Doran, B.; Koksal, H.O.; Turgay, T. Nonlinear finite element modeling of rectangular/square concrete columns confined with FRP. *Mater. Des.* **2009**, *30*, 3066–3075.
3. Wu, Y.F.; Jiang, C. Effect of load eccentricity on the stress–strain relationship of FRP-confined concrete columns.*Compos. Struct.* **2012**, *98*, 228–241.
4. Corradi, M.; Borri, A.; Vignoli, A. Strengthening techniques tested on masonry structures struck by the Umbrian-Marche earthquake of 1997–1998. *Constr. Build. Mater.* **2002**, *16*, 229–239.
5. Zhao, T.; Zhang, C.J.; Xie, J. Experimental study on earthquake strengthening of brick walls with continuous carbon fiber sheet. *Mason. Int.* **2003**, *16*, 21–25.
6. Shrive, N.G. The use of fiber reinforced polymers to improve seismic resistance of masonry. *Constr. Build. Mater.* **2006**,*20*, 269–277.
7. Eslami, A.; Ronagh, H.R. Effect of FRP wrapping in seismic performance of RC buildings with and without special detailing—A case study. *Compos. Part B* **2013**, *45*, 1265–1274.
8. Bruggi, M.; Milani, G.; Taliercio, A. Design of the optimal fiber-reinforcement for masonry structures via

topology optimization. *Int. J. Solids Struct.* **2013**, *50*, 2087–2106.

9. Luccioni, B.; Rougier, V.C. In-plane retrofitting of masonry panels with fiber reinforced composite materials. *Constr. Build. Mater.* **2011**, *25*, 1772–1788.

10. Grande, E.; Imbimbo, M.; Sacco, E. Finite element analysis of masonry panels strengthened with FRPs. *Compos. Part B***2013**, *45*, 1296–1309.

11. Corradi, M.; Grazini, A.; Borri, A. Confinement of brick masonry columns with CFRP materials. *Compos. Sci. Technol.***2007**, *67*, 1772–1783.

12. Faella, C.; Martinelli, E.; Paciello, S.; Camorani, G.; Aiello, M.A.; Micelli, F.; Nigro, E. Masonry columns confined by composite materials: Experimental investigation. *Compos. Part B* **2011**, *42*, 692–704.

13. Krevaikas, T.D.; Triantafillou, T. Masonry confinement with fiber-reinforced polymers. *J. Compos. Constr.* **2005**, *9*, 128–135.

14. Micelli, F.; de Lorenzis, L.; La Tegola, A. FRP-confined masonry columns under axial loads: Analytical model and experimental results. *Mason. Int. J.* **2004**, *17*, 95–108.

15. Aiello, M.A.; Micelli, F.; Valente, L. Structural upgrading of masonry columns by using composite reinforcements. *J. Compos. Constr.* **2007**, *11*, 650–658.

16. Marcari, G.; Oliveira, D.V.; Fabbrocino, G.; Lourenço, P.B. Shear capacity assessment of tuff panels strengthened with FRP diagonal layout. *Compos. Part B* **2011**, *42*, 1956–1965.

17. Di Ludovico, M.; D'Ambra, C.; Prota, A.; Manfredi, G. FRP confinement of tuff and clay brick columns: experimental study and assessment of analytical models. *ASCE J. Compos. Constr.* **2010**, *14*, 583–596.

18. Micelli, F.; Di Ludovico, M.; Balsamo, A.; Manfredi, G. Mechanical behavior of FRP-confined masonry by testing of full-scale columns. *Mater. Struct.* **2014**, *47*, 2081–2100.

19. Camli, U.S.; Binici, B. Strength of carbon fiber reinforced polymers bonded to concrete and masonry. *Constr. Build. Mater.* **2007**, *21*, 1431–1446.

20. Faella, C.; Camorani, G.; Martinelli, E.; Paciello, S.; Perri, F. Bond behavior of FRP strips glued on masonry: Experimental investigation and empirical formulation. *Constr. Build. Mater.* **2013**, *31*, 353–363.

21. Garbin, E.; Panizza, M.; Valluzzi, M.R. Experimental assessment of bond behavior of fiber-reinforced polymers on brick masonry. *Struct. Eng. Int.* **2010**, *20*, 392–399.

22. Carrara, P.; Ferretti, D.; Freddi, F. Debonding behavior of ancient masonry elements strengthened with CFRP sheets.*Compos. Part B* **2013**, *45*, 800–810.

23. Grande, E.; Imbimbo, M.; Sacco, E. Bond behavior of CFRP laminates glued on clay bricks: Experimental and numerical study. *Compos. Part B* **2011**, *42*, 330–340.

24. Fedele, R.; Milani, G. Three-dimensional effects induced by FRP-from-masonry delamination. *Compos. Struct.* **2011**, *93*, 1819–1831.

25. Capozucca, R. Experimental FRP/SRP–historic masonry delamination. *Compos. Struct.* **2010**, *92*, 891–903.

26. Capozucca, R. Effects of mortar layers in the delamination of GFRP bonded to historic masonry. *Compos. Part B* **2013**,*44*, 639–649.

27. Ghiassi, B.; Oliveira, D.V.; Lourenço, P.B.; Marcari, G. Numerical study of the role of mortar joints in the bond behavior of FRP-strengthened masonry. *Compos. Part B* **2013**, *46*, 21–30.

28. Kashyap, J.; Willis, C.R.; Griffith, M.C.; Inghamb, J.M.; Masia, M.J. Debonding resistance of FRP-to-clay brick masonry joints. *Eng. Struct.* **2012**, *41*, 186–198.

29. Kwiecien, A. Stiff and flexible adhesives bonding CFRP to masonry substrates—Investigated in pull-off test and Single-Lap test. *Arch. Civ. Mech. Eng.* **2012**, *12*, 228–239.

30. ČSN EN 1996–1-1 Eurocode 6: Design of Masonry Structures—Part 1–1: General Rules for Reinforced and Unreinforced Masonry Structures; Czech Office for Standards, Metrology and Testing (UNMZ): Prague, Czech Republic, 2013.

31. ČSN 731101 Design of Masonry Structures (Change B); Czech Office for Standards, Metrology and Testing (UNMZ): Prague, Czech Republic, 2013.

32. CNR-DT 200 R1/2013. *Guide for the Design and Construction of Externally Bonded FRP Systems for Strengthening Existing Structures, Revision 1*; Italian Council of Research (CNR): Rome, Italy, 2014.

33. ISO 13822:2010 Bases for Design of Structures—Assessment of Existing Structures; International Organization for Standardization: Geneva, Switzerland, 2010.

34. Abaqus Unified FEA—Simulia Package. Available online: https://www.3ds.com/fileadmin/PRODUCTS/SIMULIA/PDF/brochures/simulia-abaqus-unified-fea-brochure.pdf (accessed on 15 August 2015).

35. FEAT2000; version 3; SMARTsoft s.r.o.: Prague, Czech Republic, 2000.

CHAPTER 3

MECHANICAL AND PERMEABILITY CHARACTERISTICS OF LATEX-MODIFIED PRE-PACKED PAVEMENT REPAIR CONCRETE AS A FUNCTION OF THE RAPID-SET BINDER CONTENT

Jae-Woong Han, Ji-Hong Jeon and Chan-Gi Park

Keywords: Durability, mechanical properties, latex polymer, pavement repair concrete, rapid-set cement binder

INTRODUCTION

There is growing demand for repairs to paving concrete and to minimize disruption to traffic the road should be re-opened as soon as possible [1,2,3,4,5]. There has therefore been much interest in design standards for materials used in emergency concrete repairs [6,7,8,9,10], and rapid-set cement concrete is the most commonly used material for this purpose [1,2,3,4,5]. When rapid-set cement is used, the large hydration heat increases the rate of vapor migration and may lead to high temperatures within the concrete itself. In practice, vapor migration may be restricted, which can lead to large tensile stresses and the formation of minute cracks [5,7,8,9,10]. These cracks increase the permeability of the pavement, leading to various forms of damage [7,8,9,10]. For these reasons, the structural integrity of the pavement may be compromised, requiring further repair [5,6,7,8,9,10]. The addition of a latex polymer can increase the workability of unhardened concrete and reduce the required ratio of water to cement [7,10]. This can also prevent separation of materials due to the viscosity of the latex polymer itself, which increases the flexural strength, tensile strength, and deformation capacity following hardening. This in turn improves the adhesion, water-tightness, resistance to freezing–thaw erosion, and improves the abrasion resistance and chemical resistance of the material [7,8,9,10].

Existing rapid-set cement latex-modified concretes are typically cast by mixing coarse aggregate, sand, cement, and water. Therefore, late-modified concrete is difficult to apply without a mobile mixer truck.

Here, this study used silica sand (*i.e.*, river sand) to replace the fine aggregates. A bonding material was applied to the pre-mixed rapid-set cement and silica sand. Pre-packed concrete is a construction method whereby a bonding material is injected into the coarse aggregates. Pre-packed concrete can be applied with a rapid set binder without requiring a mobile mixer.

This study investigated the performance of latex-polymer-modified, pre-packed pavement repair concrete (LMPPRC) incorporating a rapid-set binder. The LMPPRC construction method was as follows. The pavement to be repaired was removed and coarse aggregates were applied. The rapid-set binder material was mixed with a latex polymer and the resulting mix was poured onto the concrete between the coarse aggregates using a vibration compaction method. The mixture of latex polymer and rapid-set binder was adjusted

Citation: Jae-Woong Han, Ji-Hong Jeon and Chan-Gi Park, Mechanical and Permeability Characteristics of Latex-Modified Pre-Packed Pavement Repair Concrete as a Function of the Rapid-Set Binder Content, doi:10.3390/ma8105339.

to have sufficient fluidity to exhibit self-leveling behavior. Following infusion, the surface was finished to complete the repair. By mixing the latex polymer and the rapid-set binder in this manner, concrete repairs using LMPPRC can provide the required strength in a short time, with sufficient workability and durability. We evaluated the strength, permeability, and impact resistance of several LMPPRC formulations.

MATERIALS

Materials

The rapid-set binder material used in this work was a mix of rapid-set cement and silica sands manufactured by Jung-Ang Polytec in Yangsan, Korea. The mixing ratio of rapid-set cement to silica sands was 1:1. Table 1 lists the chemical composition of the cement. Number 6 silica sand density, with a density of 2.65 g/mm^3, was used. The crushed coarse aggregates had a maximum size of 25 mm and a density of 2.62 g/mm^3. Table 2 lists the physical properties of the coarse aggregates. Table 3 lists the properties of the styrene butadiene (SB) latex polymer (Jung-Ang Polytec, Yangsan, Korea).

Table 1. Chemical compositions of the rapid set cement (Jung-Ang Polytec, Yangsan, Korea).

SiO_2 (%)	Al_2O_3 (%)	Fe_2O_3 (%)	CaO (%)	MgO (%)	K_2O (%)	SO_3 (%)
13 ± 3	17.5 ± 3	>3	50 ± 3	>2.5	0.21	14 ± 3

Table 2. Physical properties of the coarse aggregates.

Properties	Density (g/mm^3)	Absorption (%)	Fineness Modulus
Value	2.6	0.35	6.92

Table 3. Properties of the styrene butadiene latex.

Solids Content (%)	Styrene Content (%)	Butadiene Content (%)	Surface Tension (dyne/cm)	Particle Size (A)	Viscosity (cps)
8	34 ± 1.5	66 ± 1.5	30.57	1700	42

Mix Proportions

With concrete pavement repairs using the rapid-set cement, both the American Association of State Highway and Transportation Officials (AASHTO) and the Korean Highway Corporation require a maximum curing period of 4 h prior to re-opening the road to traffic [11,12]. The standards for traffic re-opening are a compressive strength of ≥21 MPa and a flexural strength of ≥3.5 MPa. Furthermore, after 28 days of curing, the compressive strength should be ≥35 MPa, the flexural strength should be ≥4.5 MPa, and the splitting tensile strength should be ≥4.2 MPa. Here we focused on concrete repairs using rapid-set cement, with the goal of achieving a compressive strength of ≥21 MPa and a flexural strength of ≥3.5 MPa after 4 h of curing, and a compressive strength of ≥35 MPa, flexural strength of ≥4.5 MPa, and a splitting tensile strength of ≥4.2 MPa after 28 days of curing. The permeability has a significant impact on the lifespan and durability of concrete pavement and we aimed for a permeability of ≤2000 C after 7 days of curing, which is recommended by the Korean Highway Corporation [12]. Chloride ion permeation tests were performed according to ASTM C1202 [13]. Table 4 lists the mixing ratios used in this work.

Preparation of Test Specimens

Cylindrical specimens were formed by packing coarse aggregates into a cylindrical mold. The rapid-set cement and silica sands were mixed together, after which the latex polymer was added and mixed for 1 min and 30 s. Following mixing, the cylindrical specimen was placed on a vibration table and the rapid set binder materials were poured into the mix while the table was vibrating. Rectangular specimens were produced by packing the coarse aggregates into a rectangular frame. The rapid-set cement and silica sands were mixed together and the latex polymer was poured into the rapid-set binder material and mixed for one minute and 30 s. Following mixing, the rectangular specimens packed with coarse aggregates were placed on a vibration table, and the rapid set binder materials were poured into it while the table was vibrating.

Table 4. Mix proportions of the latex-polymer-modified, pre-packed pavement repair concrete (LMPPRC).

Type of Mix	Rapid Set Binder Materials (kg/m^3)		Latex Polymer (kg/m^3)	Coarse Aggregate (kg/m^3)
	Rapid Set Cement	Silica Sand		
0.40	287.5	287.5	218	1450
0.33	313.5	313.5	210	1450
0.29	345	345	198	1450
0.25	366	366	184	1450

TEST METHODS

Compressive Strength Tests

Compressive strength tests were carried out in accordance with ASTM C 39 [14]. Tests were performed after curing periods of 4 h, 7 days, and 28 days. The specimens (100 mm × 200 mm) were cured in water at 23 ± 2 °C. Each variable was investigated using six specimens.

Flexural Tests

Flexural tests were carried out in accordance with ASTM C 78 [15]. Tests were carried out after curing periods of 4 h and 28 days. Specimens (100 mm × 100 mm × 400 mm) were cured in water at 23 ± 2 °C. Each variable was investigated using six specimens.

Splitting Tensile Tests

Splitting tensile tests were conducted in accordance with the ASTM C 496 standard [16]. Tests were carried out after curing periods of 4 h, 7 days, and 28 days. Specimens (100 mm × 200 mm) were cured in water at 23 ± 2 °C. Each variable was investigated using six specimens.

Chloride Ion Penetration Tests

Chloride ion penetration tests were conducted in accordance with ASTM C 1202 [13]. Specimens (150 mm × 50 mm) were cured in water at 23 ± 2 °C. Tests were carried out after curing periods of 4 h, 7 days, and 28 days. Each variable was investigated using six specimens. The test apparatus for the chloride ion penetration tests is shown in Figure 1a.

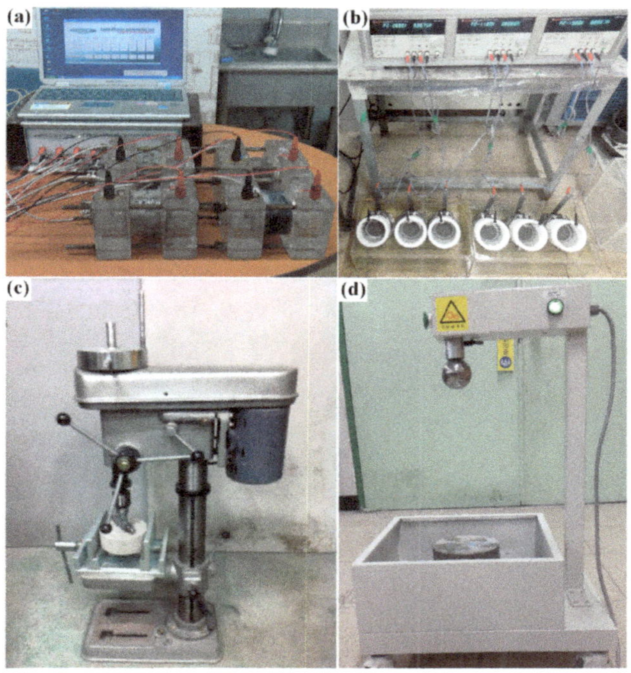

Figure 1. (**a**) Chloride ion penetration test set-up; (**b**) chloride ion diffusion test set-up; (**c**) abrasion test set-up; (**d**) impact test set-up.

Chloride Ion Diffusion Tests

To investigate the diffusion characteristics of chloride ions, diffusion tests using fast electrophoresis were carried out after 28 days of curing using the NT Build 492 test standard of Northern Europe (NT Build 492, 1999) [17]. Specimens (100 mm × 50 mm) were cured in water at 23 ± 2 °C. Each variable was investigated using six specimens. Figure 1b shows the apparatus used to measure the chloride ion diffusion coefficient.

Abrasion Tests

Abrasion tests were carried out in accordance with ASTM C 944 [18]. Specimens (150 mm × 60 mm) were tested after curing for 28 days in water at 23 ± 2 °C. Each variable was investigated using six specimens. The test apparatus for the abrasion test is shown in Figure 1c.

Impact Tests

Impact testing was carried out in accordance with the specifications of ACI Committee 544 [19]. Specimens (150 mm × 60 mm) were cured in water at 23 ± 2 °C. Tests were carried out after a curing period of 28 days. Each variable was investigated using six specimens. Figure 1d shows the test set-up.

RESULTS AND DISCUSSION

Compressive Strength

The compressive strength of LMPPRC for emergency concrete repairs is shown in Figure 2. As the mixing ratio of latex polymer and rapid-set binder decreased, the compressive strength increased. When the ratio of latex polymer and rapid-set binder material was ≤0.29, the compressive strength satisfied the standard of 21 MPa after 4 h. The required compressive strength of 35 MPa was satisfied after curing for 28 days with a mixing ratio of ≤0.29. It appears that the presence of the latex polymer retarded the manifestation of early strength, leading to a failure to satisfy the standard. Furthermore, when the quantity of rapid-set binder was small, sufficient early strength could not be obtained. Therefore, to satisfy both goals of retarding the appearance of strength using latex polymer and to acquire sufficient early strength, it was necessary to increase the quantity of rapid-set binder. When the ratio of latex polymer to rapid-set binder was ≤0.29, the mix was able to satisfy the standards for emergency concrete repair.

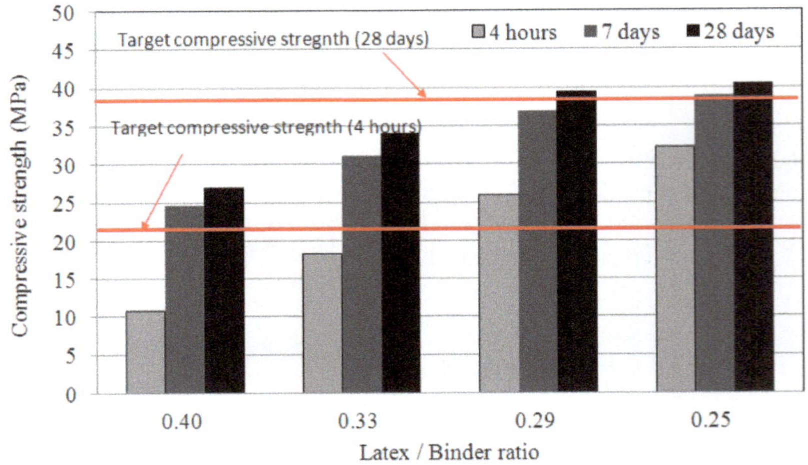

Figure 2. Compressive strength of latex-polymer-modified, pre-packed pavement repair concrete (LMPPRC).

Flexural Strength

Figure 3 shows the results of flexural strength tests for the LMPPRC for emergency repairs. The standard for flexural strength was 3.5 MPa after 4 h (for traffic re-opening) and 4.5 MPa after 28 days. When the mixing ratio of latex polymer to rapid-set binder material was ≤0.33, the concrete satisfied 3.5 MPa after 4 h and 4.5 MPa after 28 days. This shows that the addition of latex polymer to concrete had more influence on the tensile strength and flexural strength than on the compressive strength. The presence of the latex polymer increases the binding force between materials when the concrete is subjected to flexural or tensile stresses [7,8,10] Therefore (unlike the compressive strength), the flexural strength was sufficiently large to satisfy the standards for emergency concrete repair even when the ratio of latex polymer to rapid-set binder material was 0.33. Furthermore, the flexural strength increased as the ratio of latex polymer to rapid-set binder decreased.

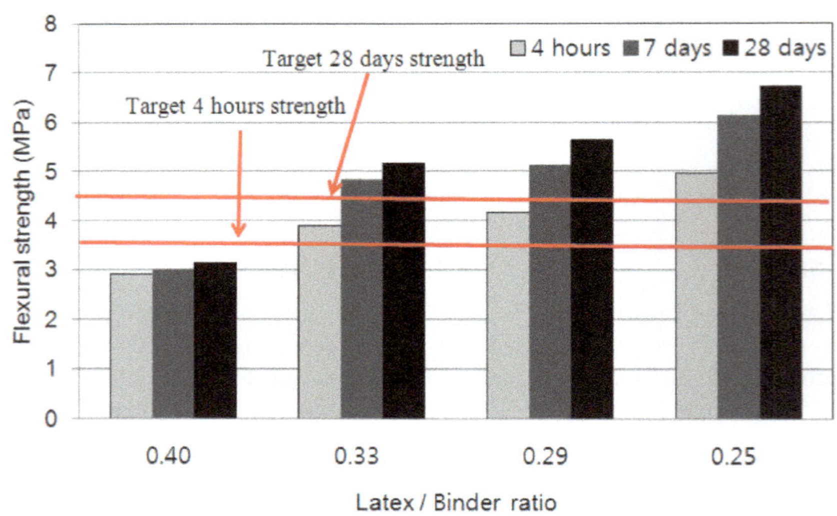

Figure 3. Flexural strength of LMPPRC.

Splitting Tensile Strength

The standard for splitting tensile strength of LMPPRC was 4.2 MPa after 28 curing for days, as defined by the Korean Highway Corporation. Splitting tensile strength increased as the ratio of latex polymer to rapid-set binder decreased (seeFigure 4). A tensile strength of 4.2 MPa after curing for 28 days was satisfied when the mixing ratio was ≤0.29, but not with mixing ratios of 0.40 and 0.33. This shows that the addition of the latex polymer to concrete had more influence on the tensile strength than on the compressive strength. The latex polymer effectively increased the binding force between the materials when the concrete was subjected to tensile stress, which enhanced the tensile strength. In general, the tensile strength of concrete is lower than its flexural strength, which is consistent with the findings of this work.

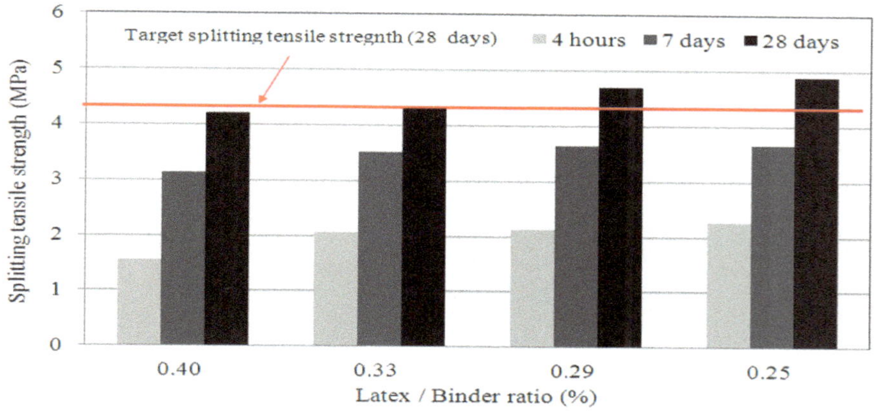

Figure 4. Splitting tensile strength of LMPPRC.

Chloride Ion Penetration

Figure 5 shows the chloride ion permeability resistance of LMPPRC. The permeability standard was ASTM C 1202 (seeTable 5) [17]. Chloride ion permeability decreased as the ratio of latex polymer to rapid-set binder decreased. Comparing these data with the ASTM C 1202 standard, we can see that the chloride ion permeability after 4 h was "high" for latex polymer/rapid-set binder ratios of 0.40 and 0.33 (*i.e.*, ≥4000 C). With latex polymer/rapid-set binder ratios of 0.29 and 0.25, the chloride ion permeability was "moderate" (*i.e.*, 2000–4000 C). After 7 days of curing, the chloride ion permeability was "low" for a latex polymer/rapid-set binder ratio of 0.40 (*i.e.*, 1000–2000 C), whereas for latex polymer/rapid-set binder ratios of 0.33, 0.29, and 0.25, the permeability was "very low" (*i.e.*, ≤1000 C). After 28 days of curing, all the specimens exhibited a "very low" chloride permeability (*i.e.*, 100–1000 C). After 7 days of curing, the target chloride ion permeability resistance of 2000 C was achieved with all the ratios of latex polymer/rapid-set binder.

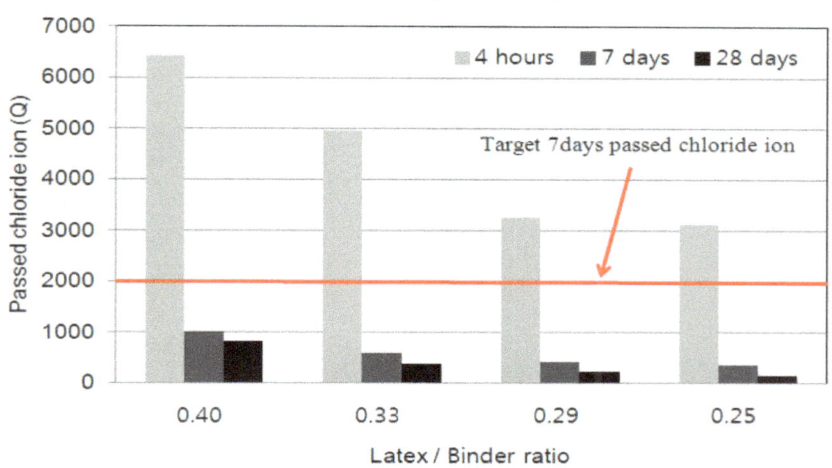

Figure 5. Chloride ion penetration test results of LMPPRC.

The addition of latex polymer reduced the permeability of chloride ions by filling pores and forming a thick latex polymer network in the concrete. Therefore, all the specimens exhibited excellent permeability resistance. Furthermore, an increase in strength generally reduces chloride ion permeability [7,8,10]. However, after curing for 4 h, the strength was insufficient. Moreover, internal cracking was initiated during the preparation of samples for chloride ion penetration tests. These cracks propagated, leading to an increase in the permeability of the LMPPRC.

Table 5. The ASTM C 1202 chloride ion permeability classifications.

Charge Passed (C)	Chloride Permeability
>4000	High
2000–4000	Moderate
1000–2000	Low
100–1000	Very low
<100	Negligible

Chloride Ion Diffusion

Chloride ion diffusion coefficients were obtained from the results of fast electrophoresis tests using the model of Tang and Nilsson (1992) [20]. Figure 6 shows the chloride ion diffusion coefficient of various LMPPRCs. The diffusion coefficient decreased as the ratio of latex polymer to rapid-set binder decreased. The chloride ion diffusion tests exhibited results similar to those of the chloride ion permeability tests. As the quantity of rapid set binder increased, the strength of the LMPPRC increased, which was attributed to a reduction in the number of internal pores. Furthermore, the increase in strength reduced permeability through improvements in the microstructure of the concrete, which appear to have led to decreases in the chloride ion diffusion coefficients.

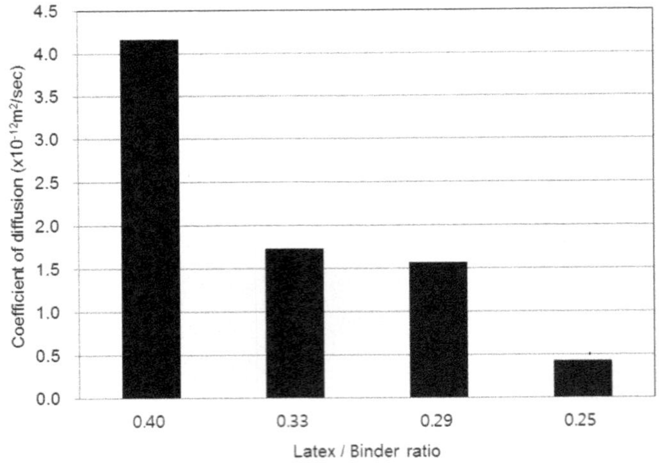

Figure 6. Chloride ion diffusion test results of LMPPRC.

Abrasion Resistance

Figure 7 shows the abrasion resistance of the LMPPRCs with varying quantities of rapid set binder for use in concrete pavement repair. Abrasion resistance increased as the quantity of rapid set binder increased. This was attributed to an increase in the strength of the surface. Additionally, the bond between materials was improved as the quantity of rapid-set binder increased. Furthermore, in increase in the quantity of binder resulted in an increase in the stiffness of the concrete, which increased the resistance to abrasion.

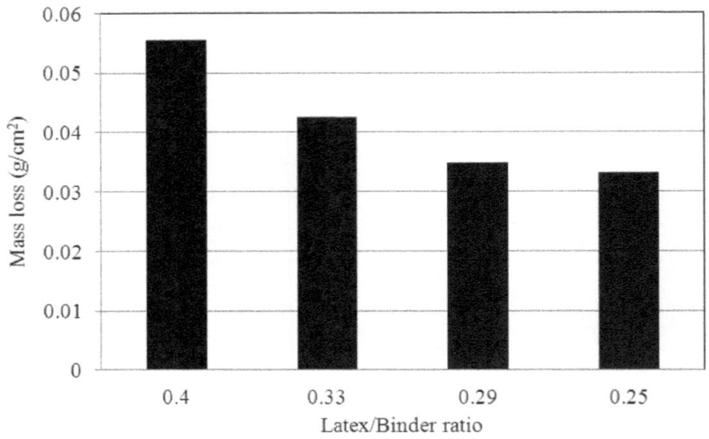

Figure 7. The results of abrasion resistance tests for LMPPRC.

Impact Resistance

Figure 8 shows the measured impact resistance of the LMPPRC for applications in the repair of concrete pavement. The number of impacts prior to the first occurrence of cracking increased as the quantity of rapid set binder increased, as did the number of impacts prior to failure. This is because the rapid set binder improved both the strength of the LMPPRC and the integrity of the bonding interface between materials, which inhibited crack formation and propagation.

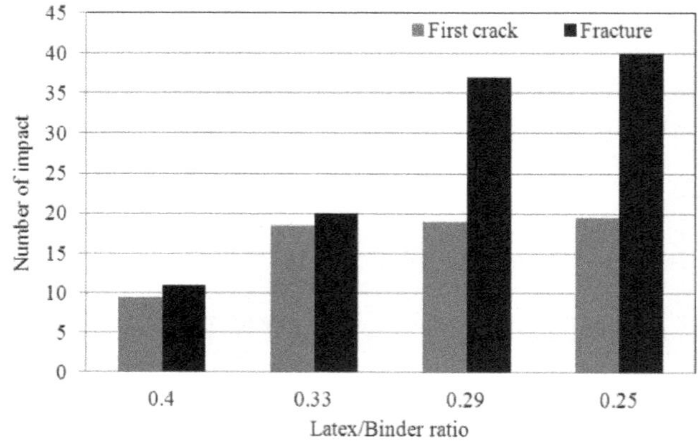

Figure 8. The results of impact tests for the LMPPRC.

CONCLUSIONS

In this study, we evaluated the strength and permeability characteristics of LMPPRCs with various ratios of latex polymer to rapid-set binder. The main results of this work can be summarized as follows.

When the ratio of latex polymer to rapid-set binder was ≤0.29, the LMPPRC exhibited a compressive strength of ≥21 MPa after curing for 4 h, and of ≥35 MPa after curing for 28 days. This satisfied the design standard for emergency repair concrete.

Flexural strength tests showed that the mixing ratio of latex polymer to rapid-set binder material should be ≤0.33 to provide a flexural strength of ≥3.5 MPa after 4 h and a flexural strength of ≥4.5 MPa after 28 days.

Splitting tensile strength tests showed that the ratio of latex polymer to rapid-set binder material should be ≤0.29 to provide a tensile strength of ≥4.2 MPa after 28 days.

Chloride ion permeability tests showed that all of the evaluated ratios of latex polymer to rapid-set binder satisfied the standard of 2000 C. As the ratio of latex polymer to rapid-set binder decreased, the permeability of chloride ions decreased.

These data show that diffusion coefficients decreased, impact resistance increased, and abrasion resistance increased as the ratio of latex polymer to rapid-set binder decreased. This was attributed to an increase in strength of the LMPPRC due to the presence of the rapid-set binder, which was in turn attributed to a decrease of the number of internal pores in the concrete. Increasing the quantity of rapid-set binder reduced the number of pores in the concrete, thereby reducing the chloride ion diffusion coefficient and increasing the impact resistance while reducing the rate of abrasion.

ACKNOWLEDGMENTS

This work was supported by the research grant of the Kongju National University in 2014.

AUTHOR CONTRIBUTIONS

Jae-Woong Han conceived and designed the experiments; Ji-Hong Jeon reviewed the references and provided useful comments; Chan-Gi Park analyzed the data and wrote the paper. All authors have read and approved the final manuscript.

Conflicts of Interest

The authors declare no conflict of interest.

REFERENCES

1. Sprinkel, M. *Latex-Modified Concrete Overlay Containing Type K Cement*; FHWA/VTRC 05-R26; Virginia Transportation Research Council: Charlottesville, VA, USA, 2005.
2. Sprinkel, M. *High Early Strength Latex Modified Concrete Overlay*; Report VTRC 88-R12; Virginia Transportation Research Council: Charlottesville, VA, USA, 1998.
3. Buch, N.; van Dam, T.J.; Peterson, K.; Sutter, L. Evaluation of high-early strength PCC mixtures used in full depth repairs. *Constr. Build. Mater.* **2008**, *22*, 162–174.
4. Won, J.P.; Kim, J.M.; Lee, S.J.; Lee, S.W.; Park, S.K. Mix proportion of high-strength, roller-compacted, latex-modified rapid-set concrete for rapid road repair. *Constr. Build. Mater.* **2011**, *25*, 1796–1800.
5. Won, J.P.; Kim, J.H.; Park, C.G.; Kang, J.W. Shrinkage cracking of styrene butadiene polymeric emulsion-modified concrete using rapid-hardening cement. *J. Appl. Polym. Sci.* **2009**, *112*, 2229–2234.
6. Zuniga, J.R. Development of Rapid, Cement-based Repair Materials for Transportation Structures. Master's Thesis, The University of Texas at Austin, Austin, TX, USA, May 2013.
7. Kim, D.H.; Park, C.G. Permeability, abrasion, and impact resistance of latex-modified fibre reinforced concrete for precast concrete pavement application. *Prog. Rubber. Plast. Recycl. Technol.* **2013**, *29*, 239–254.
8. Oh, R.O.; Kim, D.H.; Park, C.G. Durability performance of latex modified nylon fiber reinforced concrete for precast concrete pavement applications. *Indian J. Eng. Mater. Sci.* **2014**, *21*, 49–56.
9. Yun, K.K.; Choi, P. Causes and controls of cracking at bridge deck overlay with very-early strength latex-modified concrete. *Constr. Build. Mater.* **2014**, *56*, 53–62.
10. Kim, D.H.; Park, C.G. Strength, permeability and durability of hybrid fiber-reinforced concrete containing styrene butadiene latex. *J. Appl. Polym. Sci.* **2013**, *129*, 1499–1505.
11. American Association of State Highway and Transportation Officials (AASHTO). *Standard Specification of Rigid Pavement*; AASHTO: Washington, DC, USA, 1998.
12. Korea Expressway Corporation. *Construction Material Quality and Standard for Highway Construction*; Korea Expressway Corporation: Seongnam, Korea, 2005.
13. American Society for Testing and Materials (ASTM). *Standard Test Method for Electrical Indication of Concrete's Ability to Resist Chloride Ion Penetration*; ASTM C1202; American Society for Testing and Materials: Philadelphia, PA, USA, 2012.
14. American Society for Testing and Materials (ASTM). *Standard Test Method for Compressive Strength of Cylindrical Concrete Specimens*; ASTM C39; American Society for Testing and Materials: Philadelphia, PA, USA, 2015.
15. American Society for Testing and Materials (ASTM). *Standard Test Method for Flexural Strength of Concrete (Using Simple Beam with Third-Point Loading)*; ASTM C78/C78M; American Society for Testing and Materials: Philadelphia, PA, USA, 2015.
16. American Society for Testing and Materials (ASTM). *Standard Test Method for Splitting Tensile Strength of Cylindrical Concrete Specimens*; ASTM C496/C496M; American Society for Testing and Materials: Philadelphia, PA, USA, 2011.
17. NordTest Build 492. *Concrete, Mortar and Cement-*

Based Repair Materials: Chloride Migration Coefficient from Non-Steady-State Migration Experiments; NordTest Method; NordTest: Espoo, Finland, 1999.

18. ACI Committee 544. Measurement of properties of fiber reinforced concrete. *ACI Mater. J.* **1998**, *85*, 583–593.

19. American Society for Testing and Materials (ASTM). *Standard Test Method for Abrasion Resistance of Concrete or Mortar Surfaces by the Rotating-Cutter Method*; ASTM C944; ASTM: Philadelphia, PA, USA, 2012.

20. Tang, L.; Nilsson, L.O. Chloride diffusivity in high strength concrete at different ages. *Nord. Concr. Res.* **1992**, *11*, 162–171.

CHAPTER 4

MATERIAL PERFORMANCE AND ANIMAL CLINICAL STUDIES ON PERFORMANCE-OPTIMIZED HWANGTOH MIXED MORTAR AND CONCRETE TO EVALUATE THEIR MECHANICAL PROPERTIES AND HEALTH BENEFITS

Bon-Min Koo, Jang-Ho Jay Kim, Tae-Kyun Kim and Byung-Yun Kim

Keywords: Hwangtoh, health benefits, Institute for Cancer Research (ICR), material properties, mix proportion design

INTRODUCTION

A recent study by the city of Seoul, South Korea, stated that foreign and domestic ordinary Portland cement (OPC) products emitted 4.14 ppm and 13 ppm, respectively, of poisonous hexavalent chromium Cr(VI). Thus, domestic OPC is 3.25 times more hazardous than foreign OPC [1]. Since 2009 in Korea, the maximum domestic and international allowable amount of Cr(VI) in OPC cement is 20 mg/kg and 2 mg/kg, respectively. It has been reported that Hwangtoh (red clay) does not contain Cr(VI) substance [2]. Such concerns have led to a rise in the use of activated Hwangtoh as an eco-friendly construction material. As a result, much research has been conducted on Hwangtoh-mixed construction materials. However, most of it has focused on differences in the material and structural performance capacities of OPC and Hwangtoh; unfortunately, no study has yet quantified the health benefits of Hwangtoh-mixed construction materials. Published reports have considered the activation method of kaolin and activated kaolin-mixed mortar concrete as well as the development of Hwangtoh admixtures for cement mortars [3,4,5,6]. A Hwangtoh cement mortar study analyzed the effect of various firing temperatures on the compressive strength of Hwangtoh mortars and Hwangtoh concrete, suggesting an optimum firing temperature range of 500 °C–1000 °C. Moreover, the same study reported that Hwangtoh serves as an appropriate cementitious material, because its performance equals or exceeds that of other materials, such as blast furnace slag micropowders and flyash [7,8,9]. Based on mineralogical categorization, Hwangtoh is defined as a Halloysite in the kaolin or China clay group, composed mainly of SiO_2, Al_2O_3, and Fe_2O_3 [3]. Hwangtoh behaves similarly to concrete admixtures and has the pozzolanic reaction when mixed with cement and water [10,11]. To increase the usability of Hwangtoh as a construction material, activated Hwangtoh is produced by burning natural Hwangtoh at temperatures of 550 °C–1000 °C and quickly cooling it before grinding it into powder form [12]. The published studies report that partial replacement of cement with activated Hwangtoh produces Hwangtoh mortar or Hwangtoh concrete with sufficient strength for possible use as a construction material [13]. A study evaluated the workability, compressive strength, inelastic deformation, and hydration heat of concrete mixed with Hwangtoh or blast furnace slag micropowders by analyzing stress–strain relations, flexural behavior, dry shrinkage, and the creep characteristics

Citation: Bon-Min Koo, Jang-Ho Jay Kim, Tae-Kyun Kim and Byung-Yun Kim, Material Performance and Animal Clinical Studies on Performance-Optimized Hwangtoh Mixed Mortar and Concrete to Evaluate Their Mechanical Properties and Health Benefits, doi:10.3390/ma8095306.

of reinforced concrete beams mixed with Hwangtoh or blast furnace slag [12,14]. According to that study, the long-term behaviors of concrete mixed with Hwangtoh or blast furnace slag, such as creep and shrinkage, vary depending on the choice of cementitious material. The dry shrinkage increases when Hwangtoh alone is used, but decreases when both Hwangtoh and blast furnace slag are used. Furthermore, an examination of Hwangtoh mortars without cement revealed that Hwangtoh increased drying shrinkage by 20% [15]. A study of the cytotoxicity of recycled OPC mortar powder reported that the tested powders exhibit a strong and concentration-dependent cytotoxicity of 75.5% [16]. A study assessed the effect of grouting materials on pollution by examining the fish toxicity of those materials [17]. Our review of the literature showed that most studies have focused on structure and durability issues that arise from adding a Hwangtoh admixture, without consideration for any potential health benefits from its use. Meanwhile, in Japan, vigorous research has been conducted on the harmful properties of construction materials such as concrete, metal, and timber. A study found that the survival rates of experimental subjects were 7%, 41%, and 81% for concrete, metal, and timber, respectively, showing that concrete had the lowest survival rate [18]. Thus, in the present study, a preliminary experiment is conducted to determine the appropriate ratio of Hwangtoh to cement and identified the properties of Hwangtoh mixed concrete. Then, a human health benefit assessment is conducted using ICR mice in cages built with either Hwangtoh mixed mortar or cement mortar [19]. We made one cage with a cement mortar floor and four cages with Hwangtoh mixed mortar floors with varying ratios of Hwangtoh to cement (20%, 40%, 60%, and 80%) and two cages with cement mortar and Hwangtoh cement mortar walls and floor to assess the experimental subjects' choice of living environment. In addition, the subjects' weight change, food intake, water intake, residential environment preference, fertility rate, and aggression are measured to evaluate the human health benefits of the different materials. The ultimate objective of this study was to develop environmentally friendly construction materials.

MATERIAL PROPERTIES OF ACTIVATED HWANGTOH MIXED CONCRETE

Reaction Mechanism

Hwangtoh in its natural state has a very low reactivity. However, when heated at high temperatures and cooled rapidly, it is activated by the preserved crystallization energy and holds latent hydraulic properties that harden under certain conditions. The chemical composition of the Hwangtoh we used is shown in Table 1. When Hwangtoh is activated by heating at 550 °C–950 °C, it transitions into an excited state of high pozzolanic reactivity, but it does not have a hydration reactivity when in contact with water. However, Hwangtoh does initiate hydration under alkaline conditions. Concrete is a natural alkaline material; thus mixing activated Hwangtoh in concrete will induce pozzolanic reactions. The activated Hwangtoh reacts with calcium hydroxide ($Ca(OH)_2$), the hydration product of cement, and acquires latent hydraulic properties that in turn enhance the strength and water resistance of the concrete. Furthermore, mixing activated Hwangtoh into cement blocks reduces the size of internal voids by filling the spaces between the binder particles, which increases the contact area between the binder and the aggregate, reducing overall concrete bleeding and segregation. At room temperature, silicon dioxide (SiO_2), a component of activated Hwangtoh, binds with $Ca(OH)_2$ and produces a stable pozzolanic product that holds the same properties as other pozzolanic materials, such as blast furnace slag and flyash. Activated Hwangtoh is a material that under goes pozzolanic reaction. It has been reported that Hwangtoh in natural and activated state does not and does have pozzolanic activity, respectively [20]. However, it has not been completely proven that activated Hwangtoh has hydration capacity. From a published report on the topic, the measured hydration temperature of concrete mixed with activated Hwangtoh powder and OPC (41.0 °C) was slightly lower than that of 100% OPC concrete (56.4 °C), indicating that less hydration reaction has taken place by replacing a portion of OPC with activated Hwangtoh powder [21,22,23]. Therefore, a further study on this topic must be performed.

Table 1. Chemical composition of activated Hwangtoh and OPC.

Component	Unit Mass Percentage (%)	
	Hwangtoh	Cement
SiO_2	57.7	22.0
Al_2O_3	29.5	5.80
Fe_2O_3	6.32	3.30
K_2O	3.54	–
TiO_2	1.12	–
MgO	1.06	1.20
F	0.22	–
CaO	0.126	64.04
Na_2O	0.114	–
P_2O_5	0.0620	–
MnO	0.0518	–
SO_3	0.0434	–

Depending on the product formed by the reaction between activated Hwangtoh and $Ca(OH)_2$, the reaction can be classified as pozzolanic or stratlingite. A pozzolanic reaction between SiO_2 (Hwangtoh) and $Ca(OH)_2$ (cement) produces the pozzolanic product afwillite. The equation for the reaction is shown in Equation (1):

$$3Ca(OH)_2 + 2SiO_2 \rightarrow 3CaO{\cdot}2SiO_2{\cdot}3H_2O \quad (1)$$

SiO_2 and aluminum oxide (Al_2O_3) from Hwangtoh and $Ca(OH)_2$ from cement can also react to form stratlingite, which makes the activated Hwangtoh into a dense mortar. The unstable $Ca(OH)_2$ becomes stable after undergoing the stratlingite reaction, which leads to increased long-term strength, durability, and crack resistance in the concrete. The activated Hwangtoh we used was a product of the Gochang region of Korea composed of natural-state Hwangtoh sintered at 850 °C. The specific gravity and fineness of the activated Hwangtoh were 2.7 cm^2/g and 3200 cm^2/g, respectively. The chemical composition of activated Hwangtoh and OPC are shown in Table 1.

Calculating the Replacement Ratio of Activated Hwangtoh

Activated Hwangtoh absorbs more water than ordinary concrete. Therefore, the appropriate mixture proportion based on workability needed to be selected. The cement used in this study had a specific gravity of 3.16. As required by KS L ISO 679 [24], a standard aggregate produced for concrete and mortar with specific gravity of 2.6 and 2.7, respectively, was used for the fine aggregate. The coarse aggregate was a crushed stone aggregate with a maximum diameter of 25 mm and saturated surface-dry specific gravity of 2.65. In addition, to determine the appropriate mixture ratio of activated Hwangtoh, the compressive strength was measured by setting the water-to-binder (W/B) ratio at 43% and increasing the proportion of blast furnace slag from 0%–40% in 10% increments. As shown in Figure 1a, the proportion of activated Hwangtoh ranged from 15%–25%. Because concrete mixed with activated Hwangtoh has a lower compressive strength than ordinary concrete, a 20% ratio of activated Hwangtoh was chosen to secure maximal compressive strength. Figure 1b displays the changes in slump for the different replacement ratios.

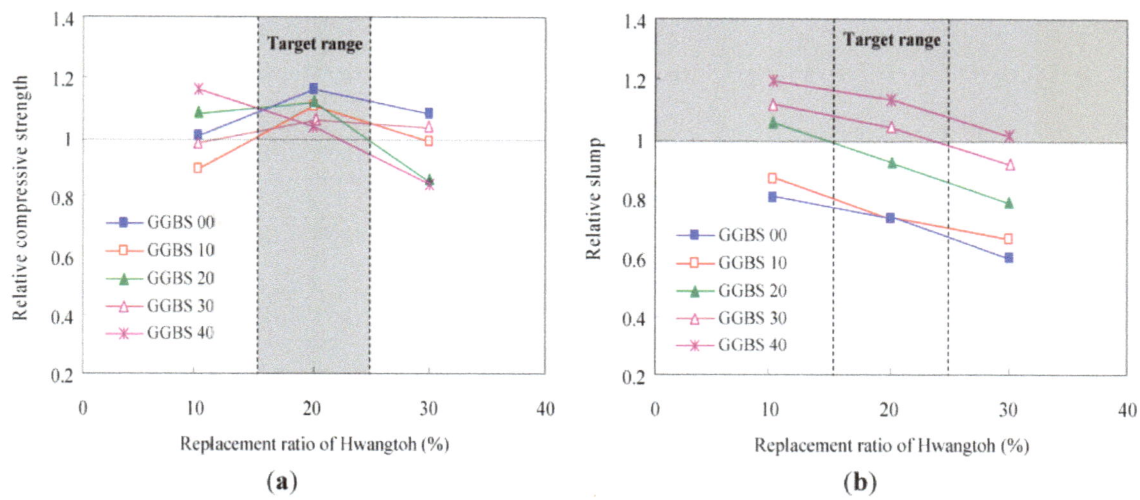

Figure 1. Relative compressive strength and slump test results on concrete samples. (**a**) Actual strength/required strength ratio and (**b**) Actual slump/required slump ratio.

Slump Test

Based on the KS F 2401 [25] and KS F 2402 [26] testing methods, the slump was measured by testing a sample of ordinary concrete and activated Hwangtoh mixed concrete. The ordinary concrete showed a slump of 10 cm–13 cm, whereas the activated Hwangtoh mixed concrete showed a slump of less than 1.0 cm and 3.5 cm for a W/B ratio of 43% and 45%, respectively. The slump was enhanced when the W/B ratio was set to 50%. However, the 28 day compressive strength of the concrete with a 50% W/B ratio and 20% activated Hwangtoh was significantly lower. To generate an appropriate slump, a polycarbonate-based superplasticizer was added at increasing ratios from 1%–2% in 0.5% increments. At 2%, the required slump was obtained, but concrete segregation occurred. The most appropriate proportion of superplasticizer for the required strength and slump was 1.5%. Table 2 shows the changes in slump for varying proportions of superplasticizer in the 20% activated Hwangtoh concrete mixture with a 45% water-to-binder (W/B) ratio.

Table 2. Slump test results.

Water-to-binder ratio (%)	Hwangtoh replacement ratio (%)	Plasticizer (%)	Slump (cm)
45	Control	–	13
	20	–	3.5
		1	6
		1.5	9–10
		2	15–18

Compressive Strength

For the preliminary test, concrete with 20% activated Hwangtoh, three W/B ratios (43%, 45%, and 50%), and varying proportions of blast furnace slag and flyash (5%–25%) were used. Table 3, Table 4 and Table 5 show the mixture proportions of Hwangtoh concrete. The air content of the fresh concrete was measured using the pressure method in KS F 2421 [27], which showed that it satisfied the criteria of 4.5%. The slump range and air content was set to 12 ± 2.0 mm and 4.5 ± 0.5%, respectively. G in Table 3 refers to the ground granulated blast furnace slag (GGBS) and F refers to the flyash. Twelve specimens were prepared for each case and wet cured each specimen at 20 °C in accordance with KS F 2403 [28]. Blast furnace slag with a specific gravity and fineness of 2.90 cm^2/g and a 4306 cm^2/g, respectively, were used. In Table 3, Hwangtoh, water, cement, fine aggregate (sand), coarse aggregate (gravel), GGBS, flyash, and fine aggregate/total aggregate ratio are abbreviated HT, W, C, S, G, GGBS, F, and S/a, respectively.

Table 3. Mix proportion of Hwangtoh concrete (W/B 43%, S/a 40%, HT 20%).

Specimen	Unit (kg/m^3)						
	W	C	S	G	HT	GGBS	F
Control		409			–	–	–
G 5F25						20.45	102.3
G10F20	176		679	1034		40.9	81.8
C15F15		204.5			81.8	61.35	61.35
G20F10						81.8	40.9
G25F 5						102.3	20.45

Table 4. Mix proportion of Hwangtoh concrete (W/B 45%, S/a 40%, HT 20%).

Specimen	Unit (kg/m^3)						
	W	C	S	G	HT	GGBS	F
Control		391			–	–	–
G 5F25						19.55	97.75
G10F20	176		681	1042		39.1	78.2
C15F15		195.5			78.2	58.65	58.65
G20F10						78.2	39.1
G25F 5						97.75	19.55

Table 5. Mix proportion of Hwangtoh concrete (W/B 50%, S/a 40%, HT 20%).

Specimen	Unit (kg/m^3)						
	W	C	S	G	HT	GGBS	F
Control		352			–	–	–
G 5F25						17.6	88
G10F20	176		694	1062		35.2	70.4
C15F15		176			70.4	52.8	52.8
G20F10						70.4	35.2
G25F 5						88	17.6

Flyash with a specific gravity and fineness of 2.19 cm^2/g and 3.490 cm^2/g, respectively, was used to mix the concrete in accordance with KS L 5405 [29]. The flyash was incorporated to improve the workability, strength, and durability of the Hwangtoh mixed concrete by reducing the heat of hydration, increasing the water permeability, and increasing the corrosion and chemical resistance. Although several types of superplasticizers are available, such as naphthalene, melamine, and aminosulfonate, the polycarbonate superplasticizer we used is the most popular option for concrete because of its excellent cement dispersing and slump maintaining characteristics. Moreover, polycarbonate superplasticizer is less vulnerable than other superplasticizers to changes in physical properties over time. Therefore, it was used to increase the workability of Hwangtoh mixed concrete. The particular superplasticizer used in this experiment had a dark brownish color with a specific gravity of 1.080 and a pH of 7.0 at 20 °C.

After casting 12,100 mm × 200 mm cylindrical specimens using activated Hwangtoh mixed concrete, their compressive strengths on the 7th, 14th, and 28th days from casting were measured in accordance with KS F 2403 [28] and KS F 2405 [30]. The lowest and highest measured values were eliminated and calculated the average compressive strength with the remaining 10 specimens. Figure 2 shows the compressive strength test results for ordinary concrete and concrete mixed with 20% activated Hwangtoh, both with a W/B ratio of 43%. Figure 3 and Figure 4 show the compressive strength test results of the same concrete with a W/B ratio of 45% and 50%, respectively. The compressive strength increased as the W/B ratio decreased. With the same W/B ratio, the differences in the compressive strengths of ordinary concrete and activated Hwangtoh mixed concrete were 7.69 MPa, 5.47 MPa and 3.48 MPa for W/B ratios of 43%, 45% and 50%, respectively. The differences in the compressive strength decreased as the W/B ratio increased. Concrete with 20% blast furnace slag and 10% flyash showed the highest compressive strength, and concrete with 25% blast furnace slag and 5% flyash had the lowest compressive strength. Further follow-up studies are required to better examine the causes of these differences in compressive strength. Figure 5 shows the 28-day compressive strength test results for ordinary concrete and Hwangtoh mixed concrete with a W/B ratio of 45%.

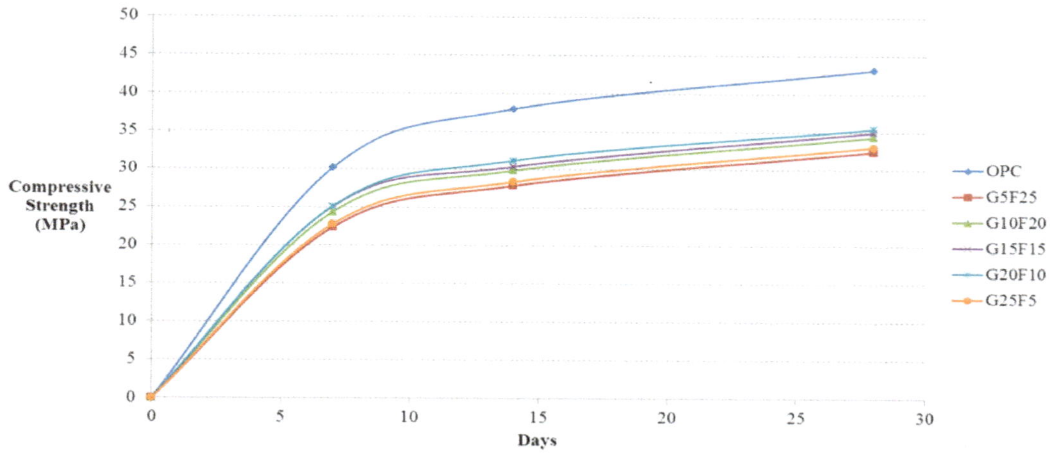

Figure 2. Compressive strength history of concrete with GGBS and flyash (W/B 43%).

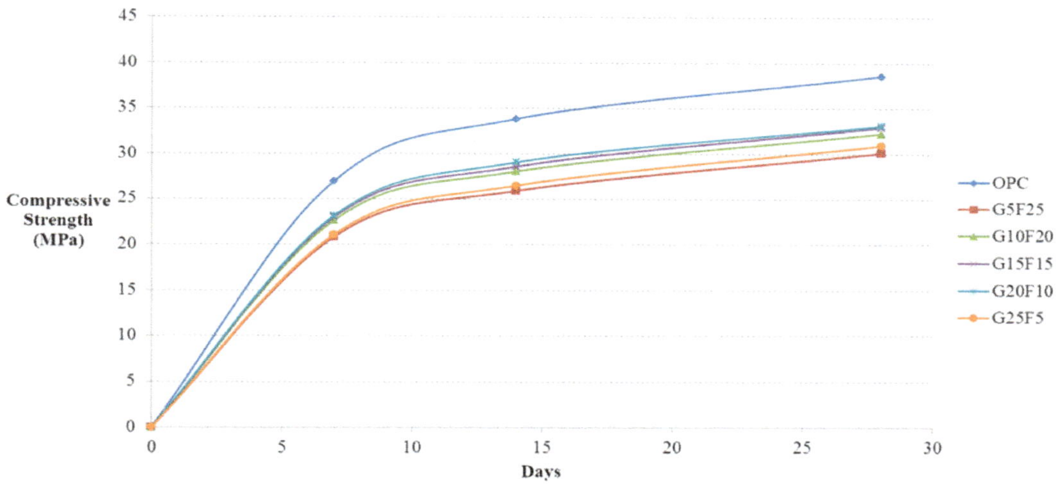

Figure 3. Compressive strength history of concrete with GGBS and flyash (W/B 45%).

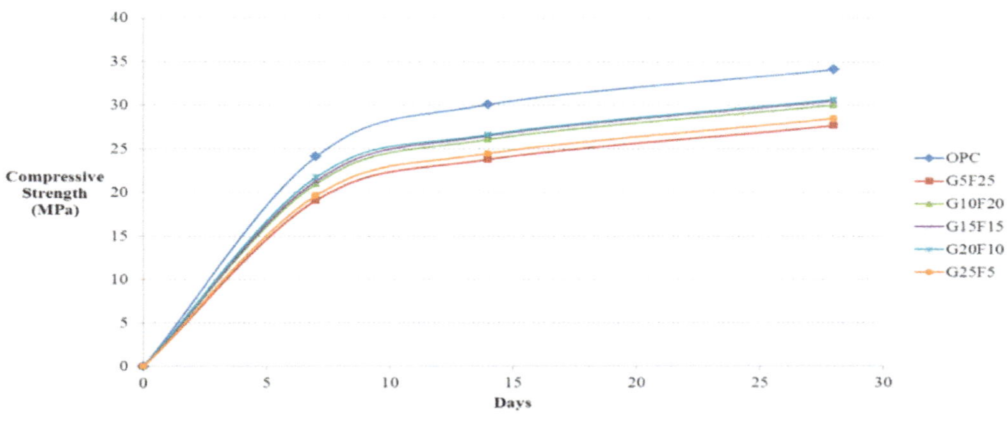

Figure 4. Compressive strength history of concrete with GGBS and flyash (W/B 50%).

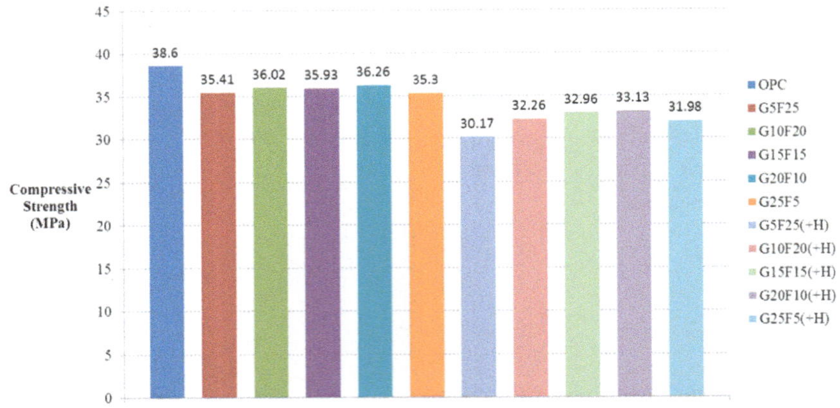

Figure 5. Compressive strength test results (W/B 45%).

pH

A pH150-C was used to measure the pH. This equipment can measure a pH range from 0–14. Flat surface electrodes were attached to the surface of the activated Hwangtoh mixed concrete specimens and measurements at three points from each specimen were obtained to calculate the mean. The pH values of the ordinary concrete, the concrete without activated Hwangtoh, and the activated Hwangtoh mixed concrete with W/B ratio of 45% were obtained and are shown as a bar graph in Figure 6. The pH of the ordinary concrete used for construction ranges from 12.5 to 13. The OPC measured in the present study had a pH of 13. The pH of the concrete without activated Hwangtoh ranged from 11.51 to 11.78, depending on the proportions of blast furnace slag and flyash. The pH of the activated Hwangtoh mixed concrete was 10.32–10.50. The $Ca(OH)_2$ produced by the hydration reaction with the cement forms a passive film that protects the rebar. However, if it loses alkalinity by binding to CO_2, it becomes carbonated and forms $CaCO_3$ with a pH of 8–9. The pH of the activated Hwangtoh mixed concrete was 1.19–1.28 lower than that of the concrete without activated Hwangtoh; the pH of the ordinary concretes was 1.25–1.54 higher than that of the activated Hwangtoh mixed concretes. These results indicate that users must pay meticulous attention not to expose activated Hwangtoh mixed concrete to substandard conditions. From a published report, poisonous and beneficial effects of OPC and Hwangtoh porous concrete on living insects, respectively, were found to be due to high and low pH property, respectively [31]. Therefore, it is safe to assume that Hwangtoh contains less harmful substances than OPC cement with regards to health issues.

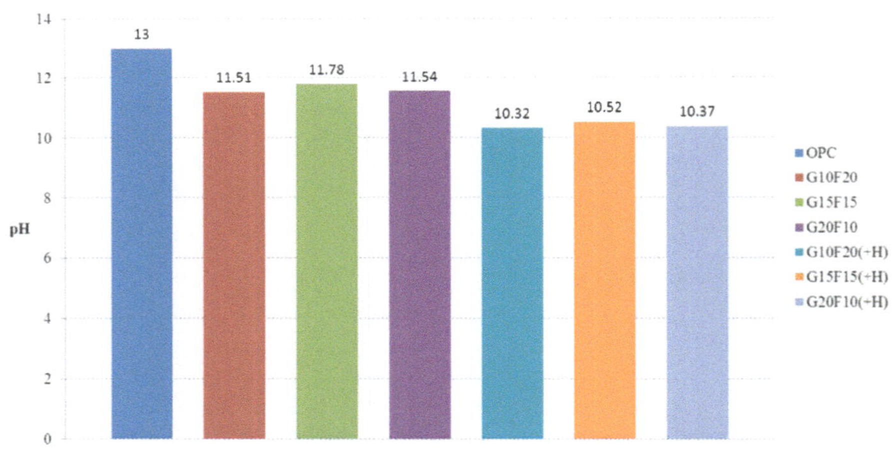

Figure 6. pH test results (W/B 45%).

HEALTH BENEFIT ASSESSMENT OF ACTIVATED HWANGTOH MIXED CONCRETE

The health benefit assessment was performed on three-week-old ICR mice, which are frequently used in animal clinical studies. All experimental procedures complied with the Institutional Animal Care and Use Committee (IACUC) mice standard operating guidelines, the Korean Food and Drug Administration guidelines for animal experimentation, and manuals on experiments involving animals [32]. The cage floors with different ratios of Hwangtoh to cement (20%, 40%, 60%, 80%) with a thickness of 3 cm were prepared, as shown in Figure 7. The titles for the floors made with 20%, 40%, 60%, and 80% Hwangtoh concrete are floor 20, floor 40, floor 60, and floor 80, respectively, and the title of OPC concrete floor is floor OPC. All of the mice were kept at the same temperature and humidity and provided with equal amounts of food and water. To alleviate their stress, exercise equipment was installed in the cage. For the comparison of residential environments, two casts with a length of 122.6 cm, a width of 51 cm, a height of 62 cm, a floor thickness of 5 cm, and a wall thickness of 3 cm were manufactured. Exactly half of each cast was poured with ordinary cement mortar, and the other half was poured with a mixture containing either 20% or 80% activated Hwangtoh, as shown in Figure 8. Therefore, the title of the cast with 20% Hwangtoh concrete and OPC concrete are H20 and OPC20, respectively. Likewise, 80% Hwangtoh and OPC concrete are titled H80 and OPC80, respectively. A passageway connecting the two casts allowed the mice to move freely to either side. The mortar's mixture ratio was: 0.5 (water):1 (cement):3 (fine aggregate). One pair of male and female ICR mice were placed in each of the five casts for the floor experiment, and five male and five female ICR mice were placed in each of the two casts for the residential selection experiment.\

(a) (b)

Figure 7. Floor cage photos. (**a**) Casted bottom surfaces and (**b**) Floor cages.

(a) (b) (c) (d) (e) (f)

Figure 8. Residence cage photos and specimen types. (**a**) Residence cage forms; (**b**) Concrete mixing; (**c**) Casting and curing; (**d**) Sleeping spaces; (**e**) 20% Hwangtoh and OPC mortar, and (**f**) 80% Hwangtoh and OPC mortar.

TEMPERATURE AND HUMIDITY CHANGES DEPENDING ON MIXTURE RATIOS

The external and internal temperatures of each cage were measured at the same time daily, as shown in Figure 9 andFigure 10, respectively. From day 21 onwards, the internal temperature surpassed the external temperature because day 21 marked the beginning of winter. Until day 21, the internal temperature was the highest in cast H20; after day 21, the internal temperature was higher in H80. By and large, the temperature was higher in the casts prepared with Hwangtoh concrete than in the OPC concrete casts. The temperature within the cast of OPC20 was higher than in that of OPC80. Under equal external temperatures, the differences in internal temperature and humidity in the activated Hwangtoh mixed environment and the OPC environment ranged from 0.1 °C to 1.5 °C and 1% to 5%, respectively. Although those differences are small, the ICR mice were sensitive to them. Moreover, the humidity of the external environment was lower than that in any internal environment throughout the duration of the study. It was the highest in the internal environment of the H80 cast, followed by the OPC20, H20, and OPC80 casts. The humidity within the OPC20 and H20 casts fell sharply between days 15 and 17, because the acrylic ceilings of the casts were damaged. The standard IACUC test guideline for live animals states that the relative humidity required for ICR mice is 80%. As shown in Figure 10, the H20 and H80 casts satisfied that requirement. However, the external relative humidity was 45%–75%, less than the required 80%. The relative humidity measurements show that the Hwangtoh environment was more effective in maintaining an appropriate relative humidity than an ordinary air environment.

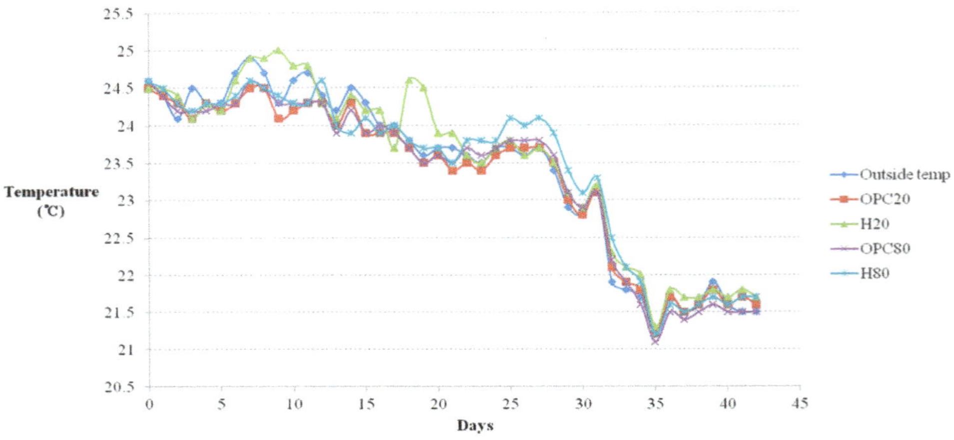

Figure 9. Temperature *versus* time curves.

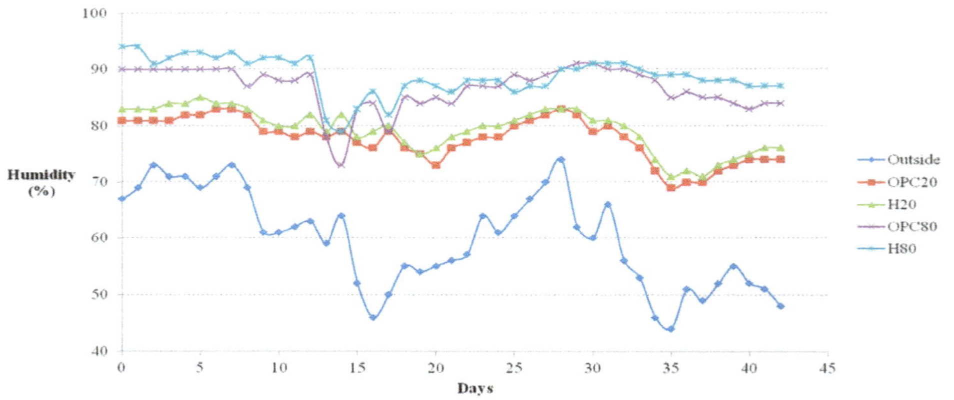

Figure 10. Relative humidity *versus* time curves.

Amount of Food and Water Intake

The amounts of food and water intake measured daily from the floor cage test are shown in Figure 11 and Figure 12, respectively. The amounts of food and water intake measured daily from the residential selection casts are shown in Figure 13 and Figure 14, respectively. The amounts were calculated by taking the difference between the initial weights of food and water and the weight of leftover amounts after 24 hours. Small food is difficult to distinguish from mouse feces; therefore, the ICR mice was fed with solid cylindrical food. As shown in Figure 11, from day 1 to 21, the amount of food intake in the floor experiment was the highest on floor 80 (80% of activated Hwangtoh), followed by floor 20, floor OPC, and floor 60. The amount of food intake initially increased moderately and reached a plateau in floor 40, possibly because the ICR mice on the floor 40 cast were infertile. From day 21 onwards, the amount of food intake was the highest on floor OPC, followed by floor 80, floor 20, floor 60, and floor 40. This trend can be attributed to the onset of colder weather on day 21, during which the mice gave birth, thereby increasing the total amount of food intake. Given the design of this study, the cause of the increased food intake in floor OPC cannot be pinpointed to a change in temperature or humidity or the increased desire for reproduction in a substandard environment. A more in-depth anatomical experiment is required to understand this trend. The ICR mice changed their food and water intake before and after parturition; the mice consumed more water and food during pregnancy and less after giving birth. The large food intake of mice on the OPC floor might be attributable to their survival instincts. The relatively lower temperature and humidity of the OPC floor might cause ICR mice to consume more food and water to supplement their energy levels and thus become stronger to overcome their harsh environment.

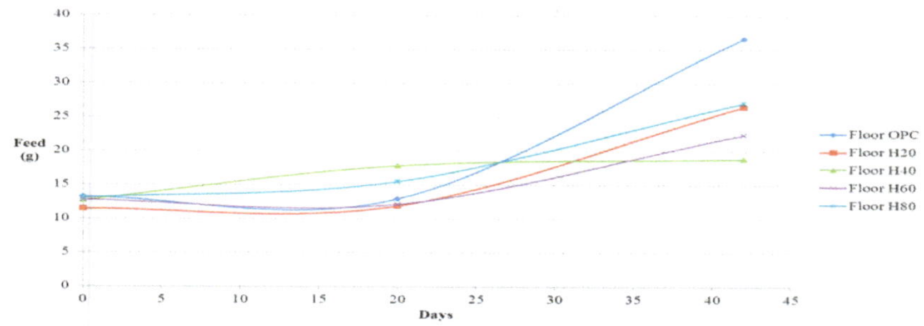

Figure 11. Food intake *versus* time curves (floor cages).

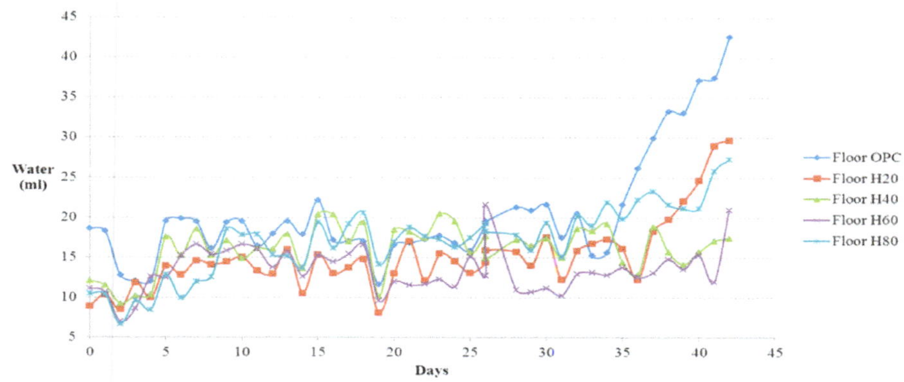

Figure 12. Water intake *versus* time curves (floor cages).

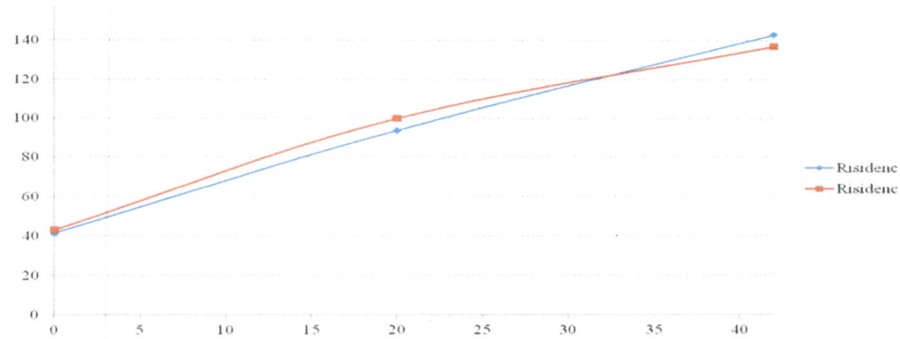

Figure 13. Food intake *versus* time curves (residence cages).

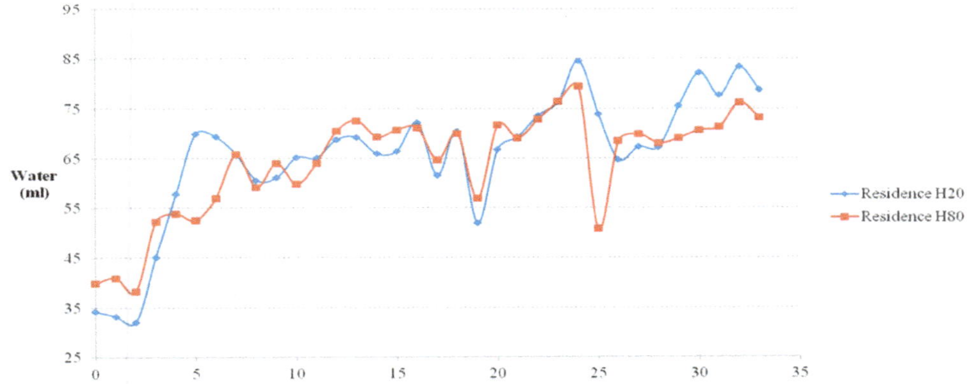

Figure 14. Water intake *versus* time curves (residence cages).

The ICR mice in the residential selection experiment showed similar upward trends in their amount of food and water intake. Before day 32, the amount of food intake increased in the mice in H80, and after day 32, the amount of food intake increased in the mice in OPC80.

In the flooring experiment, the amount of water intake was the highest in the mice on floor OPC throughout the entire study duration. The mice on floor 40 showed a consistent and repetitive trend. The next highest water intake was by the mice on floor 80, followed by floor 20, floor 40, and floor 60. In the residential selection experiment, the amount of water intake followed a nearly identical upward trend with minor differences in each cast.

Weight Gain

Weight gain at three-day intervals was measured as mandated by the regulations in the IACUC guidelines for animal experimentation. The weight changes measured from the floor cages and residence selection casts are shown in Figure 15and Figure 16, respectively. In the flooring experiment, the mice on floor 20 showed the highest weight gain; the mice on floor 80 initially showed the lowest weight gain, but they ended up with the greatest weight gain toward the end of the experiment. The weight gain by the mice on floor OPC was moderate in the initial and middle stages of the experiment but reached that of the mice on floor 20 by the end of the experiment. The mice on floor 40 and floor 60 showed identical upward trends in weight gain. Thus, the ICR mice on the concrete floor casts took in large amounts of food and water in the early and middle stages of testing to survive in their new environment. However, as they adjusted to the environment in the later stage of testing, their weight gain increased significantly. However, the ICR mice on the Hwangtoh floor casts took in less food and water than the mice on the concrete floor because of the better environmental conditions provided by the Hwangtoh environment throughout the testing. As shown in Figure 16, the residential selection experiment showed that all of the mice initially had similar weight gains, but the mice in H20 had a higher weight gain than the other mice from the middle to the end of testing.

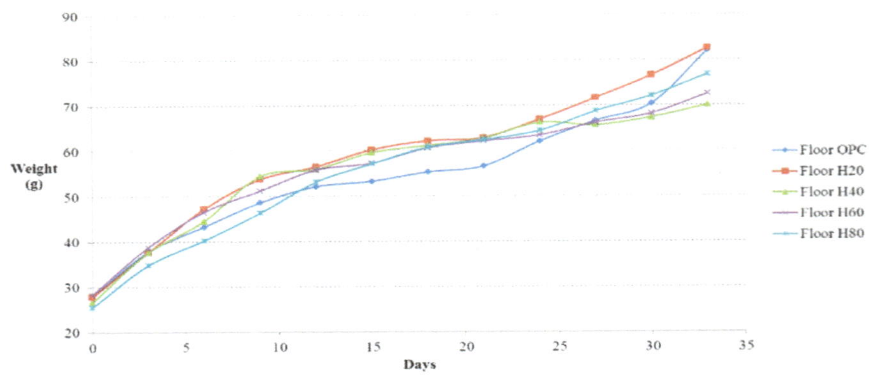

Figure 15. Weight change *versus* time curves (floor cages).

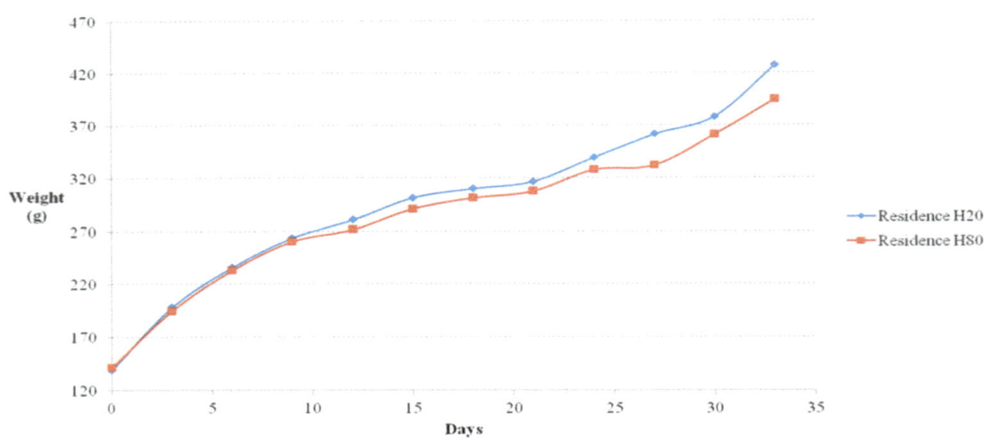

Figure 16. Weight change *versus* time curves (residence cages).

AGGRESSION

Funase (2002) reported that the level of stress and lifespan of subjects differs according to the type of construction material used in the subjects' living space, such as timber, metal, or concrete. In particular, Funase suggested that concrete can reduce subjects' lifespan by nine years as a result of the emissivity that occurs in the winter when the low external temperature is transmitted to the internal living environment through the walls [18]. Aggression is a particularly difficult behavior to quantitatively measure. In the present study, it was measured by examining the level of damage to the equipment and structures within the living spaces. The most aggressive behavior started after day 21 (the beginning of winter), and the emissivity was considered during the analysis. As shown in Figure 17, the structures within the OPC cast were more damaged than those within H80. In the residential environment selection experiment, the structures in the H20 cast were not damaged at all; however, the structures in the H80 cast sustained significant damage. Thus, as the Hwangtoh admixture ratio increased, the ICR mice felt the H80 floors and walls to be similar to soil, which brought out their natural digging and scratching instincts.

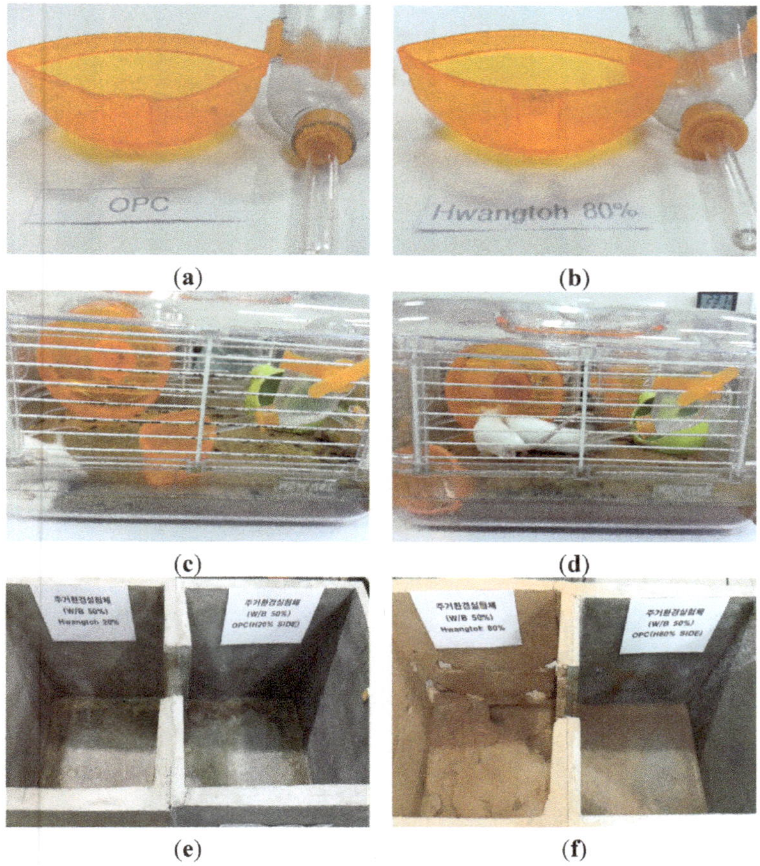

Figure 17. Damages caused by aggression. (**a**) Feeding and watering containers for OPC cage; (**b**) Feeding and watering containers for H80 cage; (**c**) Protection bars for OPC floor cages; (**d**) Protection bars for H20 floor cage; (**e**) Damaged lining of H20 residence cage, and (**f**) Damaged lining of H80 residence cage.

Fertility Rate

The ICR mice used in this experiment were fully fertile three weeks after birth, and indeed the females were pregnant around day 21 after birth. The fertility rates of the ICR mice in the five casts with different floors and two casts with different living environments were compared. Photos of the offspring from the floor cage and residence selection casts are shown inFigure 18a,b, respectively. The number of births for the floor cages and the residence selection casts are shown in Figure 19and Figure 20, respectively, as bar graphs. The regular and frequent pregnancy and reproduction shows that the amount of food and water provided was ample.

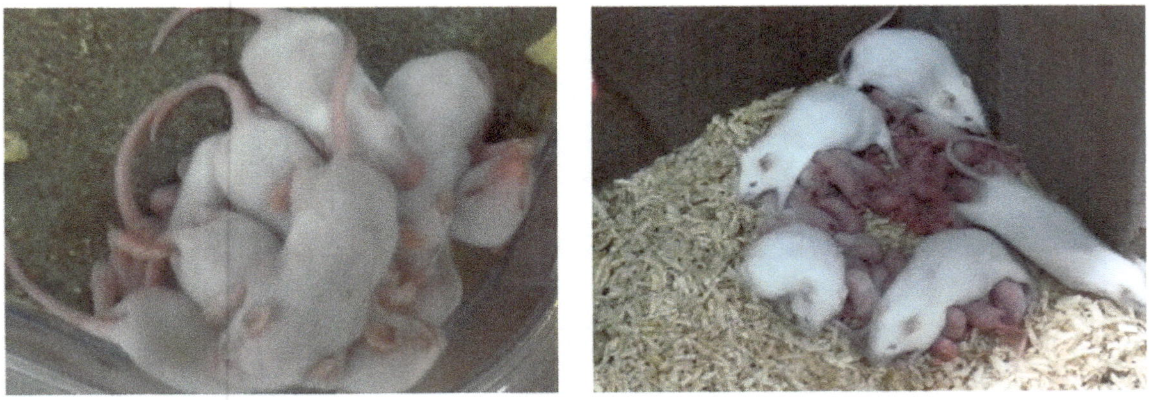

(**a**) (**b**)

Figure 18. Breeding of ICR. (**a**) Floor cages and (**b**) Residence cage.

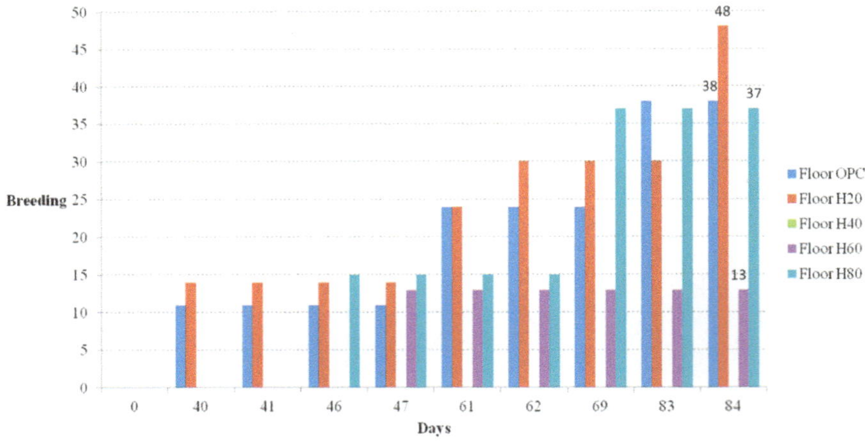

Figure 19. Breeding *versus* time curve (floor cages).

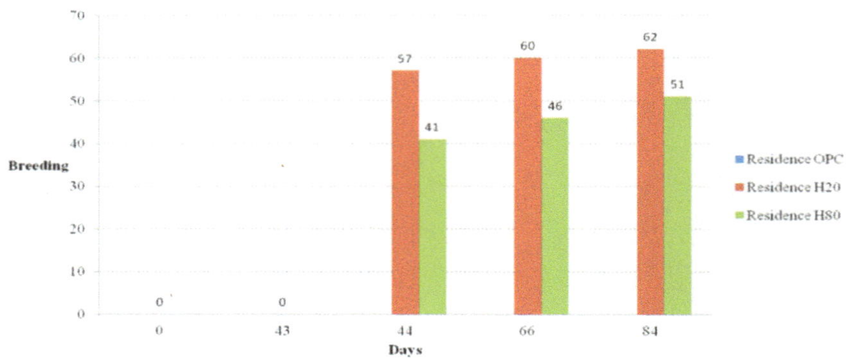

Figure 20. Breeding results curve (residence cages).

RESIDENTIAL ENVIRONMENT SELECTION

ICR mice were allowed to choose between two different living environments, one with an activated Hwangtoh mixed mortar and the other with an ordinary cement mortar to examine their choices over 12 weeks. Equal amounts of food and water were provided in each environment. As shown in Figure 21, in the first experiment, all of the mice chose the activated Hwangtoh environment within three hours of testing. They maintained their selection in repetitive experiments. Furthermore, although it initially took three hours for the mice to move to their chosen environment, the selection time grew shorter with each follow-up experiment. In the last experiment, it took less than one minute for some of the mice to move to their chosen environment. At the end of the experiment, the amount of leftover food and water in the activated Hwangtoh environment was considerably less than that in the OPC environment. Under an equal external temperature, the temperature and humidity within the activated Hwangtoh cast and OPC cast differed by 0.1 °C–1.5 °C and 1.0%–5.0%, respectively. To determine whether the differential amount of food and water intake in the two environments was caused by the differences in temperature and humidity or by hazardous components in the cement, the temperature and humidity in the OPC cast were varied to match those in the activated Hwangtoh environment. Even under identical temperature and humidity levels, the ICR mice maintained their choice of environment. This behavior indirectly indicates that the lower amount of food and water intake in the normal mortar environment was caused by harmful components in the cement rather than the difference in relative temperature and humidity. In a further study, in-depth research should be performed to examine the cement components that could pose risks to human health.

Figure 21. Residence preference. (**a**) 20% Hwangtoh residence cage and (**b**) 80% Hwangtoh residence cage.

CONCLUSIONS

The following conclusions are drawn from this study.

(1) Although the activated Hwangtoh mixed concrete showed slightly lower compressive strength than the ordinary concrete, it is a suitable construction material, except in those cases that require exceptionally high strength and durability.

(2) Because the activated Hwangtoh mixed concrete tended to absorb more water, the optimal addition rate of superplasticizer for better workability was 1.5 volume %.

(3) The pH of the activated Hwangtoh mixed concrete was lower than that of ordinary concrete.

(4) The differences in temperature and relative humidity between OPC mortar and Hwangtoh cement mortar measured from the residence selection casts were 0.1 °C–1.5 °C and 1.0%–5.0%, respectively.

(5) The amount of food and water intake by the mice in the activated Hwangtoh environment was considerably higher than that in the OPC environment. Also, the mice in the activated Hwangtoh environment had a higher fertility rate and healthier offspring.

(6) The ICR mice moved to the ordinary cement cast only for food and water. At all other times, they stayed in the activated Hwangtoh living quarter.

ACKNOWLEDGMENTS

This study was funded by the National Research foundation of Korea (NRF) grant funded by the Korea government (MSIP) (No. 2011-0030040). The authors of this study would like to extend their gratitude for the support.

AUTHOR CONTRIBUTIONS

Bon-Min Koo is a person who performed most of test and analysis works; a main writer of the paper. Jang-Ho Jay Kim is a PI of the research project, who provided the main idea of the study. Tae-Kyun Kim is a person who assisted the research and writing the paper. Byung Yun Kim is a consultant to the research work and writing of the paper.

Conflicts of Interest

The authors declare no conflict of interest.

REFERENCES

1. Hwang, J.P.; Hong, S.I.; Yang, H.J.; Ann, K.Y. Evaluation on hexavalent chromium content in loess additive as a inhibitive material to leachat of heavy metal. *Proc. Korea Concr. Inst.* **2013**, *25*, 281–282.

2. Lee, J.H.; Park, N.K.; Jung, Y.J.; Chu, Y.S.; Song, H.; Lee, J.K. Leaching Properties of Hexavalent Chromium in Sintering Condition of Clinker Material. *Proc. Korea Concr. Inst.* **2008**, *20*, 549–552.

3. Hwang, H.Z. Study on the Method of Activating Kaoline and the Mortar and Concrete Mixed with Active-Kaoline. Ph.D. Thesis, The Seoul National University, Seoul, Korea, February 1997.

4. Yang, K.H.; Hwang, H.Z.; Kim, S.Y.; Song, J.K. Development of a Cementless Mortar Using Hwangtoh Binder. *Build. Environ.* **2007**, *42*, 3717–3725.

5. Go, S.S.; Lee, H.C.; Lee, J.Y.; Kim, J.H.; Chung, C.W. Experimental Investigation of Mortars Using Activated Hwangtoh. *Constr. Build. Mater.* **2009**, *23*, 1438–1445.

6. Dinakar, P.; Sahoo, P.K.; Sriram, G. Effect of Metakaolin Content on the Properties of High Strength Concrete. *Int. J. Concr. Struct. Mater.* **2013**, *7*, 215–223.

7. Choi, H.Y.; Kim, M.H.; Hwang, H.Z. Study on the Development of Hwangtoh Admixture for the Application of Cement Mortar. *J. Archit. Inst. Korea* **2000**, *16*, 95–102.

8. Choi, H.Y.; Kim, M.H.; Hwang, H.Z.; Choi, S.W. Experimental Study on the Properties of Concrete by the Kinds of Admixture and the Replacement Ratios of Activated Hwangtoh. *J. Korea Concr. Inst.* **2001**, *13*, 672–675.

9. Becknell, N.P.; Micah-Hale, W. Effect of Slag Grade and Cement Source on the Properties of Concrete. *Int. J. Concr. Struct. Mater.* **2011**, *5*, 119–123.

10. Frisa, M.; Sanchez de Rojas, M.I.; Cabrerab, J. The Effect That the Pozzolanic Reaction of Metakaoline has on the Heat Evaluation in Metakaoline-Cement Mortars. *Cem. Concr. Res.* **2000**, *30*, 209–216.

11. Go, S.S.; Chung, C.W.; Struble, L.J.; Lee, H.C. Pozzolanic Activity of Hwangtoh Clay. *Constr. Build. Mater.* **2010**, *24*, 2638–2645.

12. Lee, N.K.; Hwang, H.Z.; Park, H.G. Flexural Performance of Activated Hwangtoh Concrete Beam. *J. Korea Concr. Inst.***2010**, *22*, 567–574.

13. Yang, K.H.; Hwang, H.Z.; Kim, S.Y.; Song, J.G. Effect of Superplasticizers and Admixtures on the Fluidity and Compressive Strength Development of Cementless Mortar Using Hwangtoh Binder. *J. Korea Concr. Inst.* **2006**, *18*, 793–800.

14. Bougara, A.; Kadri, E.; Ezziane, K. Efficiency of Granulated Blast Furnace Slag Replacement of Cement According to the Equivalent Binder Concept. *Cem. Concr. Compos.* **2010**, *32*, 226–231.

15. Kang, H.K.; Yang, K.H.; Lee, Y.H.; Hwang, H.Z.; Chung, H.S. The Stress-Strain Relationship and Compressive Strength of Concrete Containing Hwangtoh and Slag. *Korea Concr. Inst. Conf.* **2004**, *16*, 672–675.

16. Lee, S.H.; Choi, I.C.; H, S.K. Analysis on the Harmful Effect of Recycled Powder and Properties of Concrete Admixture by Recycled Powder. *J. Korea Soc. Mar. Environ. Saf.* **2008**, *14*, 289–295.

17. Chun, B.S.; Kim, J.C. Evaluation of Toxic Effect of Grouting Materials by Fish Poison Test. *J. Korea Soc. Civ. Eng.* **1998**,*18*, 531–538.

18. Funase, S. Concrete Stress. *Matibook* **2002**, 17–40.

19. Ghebrab, T.T.; Soroushian, P. Mechanical Properties of Cement Mortar: Development of Structure-Property Relationships. *Int. J. Concr. Struct. Mater.* **2011**, *5*, 3–10.

20. Lee, K.W.; Jang, J.H.; Choi, H.Y.; Goo, J.S.; Hwang, H.Z.; Kim, M.H. Experimental Study on the Resistance of Chloride Infiltration of Concrete Using Activated

Hwangtoh Admixture. *Proc. Korea Concr. Inst.* **2001**, *13*, 781–786.

21. Kang, S.S.; Lee, S.L.; Hwang, H.Z.; Cho, M.C. Hydration Heat and Shrinkage of Concrete Using Hwangtoh Binder. *J. Korea Concr. Inst.* **2008**, *20*, 549–555.
22. Ghebrab, T.T.; Soroushian, P. Mechanical Properties of Hydrated Cement Paste: Development of Structure-Property Relationships. *Int. J. Concr. Struct. Mater.* **2010**, *4*, 37–43.
23. Murmu, M.; Singh, S.P. Hydration Products, Morphology and Microstructure of Activated Slag Cement. *Int. J. Concr. Struct. Mater.* **2014**, *8*, 61–68.
24. *KS L ISO 679 Methods of Testing Cements—Determination of Strength*; Korean Agency for Technology and Standards (KATS): Seoul, Korea, 2006; pp. 1–16.
25. *KS F 2401 Method of Sampling Fresh Concrete*; Korean Agency for Technology and Standards (KATS): Seoul, Korea, 2007; pp. 1–7.
26. *KS F 2402 Method of Test for Slump of Concrete*; Korean Agency for Technology and Standards (KATS): Seoul, Korea, 2007; pp. 1–5.
27. *KS F 2421 Method of Test for Air Content of Fresh Concrete by Pressure Method*; Korean Agency for Technology and Standards (KATS): Seoul, Korea, 2006; pp. 1–17.
28. *KS F 2403 Standard Test Method for Making and Curing Concrete Specimens*; Korean Agency for Technology and Standards (KATS): Seoul, Korea, 2014; pp. 1–7.
29. *KS L 5405 Flyash*; Korean Agency for Technology and Standards (KATS): Seoul, Korea, 2009; pp. 1–6.
30. *KS F 2405 Standard Test Method for Compressive Strength of Concrete*; Korean Agency for Technology and Standards (KATS): Seoul, Korea, 2010; pp. 1–6.
31. Park, C.K.; Jeong, J.H.; Kim, H.J. pH Reduction of High Porous Concrete to Grow Plant. *Proc. Korea Concr. Inst.* **2001**,*13*, 1129–1134.
32. Korea Food and Drug Administration. *Law for Laboratory Animals*; Korea Food Drug Administration: Seoul, Korea, 2010; pp. 63–64.

CHAPTER 5

EVALUATION OF THE CHEMICAL AND MECHANICAL PROPERTIES OF HARDENING HIGH-CALCIUM FLY ASH BLENDED CONCRETE

Wei-Jie Fan, Xiao-Yong Wang and Ki-Bong Park

Keywords: High calcium fly ash, hydration, kinetic model, calcium hydroxide, strength, concrete

INTRODUCTION

Fly ash is produced in coal-burning electricity generation stations and has been widely used for producing high-performance concrete. Fly ashes can be divided into two categories according to their calcium content. The first type of fly ash, containing less than 10% analytical CaO, is generally a product of the combustion of anthracite and bituminous coals. The second type of fly ash, typically containing 15% to 40% analytical CaO, is generally a product of the combustion of lignite and sub-bituminous coals [1].

Many investigations on the physical, chemical, and durability properties of concrete incorporating low-calcium fly ash (FL, CaO less than 10%) have been performed [2,3,4,5,6]. For concrete containing high-calcium fly ash (FH, CaO more than 10%), the investigations are relatively insufficient. Due to the problems associated with the FH chemical compositions (high free CaO and sulfur contents), the suitability of FH as a mineral admixture in the concrete industry is viewed with skepticism. To overcome these problems and to apply FH blended concrete during dam construction, Tsimas and Moutsatsou-Tsima [7] put forward systematic industry methods. Untreated fly ash was cheaply upgraded by grinding at a specially designed ball mill with simultaneous hydration for the reduction of free CaO. When the free calcium oxide content in FH is less than 3.5% and the sulfur content in FH is less than 7%, the soundness, performance, and durability of FH blended concrete can meet code specifications. Antiohos and Tsimas [8] investigated the role of the reactivity of reactive silica in the hydration of FH-cement blends. They found that after the first month of the hardening process, the soluble silica of FH holds a predominant role and that the silica is increasingly dissolved in the matrix, forming additional cementitious compounds with binding properties, principally a second generation calcium silicate hydrate (CSH). Erdogdu and Turker [9] investigated the effects of FH particle size distribution on the strength of FH-cement mortars. They found the reactivity of FH will increase with the increasing fineness and that the strength of FH blended mortar relates closely to the size fractions of FH. Summarily, experimental investigations [7,8,9] show that the properties of FH blended concrete relate to cement hydration and FH reaction. To evaluate the properties of FH blended concrete, both cement hydration and FH reaction should be considered.

Compared to the abundant experimental investigations on FL blended concrete [7,8,9], theoretical models for FH-cement blended concrete are very limited. Papadakis [10,11,12] proposed a simplified scheme describing the chemical reactions of the FH in FH-cement blends and developed mathematical expressions for predicting the final chemical and volumetric composition of FH blended concrete. Schindler and Folliard [13] described the heat evolution process of FH blended cement using a three-parameter model. The hydration time parameter, hydration shape parameter, and ultimate degree of hydration parameter are adopted for evaluating the hydration of concrete. Saeki and Monteiro [14] formulated a model to

Citation: Wei-Jie Fan, Xiao-Yong Wang and Ki-Bong Park, Evaluation of the Chemical and Mechanical Properties of Hardening High-Calcium Fly Ash Blended Concrete, doi:10.3390/ma8095282.

predict the reaction between calcium hydroxide and mineral admixtures (including high-calcium fly ash, low-calcium fly ash, and slag). The parameters of the prediction model are dependent on the physical and chemical characteristics of the mineral admixtures. Conversely, there are some limits for current models of FH blended concrete. Papadakis' model is only valid for hardened concrete and cannot be used to evaluate the properties development of hardening concrete [11,12]. Schindler and Folliard's model [13] is limited to evaluating the hydration heat of FH blended concrete. The interactions between cement hydration and FH reactions are not considered. Saeki and Monteiro's model [14] is useful for evaluating the calcium hydroxide content in hydrating cement-FH blends. However, other aspects related to the hydration process, such as chemically bound water, porosity, and compressive strength, cannot be determined using Saeki and Monteiro's model.

To overcome the limitations of the current models [11,12,13,14] and to systematically evaluate the chemical and mechanical properties development of FH blended concrete, this paper proposes a numerical model for hydrating cement-FH blends. The proposed model analyzes the cement hydration, the free CaO reaction in FH, and the reaction of phases in FH other than free CaO. The properties of hardening FH blended concrete, such as the reaction degrees of cement and FH, the calcium hydroxide content, chemically bound water, porosity, and compressive strength, can be determined. To improve readability, the summary of abbreviations is shown in Table 1.

Table 1. Summary of abbreviations.

Abbreviations	Text
FH	high-calcium fly ash
FL	low-calcium fly ash
SF	silica fume
H	H_2O
C	CaO
S	SiO_2
A	Al_2O_3
F	Fe_2O_3
$\bar{S}$	SO_3
CF	free CaO
CH	$Ca(OH)_2$
C_3S	$3CaO\cdot SiO_2$
C_2S	$2CaO\cdot SiO_2$
C_3A	$3CaO\cdot Al_2O_3$
C_4AF	$4CaO\cdot Al_2O_3\cdot Fe_2O_3$
CSH	calcium silicate hydrate
A–S	aluminosilicate
C–A–S	calcium aluminosilicate
C–A–H	calcium aluminate hydrate
$C_4A_3\bar{S}$	monocalcium aluminosulfate
FHA10%	FH replaces aggregate by 10% weight of cement
FHA20%	FH replaces aggregate by 20% weight of cement
FHA30%	FH replaces aggregate by 30% weight of cement
FHC10%	FH replaces cement by 10% weight of cement
FHC20%	FH replaces cement by 20% weight of cement
FHC30%	FH replaces cement by 30% weight of cement
TG/DTA	thermogravimetry/differential thermal analysis

HYDRATION MODEL OF CEMENT-FH BLENDS

Hydration Model of Cement

In this study, the cement hydration model originally developed by Tomosawa [15] and modified by Park [16] is adopted to simulate the development of cement hydration. The kinetic processes of cement hydration, such as the formation and destruction of an initial impermeable layer, the activated chemical reaction process, and the following diffusion-controlled process, are considered in the modeling process [15,16]. This model is expressed as a single equation consisting of three coefficients: k_d, the reaction coefficient in the induction period; D_e, the effective diffusion coefficient of water through the CSH gel; and k_r, a coefficient of the reaction rate of cement, as shown in Equation (1):

$$\frac{d\alpha}{dt} = \frac{3\,(S_w/S_0)\,\rho_w C_{w\text{-free}}}{(v + w_g)r_0\rho_c}\,\frac{1}{\left(\frac{1}{k_d} - \frac{r_0}{D_e}\right) + \frac{r_0}{D_e}(1-\alpha)^{-\frac{1}{3}} + \frac{1}{k_r}(1-\alpha)^{-\frac{2}{3}}} \quad (1a)$$

$$k_d = \frac{B_{ce}}{\alpha^{1.5}} + C_{ce}\alpha^3 \quad (1b)$$

$$D_e = D_{e0}\ln\left(\frac{1}{\alpha}\right) \quad (1c)$$

where α is the degree of cement hydration; v is the stoichiometric ratio by mass of water to cement (=0.25); w_g is the physically bound water in the CSH gel (=0.15); ρ_w is the density of water; $C_{w\text{-free}}$ is the amount of water at the exterior of the CSH gel; r_0 is the radius of unhydrated cement particles ($r_0 = 3/(S\rho_c)$, in which the terms S and ρ_c stand for the Blaine surface area and the density of the cement, respectively); S_w is the effective surface area of the cement particles in contact with water and S_0 is the total surface area if the surface area develops unconstrained; B_{ce} controls the rate of the initial shell formation and C_{ce} controls the rate of the initial shell decay; and D_{e0} is the initial value of the effective diffusion coefficient.

The amount of water in the capillary pores $C_{w\text{-free}}$ is expressed as a function of the degree of hydration in the previous step, as shown in Equation (1d):

$$C_{w\text{-free}} = \frac{W_0 - 0.4 * \alpha * C_0}{W_0} \quad (1d)$$

where C_0 and W_0 are the mass fractions of cement and water in the mix proportion.

The effect of temperature on the reaction coefficients is assumed to follow Arrhenius's law [15,16]. Using the proposed Portland cement hydration model, Tomosawa [15] evaluated the heat evolution rate, chemically bound water, and compressive strength of hardening concrete. Park *et al.* [16] predicted the temperature distribution in ultra-high-strength concrete using this hydration model. A good correlation was found between the analysis results and the experimental results.

Reaction of Free CaO in FH

Cement notation chemistry is used throughout this paper, where H: H_2O, C: CaO, S: SiO_2, A: Al_2O_3, F: Fe_2O_3, S–: SO_3, CH: $Ca(OH)_2$. For FH, the majority of calcium is present in crystalline and reactive constituents, such as C_3A, C_4A_3S–, CS–, free CaO, and calcium aluminosilicate (C–A–S) glass. Free CaO has both negative and positive effects on concrete performance. Free CaO can harm the volume stability and the concrete durability [1]. Conversely, some researchers state that a moderate amount of free CaO (3.5% free CaO) is essential for the initial activation of high calcium fly ash [7]. The free CaO is hydrated rapidly to CH as follows [11]:

$$C + H \rightarrow CH \tag{2}$$

Chen [17] studied the hydration kinetic process of free CaO in systems containing high-calcium fly ash-H_2O at different curing temperatures. The reaction of free CaO in FH is a first-order reaction whose rate is directly proportional to the degree of reaction. Chen [17] proposed that the reaction degree of free CaO α_{CF} can be described as follows:

$$\alpha_{CF}(t) = 1 - \frac{1}{e^{kt}} \tag{3}$$

where k is reaction rate coefficient (k is 0.09/h when the temperature is 20 °C, and the dependence of the reaction rate coefficient on the curing temperature can be described using Arrhenius's law [17]).

Reaction of Phases in FH other than Free CaO

FH essentially consists of aluminosilicate glass modified by the presence of large amounts of calcium and magnesium. High-calcium fly ashes are pozzolans that will react with the calcium hydroxide resulting from the hydration of Portland cement to produce calcium silicate hydrate (CSH) and calcium aluminate hydrate (C–A–H). However, FH are also hydraulic in nature, as they will react directly with water to form a range of hydration products, and a mix of FH and water will set, harden, and gain a certain strength even in the absence of Portland cement (or other activators). This behavior results from the hydraulic nature of the crystalline products and the dissolution and reaction of the glass due to the presence of soluble alkalis and calcium within the FH [10,11,12]. Therefore, high-calcium fly ash has both hydraulic properties and pozzolanic properties.

Saeki and Monteiro [14] proposed that the reaction between FH and calcium hydroxide is a diffusion-controlled process. The reaction rate of FH relates to the available calcium hydroxide content in a cement-FH system. Giergiczny [18] investigated the hydraulic activity of FH using an isothermal heat evolution experiment. Similar to cement, the reaction process of an FH-water paste consists of an initial dormant period, a phase boundary reaction process, and a diffusion process. Papadakis [12] proposed that similar to the control of Portland cement, the strength of FH blended concrete relates to the calcium silicate hydrate (CSH) content. Similar to cement, the reaction products of high-calcium fly ash adhere to the surfaces of the remaining FH particles. Therefore, in this paper, similar to that of Portland cement, the reaction of FH is assumed to consist of three processes: an initial dormant period, a phase boundary reaction process, and a diffusion process. The reaction equations of FH are originally proposed as follows:

$$\frac{d\alpha_{FH}}{dt} = \frac{m_{CH}(t)}{P}\frac{W_{cap}}{W_0}\frac{3\rho_w}{v_{FH}r_{0FH}\rho_{FH}}\frac{1}{(\frac{1}{k_{dFH}} - \frac{r_{0FH}}{D_{eFH}}) + \frac{r_{0FH}}{D_{eFH}}(1 - \alpha_{FH})^{\frac{-1}{3}} + \frac{1}{k_{rFH}}(1 - \alpha_{FH})^{\frac{-2}{3}}} \tag{4a}$$

$$k_{dFH} = \frac{B_{FH}}{\alpha_{FH}{}^{1.5}} + C_{FH}\alpha_{FH}{}^{3} \tag{4b}$$

$$D_{eFH} = D_{e0FH}\ln(\frac{1}{\alpha_{FH}}) \tag{4c}$$

where α_{FH} is the reaction degree of the active part in FH other than free CaO, $m_{CH}(t)$ is the mass of calcium hydroxide, W_{cap} is the mass of capillary water, P is the quality of the FH in the mixing proportion other than the mass of free CaO, v_{FH} is the stoichiometry ratio by mass of CH to FH other than free CaO, r_{0FH} is the radius of the FH particle, ρ_{FH} is the density of FH,k_{dFH} is the reaction rate coefficient in the dormant period (B_{FH} and C_{FH} are coefficients), k_{rFH} is the reaction rate coefficient of the phase boundary reaction process, and D_{e0FH} is the initial diffusion coefficient.

Papadakis [11] proposed that FH consists of an active part and an inert part. The active part includes an aluminosilicate glass phase, a calcium phase, and a sulfate phase. The reacted ratio of the active part of FH other than free CaO can be calculated according to Equation (4a) through (4c). The crystalline part of alumina found in corundum and A_3S_2 will not react. The non-reactive part of silica is present in quartz and in crystalline A–S phases. The inert part is assumed to be chemically inert and does not react. The reacted ratio of FH other than free CaO (including both an active part and an inert part) can be obtained based on Equation (4a) through (4c) and the mass compositions of the fly ash. Considering both the active part and the inert part, the total reaction degree of FH other than free CaO can be determined as follows:

$$\alpha_{FH\text{-}total} = \gamma_{active} \times \alpha_{FH} \tag{5}$$

where $\alpha_{FH\text{-}total}$ is the reactive degree of the total FH other than free CaO and γ_{active} is the weight fraction of the active part of FH other than free CaO.

Mutual Interactions among Cement Hydration, the Reaction of Free CaO in FH, and the Reaction of Phases in FH other than Free CaO

As proposed by Papadakis [11], the chemical reactions of the mineral compounds of Portland cement can be expressed as follows:

$$2C_3S + 6H \rightarrow C_3S_2H_3 + 3CH \tag{6a}$$

$$2C_2S + 4H \rightarrow C_3S_2H_3 + CH \tag{6b}$$

$$C_3A + C\bar{S}H_2 + 10H \rightarrow C_4A\bar{S}H_{12} \tag{6c}$$

$$C_4AF + 2CH + 10H \rightarrow C_6AFH_{12} \tag{6d}$$

FH has a composition closer to Portland cement than low-calcium fly ash (FL) or silica fume (SF). The active part of FH consists of an aluminosilicate glass phase, a calcium phase, and a sulfate phase. The majority of the active part of the silica in FH is present in aluminosilicate (A–S) and calcium aluminosilicate (C–A–S) glass. Calcium silicate hydrate (CSH) is formed from the pozzolanic action of the reactive silica of FH as follows [11]:

$$2S + 3CH \rightarrow C_3S_2H_3 \tag{7a}$$

Part of alumina, found as tricalcium aluminate (C_3A) and monocalcium aluminosulfate ($C_4A_3\overline{S}$), reacts with water, CH, and/or gypsum, as in cement hydration, showing these high early strengths. This reaction in an excess of sulfate ions (as applies in this case) can be totally described as follows [11]:

$$C_3A + C\overline{S}H_2 + 10H \rightarrow C_4A\overline{S}H_{12} \quad (7b)$$

The crystalline part of alumina found in corundum and A_3S_2 will not react. The rest of the alumina present in A–S and C–A–S glass is expected to react as follows [11]:

$$A + C\overline{S}H_2 + 3CH + 7H \rightarrow C_4AH_{12} \quad (7c)$$

The Portland cement and FH can be analyzed in terms of oxides: total CaO (C), free CaO (CF), SiO_2 (S), Al_2O_3 (A), Fe_2O_3(F), SO_3 ($\overline{S}$), and other oxides or impurities denoted by R. Let $f_{i,c}$ and $f_{i,p}$ denote the weight fractions of the constituent i (i = C, CF, S, A, F, $\overline{S}$, R) in the cement and FH other than free CaO, respectively. Let γ_S and γ_A denote the active fraction of the oxides S and A in the FH, respectively. Considering the production of calcium hydroxide from cement hydration and the reaction of free CaO in FH, the consumption of calcium hydroxide from the reaction of phases in FH other than free CaO, and the content of calcium hydroxide in FH-cement blends can be determined as follows:

$$CH(t) = C_0 RCH_{CE}\alpha + 1.321CF(t) - v_{FH}\alpha_{FH}P \quad (8a)$$

$$v_{FH} = (1.851\gamma_S f_{S,p} + 2.182\gamma_A f_{A,p}) - 1.321\left(f_{C,p} - 0.7 f_{\overline{S},p}\right) \quad (8b)$$

$$CF(t) = CF_0\alpha_{CF}(t) \quad (8c)$$

where RCH_{CE} is the produced calcium hydroxide from 1 gram of cement, $CF(t)$ is the mass of the reacted free CaO at time t, and CF_0 is the mass of the free CaO content in FH. The term $C_0RCH_{CE}\alpha$ considers the production of calcium hydroxide from cement hydration, the term $1.321CF(t)$ considers the production of calcium hydroxide from the reaction of free CaO in FH, and the term $-v_{FH}\alpha_{HF}P$ considers the consumption of calcium hydroxide from the reaction of phases in FH other than free CaO.

Similarly, the mass of calcium silicate hydrate (CSH) relates to the cement hydration and FH reaction. The mass of CSH can be determined as follows:

$$CSH(t) = 2.85 f_{S,c}\alpha C_0 + \gamma_S f_{S,p}\alpha_{FH}P) \quad (8d)$$

where $2.85f_{S,c}\alpha C_0$ considers the production of CSH from cement hydration, and $2.85\gamma_S f_{S,p}\alpha_{FH}P$ considers the production of CSH from the FH reaction.

The mass of chemically bound water relates to the cement hydration, the reaction of free CaO in FH, and the reaction of phases in FH other than free CaO. The mass of chemically bound water W_{cbm} can be determined as follows:

$$W_{cbm} = v \times C_0 \times \alpha + RCW_{FH} \times \alpha_{FH} \times P + 0.321CF(t) \quad (8e)$$

where RCW_{FH} is the mass of chemically bound water from 1 gram of reacted FH (RCW_{FH} = 0.09 [11]). The term $v \times C_0 \times \alpha$ denotes the production of chemically bound water from cement hydration, the term $0.321CF(t)$ denotes the production of chemically bound water from the reaction of free CaO in FH, and the term $RCW_{FH} \times \alpha_{FH} \times P$ denotes the production of chemically bound water from the reaction of phases in FH other than free CaO.

The mass of capillary water relates to the cement hydration, the reaction of free CaO in FH, and the reaction of phases in FH other than free CaO. The mass of capillary water W_{cap} can be calculated as follows:

$$W_{cap} = W_0 - 0.4 \times C_0 \times \alpha - RCW_{FH} \times \alpha_{FH} \times P - 0.321CF(t) - RPW_{FH} \times \alpha_{FH} \times P \quad (8f)$$

where RPW_{FH} is the mass of physically bound water from 1 gram of reacted FH (RPW_{FH} = 0.15 [11]). The term $RPW_{FH} \times \alpha_{FH} \times P$ denotes the consumption of physically bound water from the reaction of phases in FH other than free CaO.

The porosity of hydrating blends is reduced due to the Portland cement hydration, the reaction of free CaO in FH, and the reaction of phases in FH other than free CaO. The porosity, can be estimated as follows:

$$\varepsilon = \frac{W}{\rho_w} - \Delta\varepsilon_c - \Delta\varepsilon_{CF} - \Delta\varepsilon_{FH} \quad (8g)$$

where $\Delta\varepsilon_c$, $\Delta\varepsilon_{CF}$, and $\Delta\varepsilon_{FH}$ are the porosity reduction due to the Portland cement hydration, the reaction of free CaO in FH, and the reaction of phases in FH other than free CaO, respectively; these terms can be determined from the amount of chemically bound water consumed in the Portland cement hydration, the reaction of free CaO in FH, and the reaction of phases in FH other than free CaO, respectively [11].

Summary of the Proposed Hydration Model of Cement-FH Blends

Papadakis [11] proposed a chemical-based steady-state model that can be used to evaluate the final chemical and volumetric composition of concrete containing high-calcium fly ash. However, on a construction site, the workers and designers are interested in not only the final properties of the FH concrete but also the evolution of the properties of the FH concrete over time. In this paper, by combining Papadakis' chemical-based steady-state model and the kinetic reaction mechanisms involved in cement hydration and FH reaction, a kinetic model is uniquely proposed to describe the hydration process of FH concrete. (The mathematical equations shown in Section 2.3 and Section 2.4 are our original work.) The FH reactions are treated separately from the Portland cement, and some interactions are taken into account through the free water content and the calcium hydroxide content. The proposed model considers the influence of the water to binder ratio, the FH to cement ratio, the mineral compositions and the particle size of cement and FH, and the curing conditions on hydration.

Based on the degree of hydration, the development of early-age properties and the durability aspect of FH concrete can be predicted.

The flow chart for a numerical simulation process is shown in Figure 1. The proposed model considers the influences of the Portland cement hydration, the reaction of free CaO in FH, and the reaction of phases in FH other than free CaO on the property developments of cement-FH blends. Using the degree of hydration of cement α, the reaction degree of free CaO in FH α_{CF}, and the reaction degree of phases in FH other than free CaO α_{FH} at each time step, the amount of calcium hydroxide, capillary water, chemically bound water, porosity, CSH, and compressive strength can be calculated.

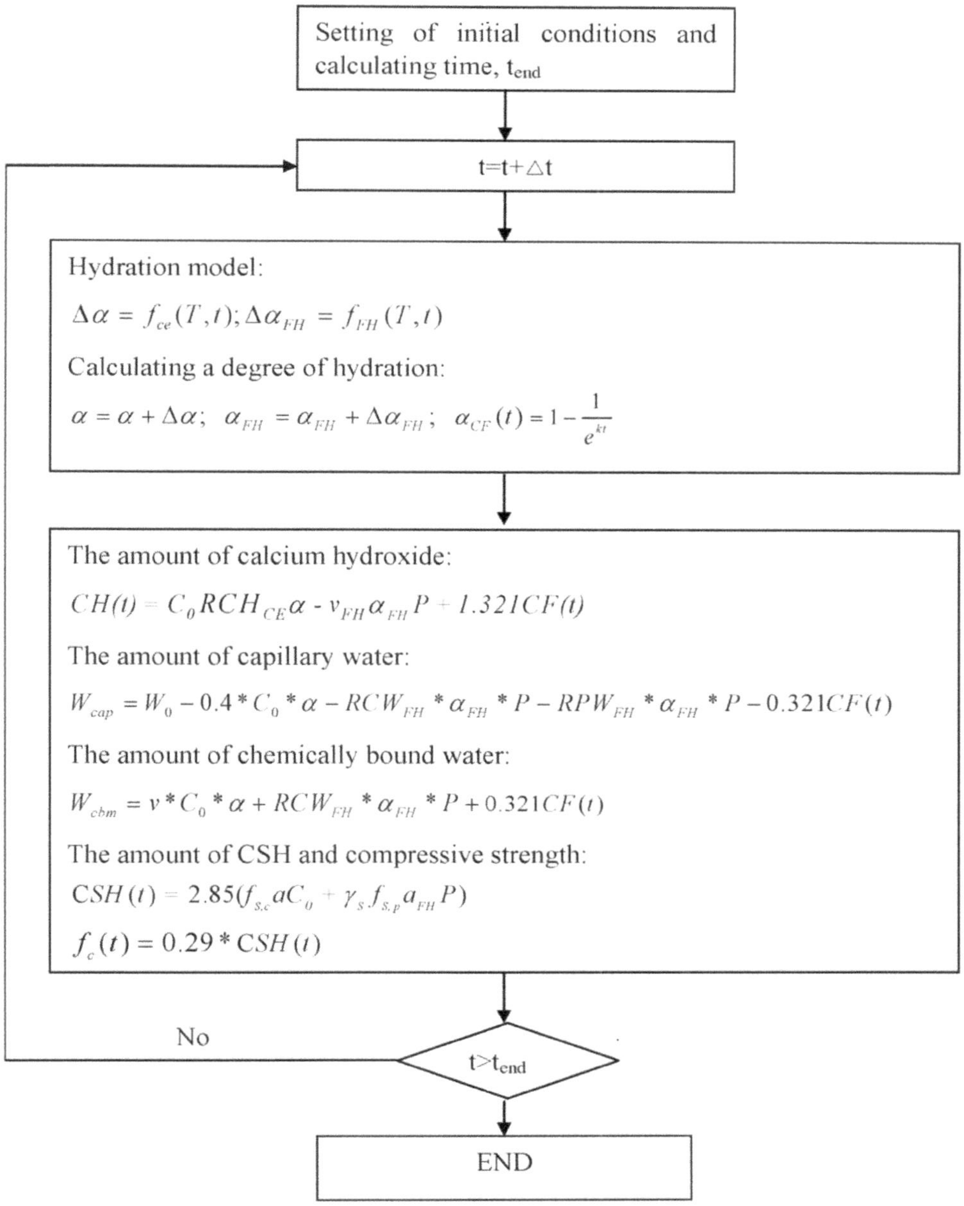

Figure 1. The flow chart for a numerical simulation process.

VERIFICATION OF THE PROPOSED MODEL

Experimental results from reference [11] are used to verify the proposed model. Papadakis [11] studied property developments of high-calcium fly ash blended mortars. A rapid-setting Portland cement was used (400 m^2/kg Blaine's fineness). Prior to use, high-calcium fly ash was pulverized to meet the cement mean particle size. The physical properties and chemical compositions of cement and FH are shown in Table 2. The total CaO content in FH is 23%, the free CaO content in FH is 5%, and the glass content in FH is 50%. The mixture proportions are given in Table 3. In the control specimen, the water-cement ratio (*W/C*) was 0.5 and the aggregate-cement ratio (*A/C*) was 3. Two different cases were studied: FH replaces either aggregate or cement. When FH replaces aggregate, three contents of FH were selected: 10%, 20%, and 30% additions to the control cement weight for specimens FHA, 10%, FHA, 20%, and FHA, 30%, respectively. When FH replaces the cement, the same FH contents were selected: 10%, 20%, and 30% replacements of the control cement weight for specimens FHC, 10%, FHC, 20%, and FHC, 30%, respectively.

The developments of strength, porosity, bound water,

and calcium hydroxide were measured [11]. The specimens for the strength measurements were prisms of 40 × 40 × 160 mm, cured under lime-saturated water at 20 °C and tested after 3, 14, 28, 49, 91, 182, and 364 days. FH-cement pastes were also prepared, representing the paste matrix of the mortar specimens, and analyzed for CH content, chemically bound water content, and porosity at 3, 7, 14, 28, 49, 112, 182, and 364 days after casting and moist curing [11]. For each test, the moist-cured paste specimen was stripped, placed in a pre-weighed glass mortar, and pulverized. The mortar with the material was placed in an oven at 105 °C until the weight was constant. These weight indications were used to determine the chemically bound water content and porosity. To determine the CH content, a paste sample of the oven-dried powder was examined by a combination of quantitative X-ray diffraction analysis and TG/DTA (thermogravimetry/differential thermal analysis). The mass loss corresponding to the decomposition of $Ca(OH)_2$ occurs between 440 °C and 520 °C. The paste properties were reduced to the corresponding mortar specimen volume by multiplying the appropriate paste volume/mortar volume fraction [11].

Table 2. Physical and chemical characteristic of cement and FH.

Properties	Cement	FH
Physical properties	-	-
BET specific surface (m^2/g)	1.3	6.2
Particle mean diameter (μm)	14.0	12.6
Density (kg/m^3)	3130	2660
Chemical analysis (%)	-	-
SiO_2	20.10	39.21
Al_2O_3	4.25	16.22
Fe_2O_3	3.49	6.58
CaO	63.20	22.78 (5.18 free)
SO_3	2.88	4.3
LOI	0.86	2.10

Table 3. Mixing proportions of the specimens.

Specimens	Cement (kg/m^3)	Water (kg/m^3)	FH (kg/m^3)	Aggregate (kg/m^3)	Water to Binder Ratio	FH to Cement Ratio	Aggregate to Cement Ratio
control	514.6	257.4	0	1543.8	0.5	0	3
FHA, 10%	514.6	257.4	51.5	1492.1	0.46	0.1	2.9
FHA, 20%	514.6	257.4	102.9	1440.5	0.42	0.2	2.8
FHA, 30%	514.6	257.4	154.4	1388.8	0.39	0.3	2.7
FHC, 10%	463.1	257.4	51.5	1536.0	0.5	0.11	3.32
FHC, 20%	411.7	257.4	102.9	1528.3	0.5	0.25	3.71
FHC, 30%	360.2	257.4	154.4	1520.5	0.5	0.43	4.22

Evaluation of Calcium Hydroxide Contents

In the proposed model, the reaction of free CaO and other phases in FH are simulated separately. Equation (3) is used to model the rapid reaction of free CaO in FH, and Equation (4) is used to model the reaction of phases in FH other than free CaO. Using the chemical analysis of FH shown in Table 2, we can determine the weight fractions of S, A, C, and $\overline{S}$ in the FH other than free CaO. Furthermore, the active part of FH other than free CaO, γ_{active}, which includes an aluminosilicate glass phase, a calcium phase, and a sulfate phase, can be calculated. The weight fractions of FH used for modeling are shown inTable 4.

Using the experimental results of calcium hydroxide from Portland cement and a predictor-corrector algorithm [19], the reaction coefficients of Portland cement B_{ce}, C_{ce}, D_{e0}, and k_r can be determined. Using the experimental results of calcium hydroxide from cement-FH paste, the FH reaction coefficients B_{FH}, C_{FH}, D_{e0FH}, and k_{rFH} can be determined. The values of the reaction coefficients are shown in Table 5. The fit parameters for a material are not changed from one mix to the other. For concrete with different water to binder ratios or FH replacement ratios, the reaction coefficients of cement and FH do not change.

Table 4. Weight fractions of FH.

γ_S	$f_{S,p}$	γ_A	$f_{A,p}$	$f_{\bar{S},p}$	$f_{C,p}$	CaO_{free}
0.7	0.413	0.7	0.1711	0.0453	0.1856	0.0518
$\gamma_{active} = \gamma_S f_{S,p}(1-CaO_{free}) + \gamma_A f_{A,p}(1-CaO_{free}) + f_{C,p}(1-CaO_{free}) + f_{\bar{S},p}(1-CaO_{free})$						0.61

Table 5. Coefficients of the reaction model.

B_{ce} (cm/h)	C_{ce} (cm/h)	k_r (cm/h)	D_{e0} (cm²/h)
1×10^{-8}	0.81	2.33×10^{-6}	3.79×10^{-10}
B_{FH} (cm/h)	C_{FH} (cm/h)	k_{rFH} (cm/h)	D_{e0FH} (cm²/h)
2.22×10^{-11}	0.08	5.45×10^{-7}	2×10^{-8}

After calibrating the coefficients of the reaction model, multiple checks, such as the evolution of chemically bound water, porosity, CSH contents, and compressive strength, were performed to verify the proposed model. The procedure of the proposed model is similar to that of Tomosawa *et al.* [15,19]. In Tomosawa's model, the heat evolution of Portland cement was adopted to calibrate the coefficients of the hydration model. Furthermore, the evolution of the mechanical properties in early-age concrete was described using functions of the degree of hydration [15,19].

The evaluation results of the calcium hydroxide contents are shown in Figure 2. The CH content of the control specimen increases with time until a steady state is attained (Figure 2a). The CH content of FHA specimens presents a rather complicated picture due to the simultaneous CH production from the reaction of cement and the free CaO and CH consumption from other phases of the FH. In the first week, due to the rapid free CaO hydration, the CH contents are higher in the fly ash specimens (Figure 2b–d) than that of the control specimen (Figure 2a), These CH contents pass through a maximum due to higher CH production but decrease afterward as the FH-CH reaction proceeds at higher rates. At a late age, with the FH replacing ratio increasing from 10% to 30%, the CH contents decrease correspondingly. For concrete with 10% FH (Figure 2c) and 20% FH (Figure 2b), at the age of one year, the calcium hydroxide will slightly increase. This result may be due to the depletion of the active part of FH. Because the proposed model has modeled the rapid production of calcium hydroxide from the free CaO reaction in FH, the production of calcium hydroxide from cement hydration, and the consumption of calcium hydroxide from other phase reactions in FH, the proposed model can describe the complicated evolution process of cement-FH blends. Figure 2e shows the comparison between the experimental results and the analysis results for different mixing proportions. The correlation coefficient between the experimental results and the analysis results is 0.95.

In addition, Figure 2 shows the analysis results of calcium hydroxide from Papadakis' model [11]. Papadakis assumed that at the age of 365 days, the reaction rate of cement and FH is very slow. The age of 365 days can be approximately regarded as the steady state age of FH blended concrete. As shown in Figure 2, Papadakis' model can calculate the final CH content. However, because Papadakis' model does not consider the kinetic reaction process of FH, this model cannot calculate the evolution of CH in cement-FH blends. The calculation result of Papadakis' model is a point, not a curve. Conversely, the calculation result from our kinetic hydration model is a curve.

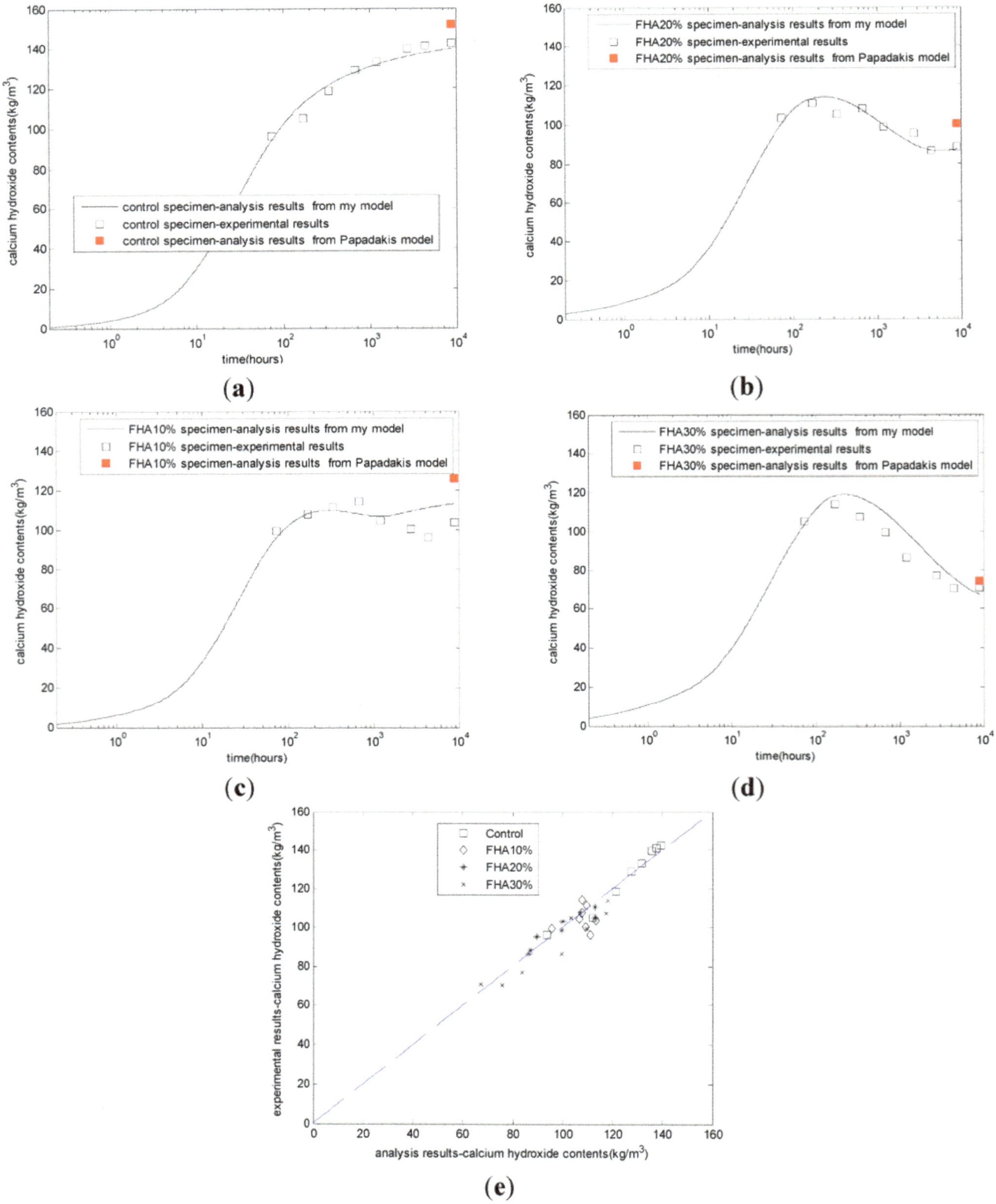

Figure 2. Development of calcium hydroxide contents. (**a**) Control concrete; (**b**) FHA, 20% concrete: FH replaces aggregate by 20% weight of cement; (**c**) FHA, 10% concrete: FH replaces aggregate by 10% weight of cement; (**d**) FHA, 30% concrete: FH replaces aggregate by 30% weight of cement; (**e**) comparison between analysis results and experimental results.

The evaluation of the reaction degree of fly ash $\alpha_{FH\text{-}total}$ is shown in Figure 3. As shown in this figure, given a certain water to binder ratio, with an increasing replacement level of fly ash, the alkaline-activating effect of the cement would be weaker, so that the reactivity of the fly ash decreases [20]. Conversely, at a late age, for concrete with 10% and 20% FH, the active part of FH has totally reacted (the ultimate reaction degree equals the weight fraction of the active part γ_{active}). For concrete with 30% FH, until the age of one year, part of FH is still un-reacted.

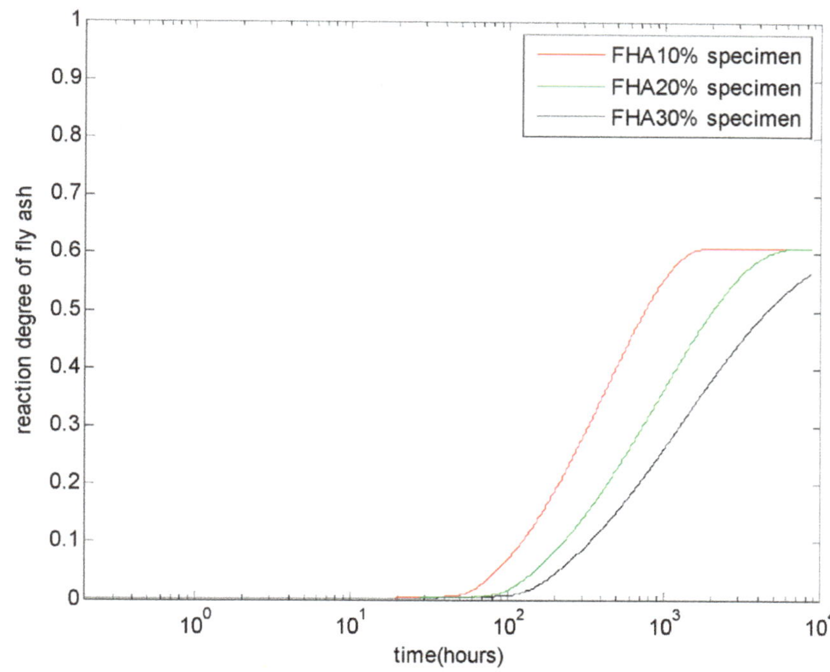

Figure 3. Reaction degree of high-calcium fly ash with different FH contents: FH replaces aggregate by 10%, 20%, and 30% weight of cement and a water to binder ratio of 0.5.

Evaluation of Chemically Bound Water and Porosity

The evaluation results of chemically bound water and porosity contents are shown in Figure 4 and Figure 5, respectively. A higher chemically bound water (H) content (Figure 4) and a lower porosity (Figure 5) is observed for all of the FHA specimens compared to the corresponding control values from the initiation of the cement hydration. This increase in H values or the similar decrease in porosity values is almost proportional to the fly ash content in the specimen. This behavior can be partly explained as being due to the presence of free CaO in the FH and the hydration of the tricalcium aluminate present in FH. The early formation of water-rich and pore-filling ettringite and its subsequent transformation to other hydrates contributes significantly to the increased hydration and decreased porosity [11].

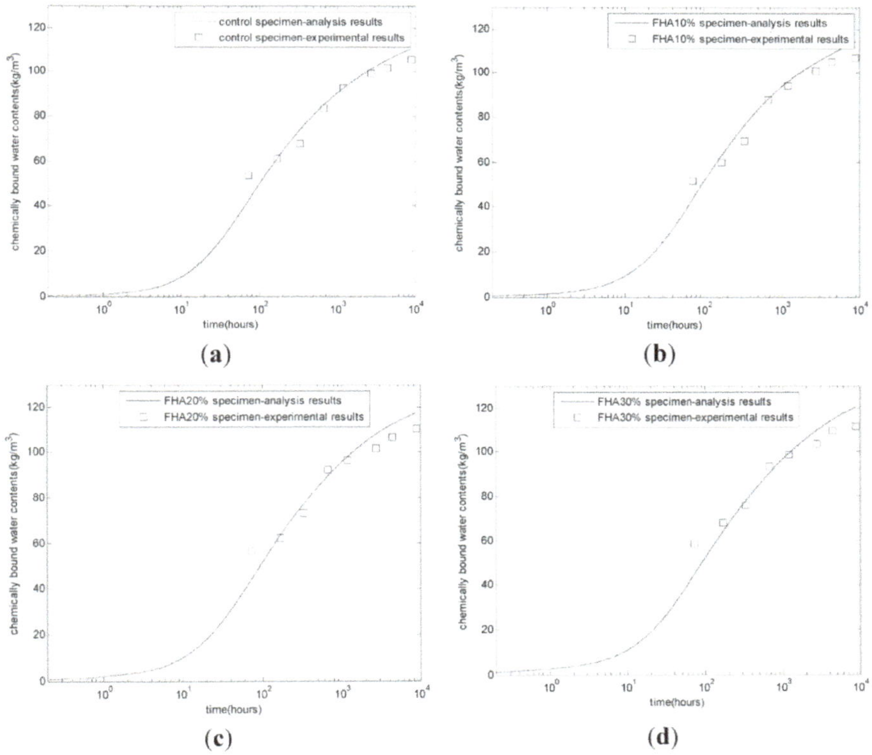

Figure 4. Development of chemically bound water contents. (**a**) Control concrete; (**b**) FHA10% concrete: FH replaces aggregate by 10% weight of cement; (**c**) FHA20% concrete: FH replaces aggregate by 20% weight of cement; (**d**) FHA30% concrete: FH replaces aggregate by 30% weight of cement.

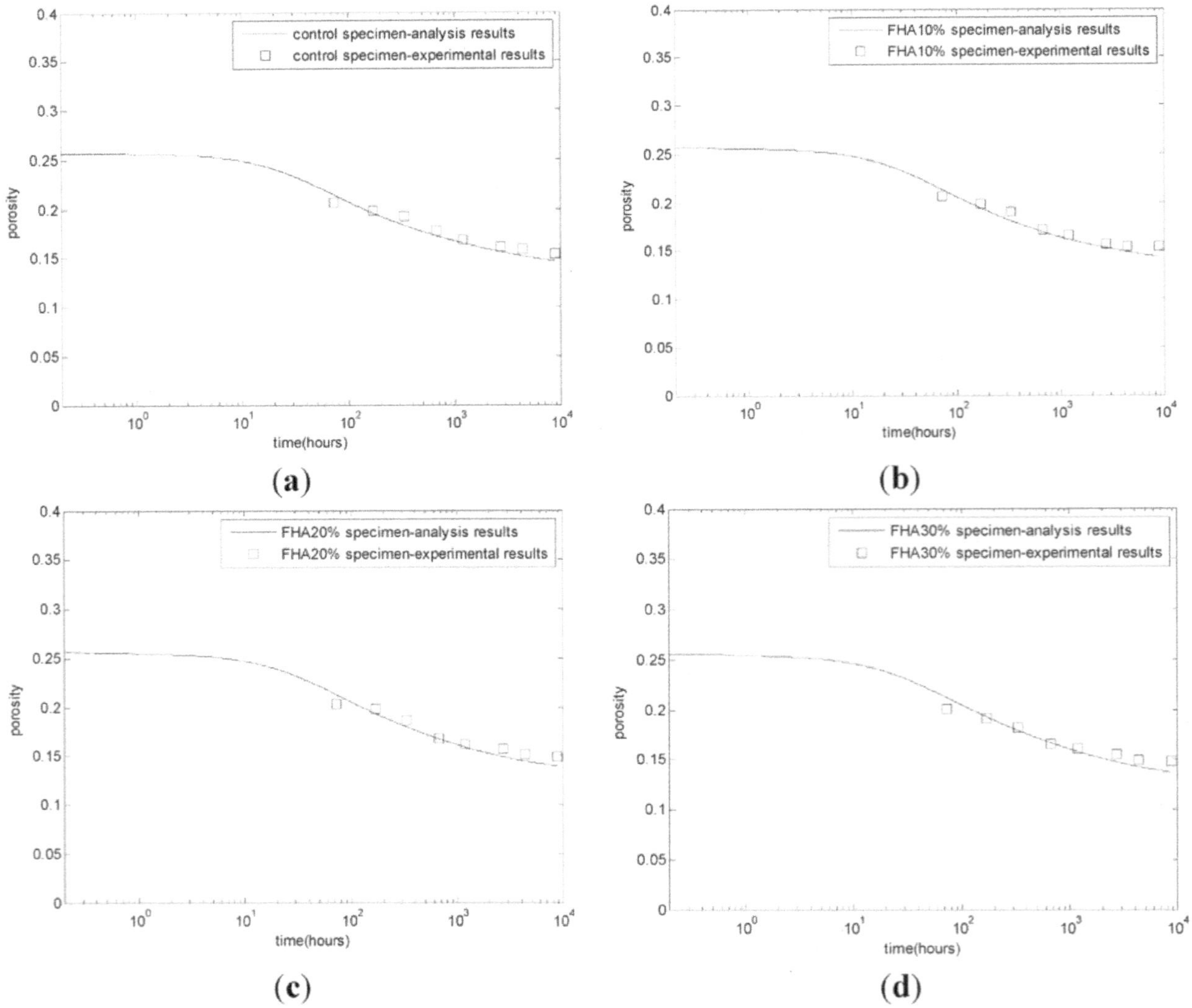

Figure 5. Development of porosity contents. (**a**) Control concrete; (**b**) FHA10% concrete: FH replaces aggregate by 10% weight of cement; (**c**) FHA20% concrete: FH replaces aggregate by 20% weight of cement; (**d**) FHA30% concrete: FH replaces aggregate by 30% weight of cement.

Evaluation of Compressive Strength

Papadakis [11,12] proposed that the compressive strength of cement-based materials, such as concrete containing low-calcium fly ash, high-calcium fly ash, and slag, can be estimated from the CSH content. Using the proposed hydration model considering both cement hydration and FH reaction (Equation (8d)), we can calculate the CSH contents for mortars with different FH contents, such as the control specimen, FHA, 10% (FH replaces aggregate by 10% weight of cement), FHA, 20% (FH replaces aggregate by 20% weight of cement), FHA30% (FH replaces aggregate by 30% weight of cement), and FHC, 20% (FH replaces cement by 20%). Figure 6 presents the compressive strength of the FH blended mortars as a function of the calculated CSH contents. As shown in Figure 6, for concrete with different FH contents (control, FHA, 10%, FHA, 20%, FHA, 30%, and FHC, 20%) and at different ages (3, 14, 28, 49, 91, 182, and 364 days), a single linear relationship exists between the compressive strengths and the calculated CSH amounts, *i.e.*, the compressive strength = 0.29 × CSH.

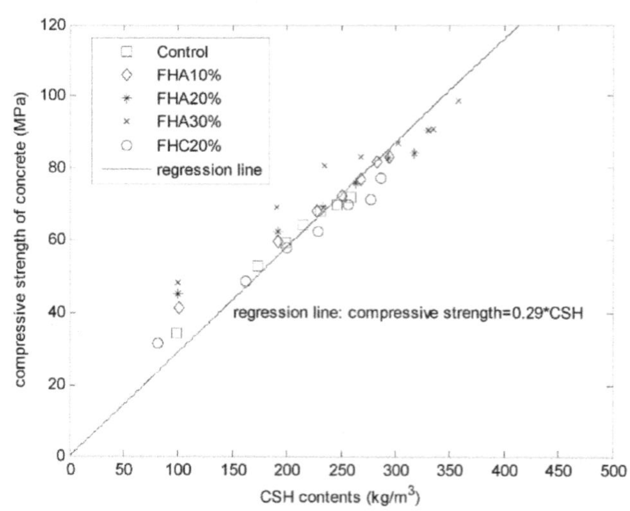

Figure 6. Compressive strength of mortars as a function of calculated (CSH) content.

Using the regressed linear relationship between the compressive strength and the CSH content, we can evaluate the development of the compressive strength of FH-cement blends. Figure 7 shows the comparison between the experiment results and the simulation results of the compressive strength. For mortar containing FH-replacing aggregate (FHA mortars, as shown in Figure 7b–d), the compressive strength of FHA specimens is higher than that of the control specimen (Figure 7a) since the early ages. However, for mortar containing FH-replacing cement (FHC mortar, as shown in Figure 7e), at early ages, the compressive strength of the FHC specimen is lower than that of the control specimen (Figure 7a), and at late ages, due to the evolution of the FH reaction, the compressive strength of the FHC specimen surpasses that of the control specimen. Conversely, in the early age of three days, the simulation value is slightly lower than the experimental value. This result may be due to the omission of some factors, such as the contribution of ettringite from the FH reaction [11] and the nucleation effects of FH [11]. For FHA, 30% concrete (Figure 7d), due to the abundant ettringite produced from the FH reaction, at the early ages of three and 14 days, the analysis results show more deviations from the experimental results than for the other mixing proportions.

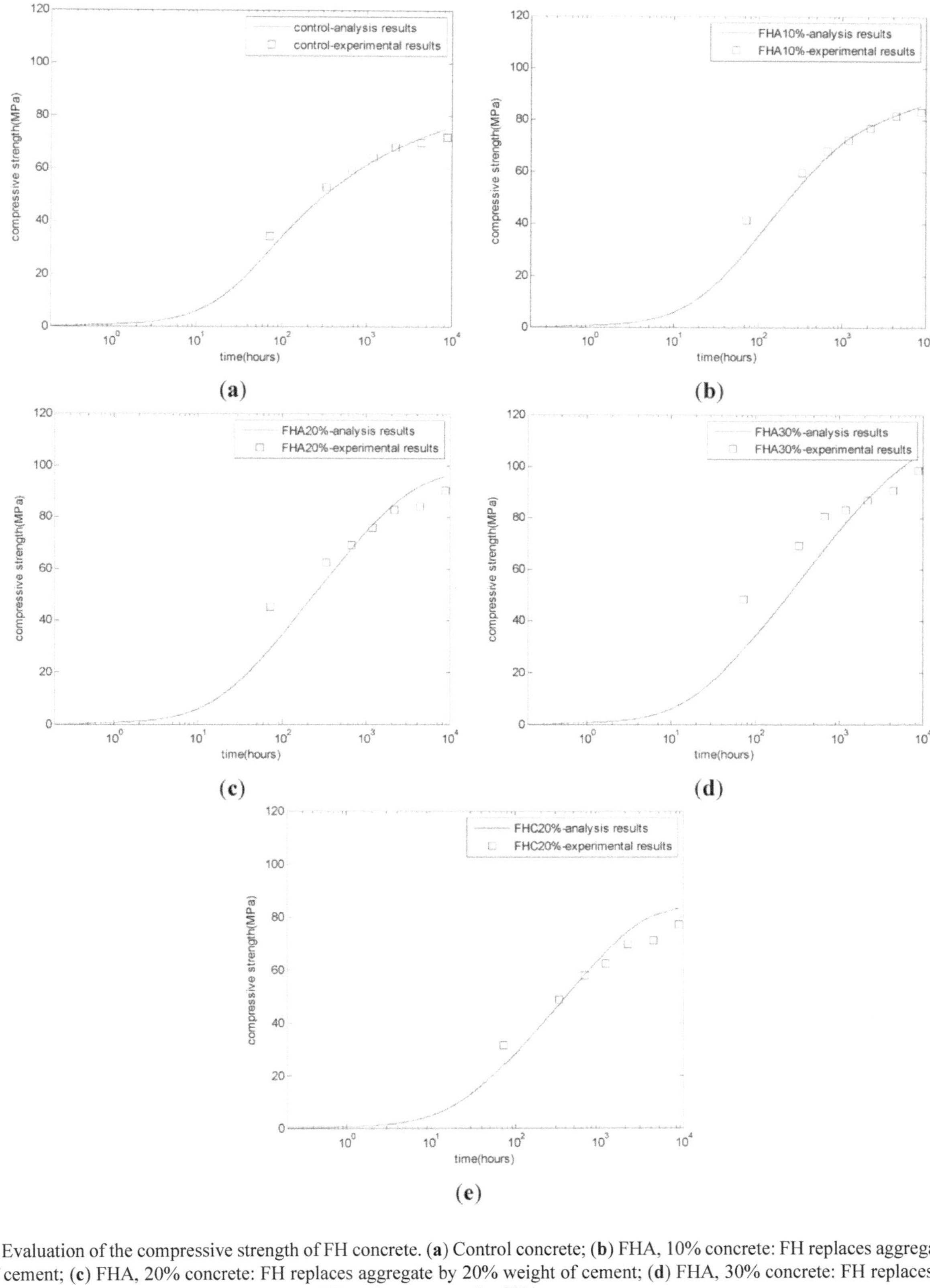

Figure 7. Evaluation of the compressive strength of FH concrete. (**a**) Control concrete; (**b**) FHA, 10% concrete: FH replaces aggregate by 10% weight of cement; (**c**) FHA, 20% concrete: FH replaces aggregate by 20% weight of cement; (**d**) FHA, 30% concrete: FH replaces aggregate by 30% weight of cement; (**e**) FHC, 20% concrete: FH replaces cement by 20%.

In addition, the free CaO in FH also affects the compressive strength developments of hardening cement-FH blends. Tsimas and Moutsatsou-Tsima [7] investigated the influence of free CaO contents on the compressive strength development. When the free CaO content increases from 1.8% to 3.5%, the compressive strength will increase correspondingly. This result is mainly due to the stimulus of the free CaO on the initial FH reaction [7]. However, when the free CaO content further increases from 3.5% to 6.8%, the soundness of the concrete will be impaired and the compressive strength will decrease. Therefore, the free CaO presents double-edged effects on the strength development. For the model proposed in this paper, further investigations should be performed regarding FH with various chemical compositions and free CaO contents.

In the concrete industry, considering the economic and environmental effects, high-calcium fly ash is generally used as a mineral admixture to replace cement. Using the hydration model, we can calculate the CSH contents. Furthermore, using the regressed linear relationship between the compressive strengths and CSH amounts, we can calculate the compressive strength development of FH concrete. Figure 8 shows the analysis results of the compressive strength of FH blended concrete with a water to binder ratio of 0.5 and different FH contents: FH replacing cement by 10%, 20%, and 30%. The proposed model can reproduce the compressive strength crossover phenomenon between the control Portland cement concrete and the FH blended concrete. With increasing FH replacement ratios, the reactivity of FH will decrease (as shown in Figure 3) and the age corresponding to the crossover of the compressive strength will be postponed.

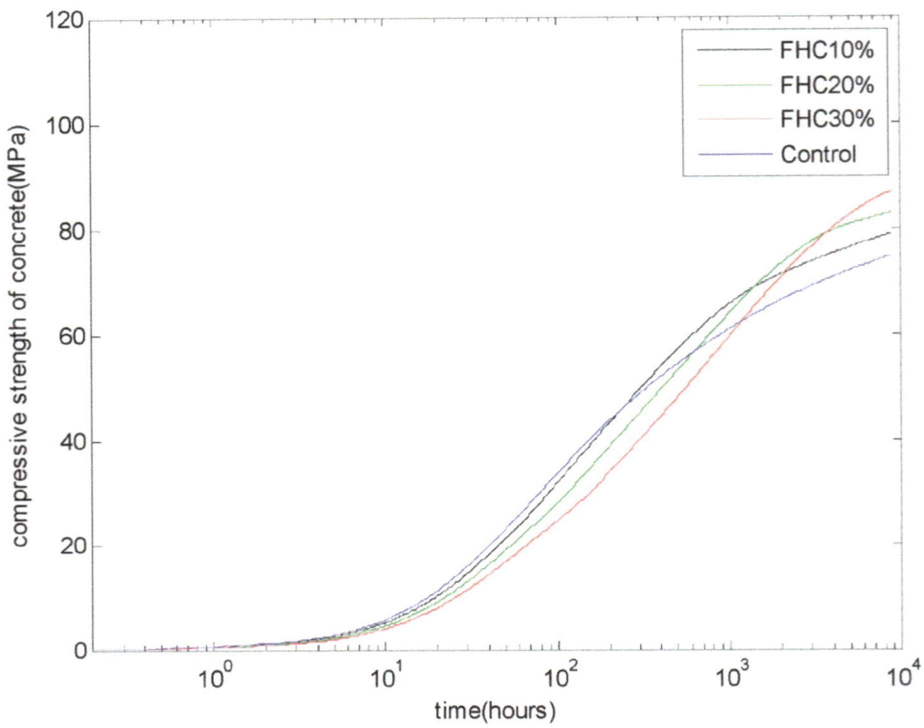

Figure 8. Compressive strength development of FH concrete: water to binder ratio of 0.5 and FH replaces cement by 10%, 20%, and 30%.

Figure 9 shows the compressive strength development of FH blended concrete with different FH contents (FH replaces cement by 10% to 70%) at different ages of seven days, 28 days, 90 days, and 365 days. At an early age of seven days, compared to the control concrete, the compressive strength of FH concrete almost linearly decreases with increasing FH content. With increasing curing age, due to the development of the FH reaction, the compressive strength of FH concrete will surpass that of the control concrete. However, for concrete incorporating larger FH contents (more than 50% cement is replaced by FH), due to the reduction of FH reactivity, at the age of one year, the compressive strength of FH concrete is still lower than that of the control concrete. Therefore, 50% can be regarded as the maximum replacement ratio of FH. Over this maximum replacement ratio, the compressive strength of FH blended concrete will be lower than that of the control concrete. Conversely, at the age of one year, when the FH replacement ratio is higher than 30%, the compressive strength of concrete will decrease. Therefore, 30% FH can be regarded as an optimum FH content for concrete incorporating FH as a cement-replacing mineral admixture. The optimum FH content can be explained as follows: according to Equation (8d), the CSH content produced from the FH reaction, $2.85\gamma_S f_{S,p}\alpha_{FH}P$, relates to the FH content P and the reaction degree of FH α_{FH}. With increasing FH content P, the reactivity of FH will decrease (as shown in Figure 3), and therefore, there is a peak value for the product of FH content P and the reaction degree of FH α_{FH}. The FH content corresponding to the peak value can be regarded as the optimum replacement of FH.

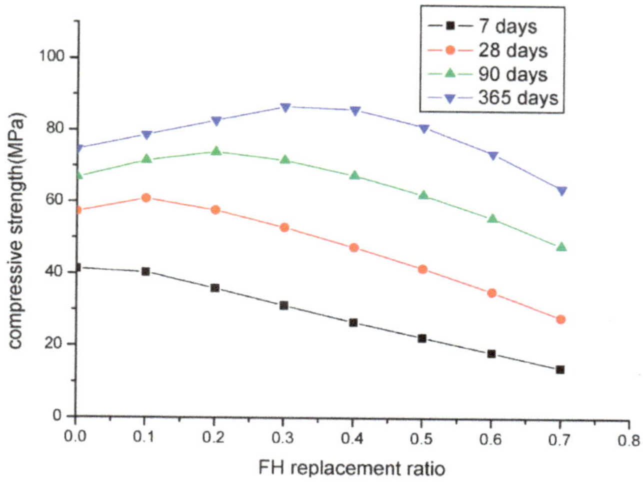

Figure 9. Parameter analysis of the compressive strength of concrete: a water to binder ratio of 0.5 and FH replaces cement by 10% to 70%.

CONCLUSIONS

(1) This paper proposes a numerical procedure to simulate the hydration of cement and high-calcium fly ash (FH) blends. The hydration of cement, the reaction of free CaO in FH, and the reaction of phases in FH other than free CaO are modeled. The interactions among the hydration of cement, the reaction of free CaO in FH, and the reaction of phases in FH other than free CaO are considered through the available calcium contents and capillary water contents in the system. The reaction of free CaO in FH is described as a first-order reaction. The reaction of other phases in FH is divided into three processes, *i.e.*, an initial dormant period, a phase boundary reaction process, and a diffusion process. The reaction coefficients of other phases in FH are obtained from experimental results of calcium hydroxide in cement-FH blends. The fit parameters for a material are not changed from one mix to the other. For concrete with different water to binder ratios or FH replacement ratios, the reaction coefficients of cement and FH do not change.

(2) Using the numerical procedure, the hydration degree of cement, the reaction degree of free CaO in FH, and the reaction degree of phases in FH other than free CaO are calculated. The amount of calcium hydroxide, bound water, porosity, and compressive strength of FH blended concrete are determined considering the contributions from the hydration of cement, the reaction of free CaO in FH, and the reaction of phases in FH other than free CaO. At different curing ages, for FH blended concrete with different FH contents, a single linear relationship exists between the compressive strengths and the calculated CSH amounts. The proposed model can reproduce the compressive strength crossover phenomenon between control Portland cement concrete and FH blended concrete. The proposed model also can be used to determine the optimum content of FH.

ACKNOWLEDGMENTS

This research was supported by Basic Science Research Program through the National Research Foundation of Korea (NRF) funded by the Ministry of Science, ICT & Future Planning (No. 2015R1A5A1037548).

AUTHOR CONTRIBUTIONS

Wei-jie Fan conducted the programming and wrote the initial draft of the manuscript. Xiao-Yong Wang designed the project and analyzed the data. Ki-Bong Park wrote the final manuscript. All of the authors contributed to the analysis of the data and read the final paper.

Conflicts of Interest

The authors declare no conflict of interest.

REFERENCES

1. Metha, P.K.; Monteiro, P.J.M. *Concrete: Microstructure, Properties and Materials*; McGraw-Hill: New York, NY, USA, 2006.
2. Wang, Q.; Feng, J.J.; Yan, P.Y. An explanation for the negative effect of elevated temperature at early ages on the late-age strength of concrete. *J. Mater. Sci.* **2011**, *46*, 7279–7288.
3. Hwang, K.; Noguchi, T.; Tomosawa, F. Prediction model of compressive strength development of fly-ash concrete. *Cem. Concr. Res.* **2004**, *34*, 2269–2276.
4. Maekawa, K.; Ishida, T.; Kishi, T. *Multi-Scale Modeling of Structural Concrete*; Taylor & Francis: London, UK, 2009.
5. Wang, X.Y.; Lee, H.S. Modeling the hydration of concrete incorporating fly ash or slag. *Cem. Concr. Res.* **2010**, *40*, 984–996.
6. Yu, Z.Q.; Ye, G. The pore structure of cement paste blended with fly ash. *Constr. Build. Mater.* **2013**, *45*, 30–35.
7. Tsimas, S.; Moutsatsou-Tsima, A. High-calcium fly ash as the fourth constituent in concrete: Problems, solutions and perspectives. *Cem. Concr. Compos.* **2005**, *27*, 231–237.
8. Antiohos, S.; Tsimas, S. Investigating the role of reactive silica in the hydration mechanisms of high-calcium fly ash/cement systems. *Cem. Concr. Compos.* **2005**, *27*, 171–181.
9. Erdoğdu, K.; Türker, P. Effects of fly ash particle size on strength of portland cement fly ash mortars. *Cem. Concr. Res.* **1998**, *28*, 1217–1222.
10. Papadakis, V.G.; Pedersen, E.J.; Lindgreen, H. An AFM-SEM investigation of the effect of silica fume and fly ash on cement paste microstructure. *J. Mater. Sci.* **1999**, *34*, 683–690.
11. Papadakis, V.G. Effect of fly ash on Portland cement systems: Part II. High-calcium fly ash. *Cem. Concr. Res.* **2000**, *30*, 1647–1654.
12. Papadakis, V.G.; Tsimas, S. Supplementary cementing materials in concrete Part I: Efficiency and design. *Cem. Concr. Res.* **2002**, *32*, 1525–1532.
13. Schindler, A.K.; Folliard, K.J. Heat of hydration models for cementitious materials. *Am. Concr. Inst. Mater. J.* **2005**, *102*, 24–33.

14. Saeki, T.; Monteiro, P.J.M. A model to predict the amount of calcium hydroxide in concrete containing mineral admixture. *Cem. Concr. Res.* **2005**, *35*, 1914–1921.

15. Tomosawa, F. Development of a kinetic model for hydration of cement. In Proceedings of the Tenth International Congress Chemistry of Cement, Gothenburg, Sweden, 2–6 June 1997; pp. 43–50.

16. Park, K.B.; Jee, N.Y.; Yoon, I.S.; Lee, H.S. Prediction of temperature distribution in high-strength concrete using hydration model. *Am. Concr.Inst. Mater. J.* **2008**, *105*, 180–186.

17. Chen, R.; Chen, Z.Y. Study on the hydration kinetics of free CaO in high calcium fly ash. *J. Build. Mater.* **2000**.

18. Giergiczny, Z. The hydraulic activity of high calcium fly ash. *J Therm. Anal. Calorim.* **2006**, *83*, 227–232.

19. Tomosawa, F.; Noguchi, T.; Hyun, C. Simulation model for temperature rise and evolution of thermal stress in concrete based on kinetic hydration model of cement. In Proceedings of the Tenth International Congress Chemistry of Cement, Gothenburg, Sweden, 2–6 June 1997; pp. 72–75.

20. Chen, W.; Brouwers, H.J.H.; Shui, Z.H. Three-dimensional computer modeling of slag cement hydration. *J. Mater. Sci.***2007**, *42*, 9595–9610.

CHAPTER 6

EFFECT OF INDUSTRIAL BY-PRODUCTS ON UNCONFINED COMPRESSIVE STRENGTH OF SOLIDIFIED ORGANIC MARINE CLAYEY SOILS

Chan-Gi Park, Sung-Wook Yun, Phillippe C. Baveye and Chan Yu

Keywords: Organic marine clayey soils, solidification agents, admixture, steelmaking slag, unconfined compressive strength

INTRODUCTION

Coastal mudflats (or estuarine sediments) are deposited in estuarine or river-mouth areas and are generally composed of clay, silt, calcareous shell debris, and organic matter. They have been recognized as weak grounds and are therefore not appropriate for building structures [1]. However, there is currently a need to develop harbors and other civil works, such as roads or railroads, upon such weak organic marine clayey materials.

Traditional treatment strategies for these areas tended to be overly time-consuming and expensive, and inevitably disturbed the ecology of the tidal lands. Solidification methods have recently become more popular because they improve the mechanical properties of the weak ground in a short time period, while causing minimal damage to the ecological systems at the sites [2,3,4,5].

The solidification method was developed in the late 1960s in Sweden and Japan, and involves mixing solidification agents with weak marine sediments (mainly organic clayey soils) [6]. ASTM Type I cement (ordinary Portland cement (OPC)) is now commonly used as a solidification agent in a slurry form (wet method) [6,7]. The higher the solidification agent content, the greater the amounts of calcium silicate hydrates (C–S–H) and calcium hydroxide ($Ca(OH)_2$) that are available and, hence, the greater the amount of calcium carbonate ($CaCO_3$) that eventually results from carbonation. $CaCO_3$ forms from the reaction of carbon dioxide (CO_2) with $Ca(OH)_2$ during the pozzolanic reaction and hardening process.

Organic marine clayey soils contain very fine particles, have considerable impurities and have high moisture contents. It is accepted that a high OPC addition ratio (over 20% based on the sediment dry weight) is required to achieve the treatment goal. Admixtures, such as silica or lime, can be used to reduce the addition ratio of OPC and enhance the treatment efficiency [8,9,10]. Other industrial by-products with high calcium contents and low cost could also be attractive alternative admixtures [11,12,13,14]. Additionally, significant economic benefits could be gained by replacing a substantial part of the OPC with inexpensive industrial by-products; greenhouse gas emissions associated with OPC production could be alleviated, and large amounts of industrial or societal waste could be recycled into durable underground construction materials [15].

In this general context, the primary objective of this study was to examine the potential of industrial by-products as alternative solidification agents for organic marine clayey soils. Unconfined compressive strength (UCS) tests were used to evaluate solidification effects with industrial by-product use. Changes in chemical properties and microstructural development during curing were also investigated.

MATERIALS AND METHODS

Materials

A representative sample was selected at Bosung County, Chunnam Province, approximately in the middle between

Citation: Chan-Gi Park, Sung-Wook Yun, Phillippe C. Baveye and Chan Yu, Effect of Industrial By-Products on Unconfined Compressive Strength of Solidified Organic Marine Clayey Soils, doi:10.3390/ma8085098.

the western and eastern coasts of Korea. Because estuary sediments in Korea have similar physical and chemical characteristics due to having the same geological depositional units, the succession as a typical Holocene to pre-Holocene tidal-flat sedimentary up to 20 m depth [15,16]. The organic matter content of the sample was 5.87% (Table 1). The natural water content of the sample was not much lower than its liquid limit (LL) water content. The ratio of the LLs between the as-sampled and the oven-dried sample was <0.75 and, hence, it was classified as organic high plasticity soil by Unified Soil Classification System. The pH and electrical conductivity (EC) were determined in 1:5 (w/v) soil water suspensions using a Orion 550A pH meter (Thermo Fisher Scientific, Inc., Waltham, MA, USA). The pH value of hydration reaction of OPC concrete mass is above 12 even though it must be lower than that for soils.

Table 1. Physical and chemical properties of organic marine clayey soil.

Item	Value	Item	Value
Gs [1]	2.44	Plasticity Index [2] (%)	16.67
Natural moisture content (% by mass)	60.2	Organic Content (%)	5.87
Liquid Limit (%)	68.89	pH	8.3
Plasticity Index (%)	21.1	Electric Conductivity (dS/m)	2.46
Liquid Limit [2] (%)	51.29	-	-

[1] specific gravity of soil sample; [2] result after oven drying.

In this study, the industrial by-product materials of oyster-shell powder (OSP), steelmaking slag dust (SMS) and flue-gas-desulfurized gypsum powder (FGD) were explored as alternative admixtures to OPC. The OSP was produced through processes of "calcining (at 900 °C)" and "grinding." The former consists of burning out impurities remaining on the surface of oyster-shell. SMS was produced during the separation of molten steel from impurities in steel-making electrical arc furnace (EAF). New SMS was stored and "cured" for about six months, and then was crushed, sized, and screened for sale for many different uses, *i.e.*, subbase of road, fill material, surcharge load in earth work, *etc.* During these procedures, the powder (or dust) was collected. FGD is a unique synthetic product derived from flue gas desulfurization systems at coal-fired furnace system.

Particle size distributions of the various materials were determined with a CILAS (Orleans, France) laser diffraction particle size analyzer (Figure 1). FGD was finer than the sediment and OSP was coarser than the sediment; however, the particle size distributions of OPC and the sediment were similar.

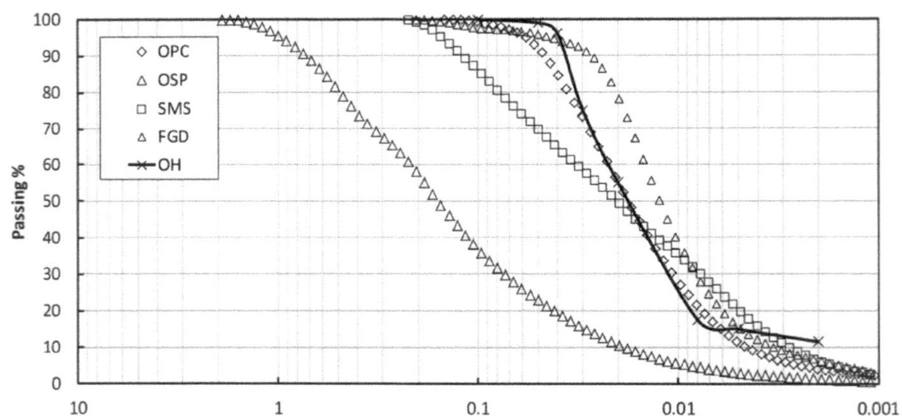

Figure 1. Particle size distribution of the different materials.

X-ray fluorescence (XRF; Model XRF-1800/WD, Shimadzu Co., Kyoto, Japan) analyses of the OPC and admixtures (Table 2) suggests that FGD had a high Ca content (> 64.9%) while that of OSP was lower (46.4%). Processed OSP is generally a component of CaO, with about 97% total CaO composition, and often applied as an additive of OPC [17]. SMS had a much lower Ca content (24.7%) but had high Si and Fe and relatively high Al and Mn contents. The mineralogical contents of the soil and admixtures were determined by XRD (Model D8 Advance, Bruker AXS, Karlsruhe, Germany). The analyzed material was ground to a powder so that it completely passed through a No. 200 (75 μm) sieve.

Table 2. Chemical properties and X-ray fluorescence (XRF) analysis results of admixtures. OPC: ordinary portland cement; OSP: oyster-shell powder; SMS: steelmaking slag dust; FGD: fuel-gas-desulfurized.

Materials	pH	EC (dS·m^{-1})	Blaine Finess (cm^2/g)	Main Elements Contents (%) by XRF								
				Ca	O	Si	Fe	Mg	Al	S	Cl	LOI ♯
OPC	12.1	7.50	2800	44.99	35.22	10.37	2.03	2.01	1.86	1.51	-	1.96
OSP	8.7	0.61	3843	46.43	48.08	1.77	0.31	-	0.56	0.30	0.83	2.12
SMS	12.0	4.95	4500	24.74	33.92	8.63	21.02	2.81	3.06	0.44	0.15	1.88
FGD	12.7	7.98	3800	64.84	-	1.45	1.65	1.37	0.38	16.17	11.53	2.49

♯ Loss on ignition.

In the XRD diffractograms for the sediment sample (Figure 2), illite and kaolinite appeared to be the dominant clay minerals; quartz (SiO_2) was also identified. OPC consisted primarily of four minerals, *i.e.*, alite (C_3S), belite (C_2S), celite (C_3A) and ferrite (C_4AF) [18,19]. The XRD diffractogram of OSP (Figure 3b) indicates that it contained ferrite, calcite with alite and belite. The SMS contained alite, belite, celite and other minerals; the latter were similar to those found in OPC (Figure 3c). The FGD contained portlandite and calcite with alite, belite and celite. On the basis of the XRD data, it was expected that SMS would be a good solidification material because of its finer particle distribution and similar chemical constituents to OPC.

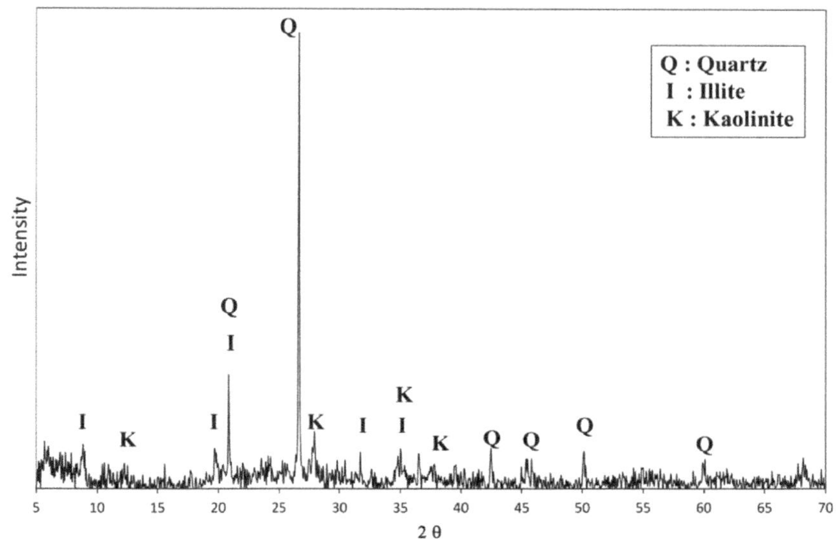

Figure 2. X-ray diffraction (XRD) diffractograms of organic marine clayey soil.

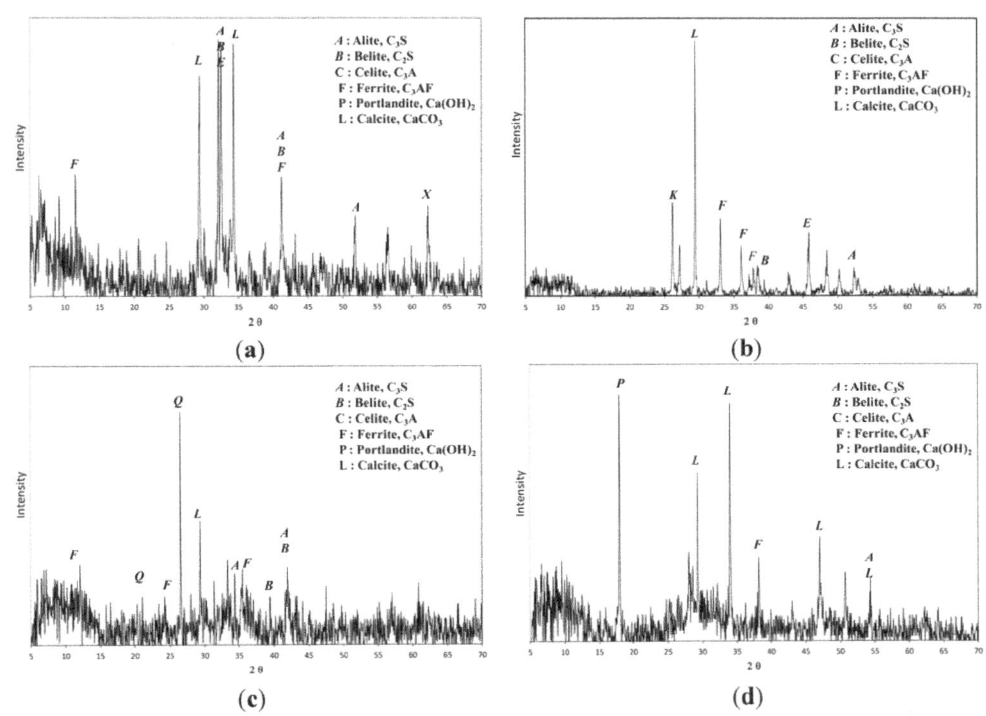

Figure 3. XRD diffractograms of ordinary Portland cement (OPC) and admixtures (**a**) OPC; (**b**) Oyster-shell powder (OSP); (**c**) Steelmaking slag dust (SMS); (**d**) Fuel-gas-desulfurized (FGD).

Extraction tests were also carried out to verify the environmental safety of the admixture materials according to a Korean standard [20]. Extraction solution (diluted HCl solution by distilled water at pH = 5.8–6.3) was added to samples (<5 mm size) to collect soluble contaminants and was shaken for 6 h at 2000 rpm. The extracted solution was then filtered through a 0.45 μm filter and concentrations were determined using an inductively coupled plasma optical-emission spectrometer (ICP-OES, ELAN DRC II, Perkin Elmer, Waltham, MA, USA).

Table 3 shows the test results on admixtures as well as soil sample (OH). None of the materials presented concentrations that exceeded permitted levels.

Table 3. Results of heavy metal concentration test.

Material	Heavy Metal Concentration (mg/L)				
	Cu	Cd	Pb	Cr	As
Standard of Korea	1.0	0.3	3.0	1.5	1.5
OH	0.031	N.D.	0.003	0.004	N.D.
OPC	0.001	N.D.	N.D.	0.729	N.D.
OSP	0.085	N.D.	0.004	0.004	0.007
SMS	0.004	N.D.	N.D.	0.007	N.D.
FGD	N.D.	N.D.	N.D.	0.013	N.D.

N.D.: Non detected.

Mixing and Manufacturing of Test Specimens

Test specimens were prepared by mixing the organic marine clayey soil with OPC and the admixture materials. Admixture materials were first dried and then added to OPC based on the dry weight at a ratio of 5% to produce the solidification agents. Most patented solidification agents, which are OPC-based systems, have been used at the 5% mixing ratio [21]. This mixing ratio was also used in this study to facilitate comparisons with previous work. The solidification agent was mixed with organic marine clayey soils but generally not directly; *i.e.*, dry-type mixing. Rather, it was mixed with organic marine clayey soils as a slurry; *i.e.*, wet-type mixing. A slurry of the solidification materials was prepared by mixing it with fresh water at a desired ratio.

The water-mixture ratio typically used is 0.8–1.2 for deep mixing, but is reduced for shallow mixing [21]. A water-mixture ratio of 1:1 was used in this study. The dosage ratio of the solidification materials to organic marine clayey soil was 5%, 10% or 15% on a dry weight basis. The mixtures (solidification agents + organic marine clayey soil + water) were mixed by hand for *ca.* 5 min. After mixing of the mixture, the duplicate specimens were molded for UCS test with curing day and mixing ratio using a cylinder mold (cylindrical PVC mold, 50 mm in diameter with an aspect ratio of 2.0). The mixture was cast into the mold and cured in a humidity chamber (23 ± 1 °C and relative humidity >96%) for 24 h and then extruded from the mold using a hydraulic jacking device; the samples, then, were sunk in the sea water tube for 7, 14 or 28 days (23 ± 1 °C and seawater) prior to strength testing. This curing condition was appropriate as they reflect conditions typically employed on local sites where treated soils are exposed under the seawater level.

Test Methods

Unconfined Compressive Strength

UCS test was conducted at a strain rate of 1 mm/min according to the ASTM C 39–96 standard [22]. Each test was performed after 7, 14 and 28 days of curing.

Optical Observations

Test involving pH, XRD and SEM (Model JEM–2010; JEOL Ltd., Tokyo, Japan) analyses, which were carried out to investigate the differences between the solidification materials with different types of admixture materials.

The Content of $CaCO_3$

The calcium content of the mixtures was assumed to derive from the products associated with the hydration reaction. The EDTA titration method was used as follows [23]. Subsamples taken from the mixtures were dried at 105 °C for 24 h, cooled and weighed. The dry mixtures were then rinsed with a 0.1 N H_2SO_4 solution to allow the precipitates to dissolve into the liquid phase. The sample plus solution was mixed by hand for 10 min and filtered through 0.45 μm filter paper, after which the calcium content was determined by EDTA titration.

RESULTS AND DISCUSSION

Unconfined Compressive Strength

The experimental results are summarized in Table 4 and shown graphically in Figure 4, Figure 5, Figure 6. The UCSes of OPC + SMS and OPC + OSP were higher than the UCS of OPC alone and OPC + FGD at 28 curing days. The UCSes of OPC alone were higher at initial curing stage than other cases, but at a mixing ratio of 15% content, the situation was reversed at 28 days of curing. It is effective in the strength development when added to an appropriate amount of pozzolanic material. Insufficient or excessive pozzolanic material would limit strength development [24,25]. The pozzolan reaction was activated by the increased OPC + OSP and OPC + SMS content at 15% mixing ratio at 28 days of curing.

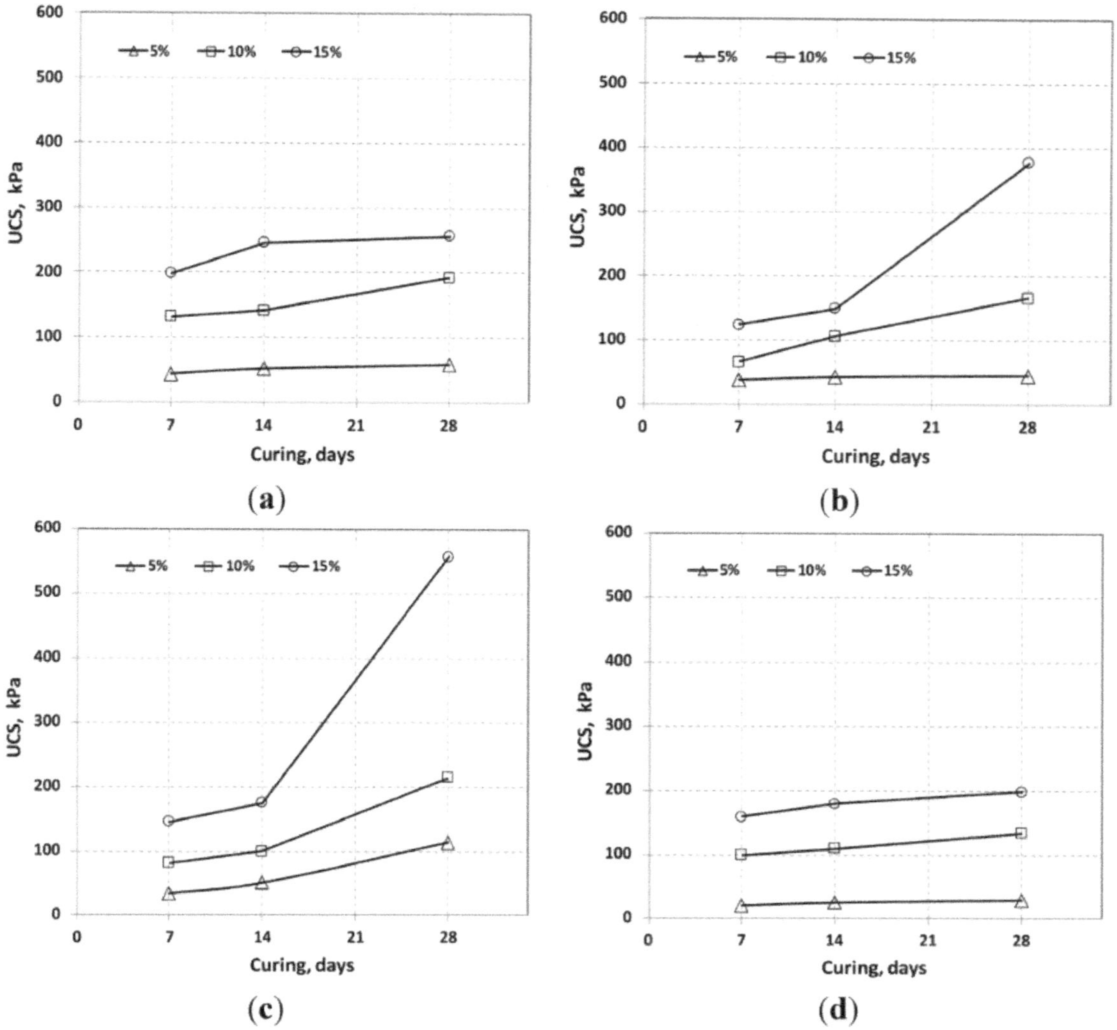

Figure 4. Relationship between curing days and unconfined compressive strength (UCS) of mixtures with admixtures. (**a**) OPC alone; (**b**) OPC + OSP; (**c**) OPC + SMS; (**d**) OPC + FGD.

The performance with OPC + FGD was mediocre: the UCS at a mixing ratio of 15% was <200 kPa at 28 curing days. It is presumed that some interference prevented the FGD reaction from occurring with the OPC or sediment particles. On the other hand, FGD decreases the early hydration rate of cement-based materials and prolongs the setting time, thus requiring long curing times to achieve a high UCS [26,27].

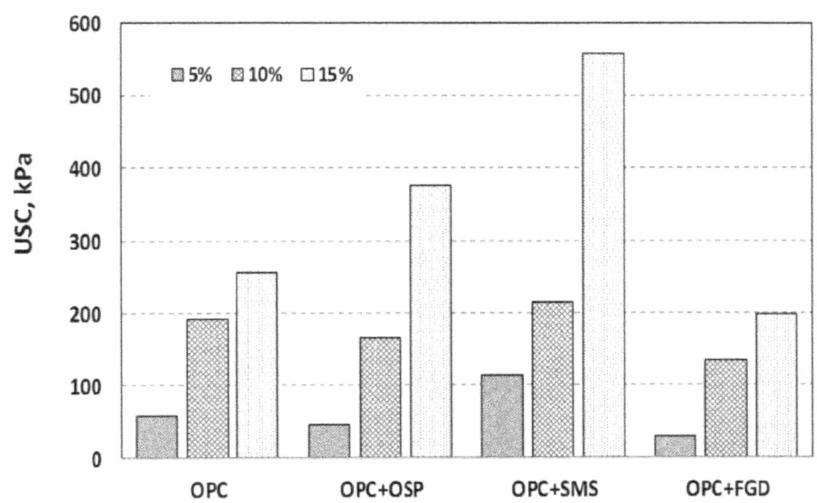

Figure 5. UCS with mixture with type of admixtures at 28 curing days.

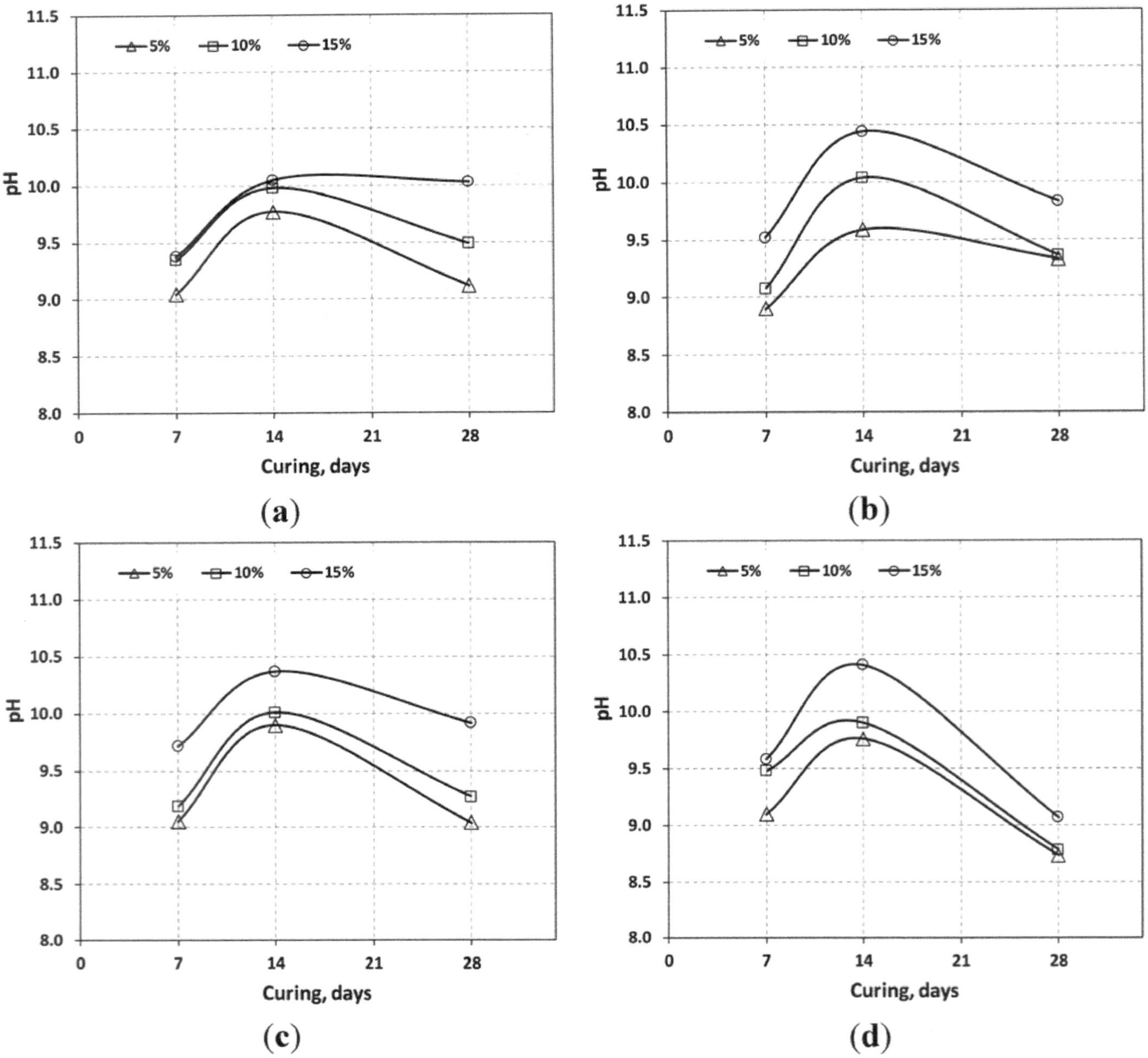

Figure 6. The pH values of mixtures with the curing time (**a**) OPC; (**b**) OPC + OSP; (**c**) OPC + SMS; (**d**) OPC + FGD.

The addition of quicklime to OPC is generally acknowledged to help counteract the high moisture content of sediments that affects the hardening process of OPC. Quicklime added to sediments modifies the fine particles, and OPC binds the modified peds of fine particles and thereby provides a degree of long-term strength gain because of enhanced pozzolanic reactions [26,27]. However, such phenomena were not clearly observed in this study, especially in the case of OPC + FGD. Furthermore, evidence in this respect was not strong at a low mixing ratio. A measureable improvement in UCS was observed under these conditions only with OPC + SMS at the 5% mixing ratio after 28 curing days.

Table 4. Results of experiment with the curing period of test specimens. UCS: unconfined compressive strength.

Curing (days)	Additive	Mixing Ratio	pH	Dry Density (g/cm^3)	UCS (kPa)
7	OPC	5%	9.05	1.06	42.6
		10%	9.35	1.06	131.2
		15%	9.38	1.01	198.8
	OPC + OSP	5%	8.90	1.02	36.9
		10%	9.08	1.01	65.1
		15%	9.52	0.99	123.2
	OPC + SMS	5%	9.06	1.02	33.9
		10%	9.19	1.00	81.6
		15%	9.72	1.03	145.6
	OPC + FGD	5%	9.10	1.04	20.0
		10%	9.48	1.04	98.5
		15%	9.58	1.04	158.1
14	OPC	5%	9.77	1.04	51.0
		10%	9.98	1.05	140.9
		15%	10.05	1.03	245.9
	OPC + OSP	5%	9.59	1.02	41.8
		10%	10.04	1.02	106.2
		15%	10.44	1.01	148.8
	OPC + SMS	5%	9.90	1.02	50.6
		10%	10.01	1.03	99.3
		15%	10.37	1.05	174.4
	OPC + FGD	5%	9.76	1.03	24.6
		10%	9.90	1.02	109.2
		15%	10.41	1.05	179.0
28	OPC	5%	9.12	1.02	57.8
		10%	9.49	1.02	191.9
		15%	10.03	1.04	256.7
	OPC + OSP	5%	9.33	1.01	44.5
		10%	9.36	1.04	165.8
		15%	9.83	1.04	376.1
	OPC + SMS	5%	9.04	1.02	113.5
		10%	9.27	1.05	214.1
		15%	9.92	1.02	556.7
	OPC + FGD	5%	8.73	1.05	29.0
		10%	8.78	1.03	133.3
		15%	9.07	1.05	198.4

The UCS results for specimens mixed with OPC + OSP and OPC + SMS were expected, showing UCS increase rate values that were 50%–100% higher than the results demonstrated by OPC after 28 days of curing. This enhancement was attributed to the addition of the calcium component (mainly lime) in OSP and SMS, which led to a relatively slow strengthening of the mixtures because of the slow lime reaction kinetics. The higher UCS with OPC + SMS than with OPC + OSP was also attributed to the finer particle distribution of the former mix.

The UCS for the 15% mixing ratio of OPC + FGD was 198.4 kPa at 28 curing days. The Japanese Cement Association (1994) requires a surcharge of 150 kPa to improve soft ground. Our data indicate that the addition of 10% OPC, OPC + OSP or OPC + SMS is sufficient to meet the UCS requirement for the movement of heavy equipment in fields. Additionally, the higher UCS with OPC + SMS means that it could be used with the pavement subgrade method.

pH Variation

The pH test results are shown in Figure 6. The pH increased up to 14 curing days and then decreased afterwards until the 28th curing day. This happened in most cases, with the exception of the sample involving 15% OPC; in this case, the pH did not change after 14 days. With OPC + OSP and OPC + SMS, the pH increased up to 14 curing days and then decreased until the 28th day. OSP and SMS mixtures had a similar range of pH values, *i.e.*, 9.3–9.8 and 9.0–9.9, respectively. The reduction in pH after 14 days of curing was more pronounced for the OPC + FGD mixture. When cementitious material is immersed in water, the alkaline component leaches out and the pH increases. However, the pH decreases with increasing age so that the decreased alkaline component leaches out and an increase in carbonate occurs [28].

Optical Observations

Optical analysis was conducted on the specimens made with SMS that had the highest UCS. This technique is often applied to solidified samples to assess changes in the mixture components with solidification materials or as a function of curing time. XRD and SEM analyses were carried out to compare the changes in the components of the 15% mixing ratio mixtures, which were the most greatly affected by the admixture materials (Figure 7).

The XRD diffractograms of all mixtures displayed prominent reflections for C–S–H, calcite and ettringite for the OPC + OSP (Figure 7b) and OPC + FGD (Figure 7d) mixtures. The diffractogram of the OPC + SMS mixture was similar to that of OPC and could be predicted from the XRD data of the individual admixture materials (see Table 2). Even at the lowest calcium oxide content among admixture materials, the observed reflections for the OPC + SMS system are evidence that hydration and the pozzolanic reaction were more active in the case of this combined system than in the other examples.

Alkalis act as catalysts in the formation of calcium silicate hydrate from calcium oxide and silica. The calcium contents of the OPC alone, OPC + OSP and OPC + SMS mixtures after 28 days of curing were measured by EDTA titration (Figure 8). There was a good correlation between the UCS and the calcium content. C–S–H gel and portlandite amount to >75% of the hydrated OPC paste during the setting process and lead to either development of strength or its enhancement.

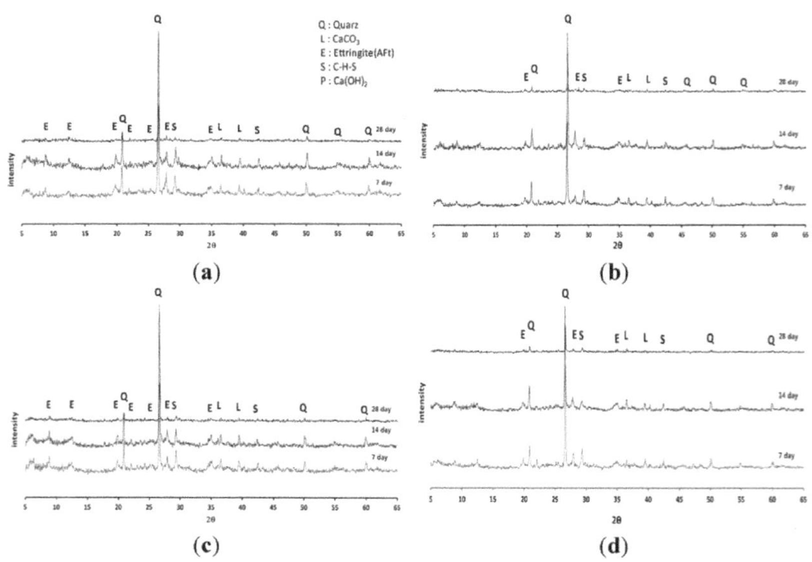

Figure 7. Comparison of XRD diffractogram of mixtures with the types of admixtures at 15% mixing ration of solidification agents. (**a**) OPC; (**b**) OPC + OSP; (**c**) OPC + SMS; (**d**) OPC + FGD.

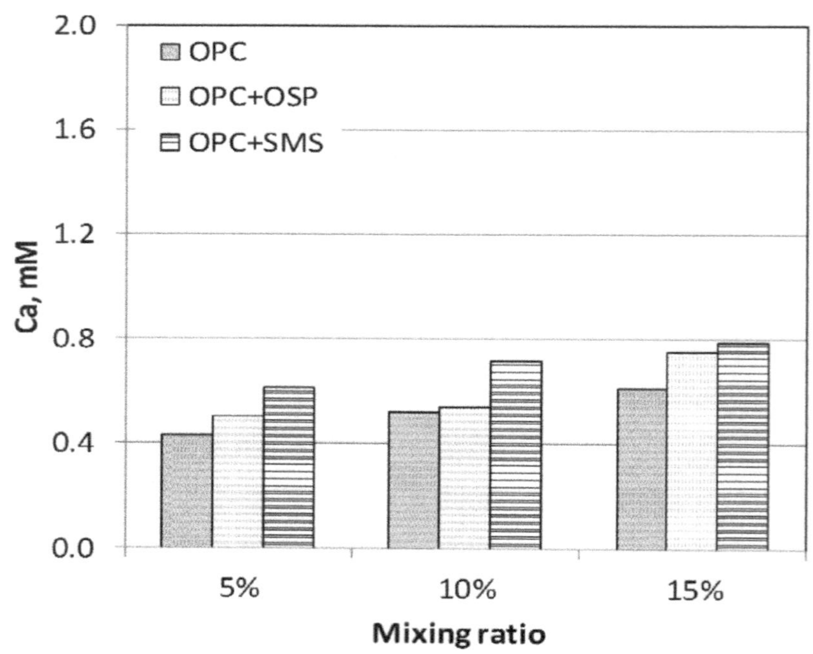

Figure 8. Ca content of mixtures with type of admixtures after 28 days.

Ettringite also forms during the pozzolanic reaction and fills the pore spaces between particles. Therefore, the higher the calcium concentration, the higher the USC that can be expected for the mixture. However, the UCS for 5% OPC alone was larger than that for 5% OPC + OSP (Figure 5), but the calcium content for 5% OPC was smaller than that for 5% OPC + OSP (Figure 7). These results were attributed to insufficient mixing and possible compaction; however, this was a relatively small effect. SEM images (Figure 9) revealed fine platelets in the sediment (organic clay) that contained the Si and Al atoms corresponding to the kaolin detected by XRD analysis; the Si, Al and K corresponded to illite. There were no large or round minerals. There were some voids between the peds' hydrated sediments for the OPC and OSP mixtures, but almost all of the voids were filled in the SMS case, giving the appearance of higher density.

Figure 9. Scanning electron microscopy (SEM) photograph of organic clayey soil and mixtures with different types of admixtures. (**a**) Sediment; (**b**) OPC; (**c**) OPC + OSP; (**d**) OPC + SMS.

CONCLUSIONS

The utility of industrial by-product materials as alternative solidification materials to OPC to improve the UCS of organic marine clayey sediments was explored. OSP, SMS and FGD were evaluated as alternative admixture materials.

The highest strengths were achieved with a 15% mixing ratio of OPC + SMS; the UCS exceeded 500 kPa after 28 days of curing. This compared with *ca.* 200 kPa for OPC alone and 300 kPa with the OPC + OSP mixture. OSP and SMS could therefore be used as alternative admixture materials to OPC to improve the solidification efficacy of organic marine clays instead of the expensive conventional admixture materials.

ACKNOWLEDGMENTS

The authors are thankful to Sanha Engineering & Construction Co., Ltd. (http://www.sanhaenc.com), Seongnam-si, Korea, for financial support.

Author Contributions

Chan Yu conceived and designed the experiments; Sung-Wook Yun performed the experiments; Chan Yu and Chan-Gi Park analyzed the data; Phillippe C. Baveye provided input on data analysis and edited the English; Chan-Gi Park wrote the paper.

Conflicts of Interest

The authors declare no conflict of interest.

REFERENCES

1. Jauberthie, R.; Rendell, F.; Rangeard, D.; Molez, L. Stabilisation of estuarine silt with lime and/or cement. *Appl. Clay Sci.* **2010**, *50*, 395–400.
2. Tremblay, H.; Duchesne, J.; Locat, J.; Leroueil, S. Influence of the nature of organic compounds on fine soil stabilization with cement. *Can. Geotech. J.* **2002**, *39*, 535–546.
3. Paria, S.; Yuet, P.K. Solidification-stabilization of organic and inorganic contaminants using portland cement: A literature review. *Environ. Rev.* **2006**, *14*, 217–255.
4. Porbaha, A. Recent advances in deep mixing research and development in the USA. *Ground Modif. Seism. Mitig.* **2006**,*152*, 45–50.
5. Masher, A.; Douglas, W.S.; Yang, D.; Jafari, F.; Schaefer,

V.R. Cement deep soil mixing (CDSM) for solidification of soft estuarine sediments. *Mar. Georesour. Geotechnol.* **2007**, *25*, 221–235.

6. Bobet, A.; Hwang, J.; Johnston, C.T.; Santagata, M. One-dimensional consolidation behavior of cement-treated organic soil. *Can. Geotech. J.* **2011**, *48*, 1100–1115.
7. Oliveira, P.J.V.; Correia, A.A.; Garcia, M.R. Effect of organic matter content and curing conditions on the creep behavior of an artificially stabilized soil. *J. Mater. Civ. Eng.* **2012**, *24*, 868–875.
8. Al-Tabbaa, A.; Evans, C.W. Medium term performance of stabilised/solidified contaminated soil-grout materials. In Proceedings of the 14th International Conference on Soil Mechanics and Foundation Engineering, Hamburg, Germany, 6–12 September 1997; pp. 1941–1946.
9. Al-Tabbaa, A.; Evans, C.W. Pilot *in situ* auger mixing treatment of a contaminated site. Part 1: Treatability study. *Proc. ICE Geotech. Eng.* **1998**, *131*, 52–59.
10. Al-Tabbaa, A.; Tabbaa, A.; Evans, C.W.; Wallace, C.J. Pilot *in situ* auger mixing treatment of a contaminated site. Part 2: Site trial. *Proc. ICE Geotech. Eng.* **1998**, *131*, 89–95.
11. Gidley, J.S.; Sack, W.A. Environmental aspects of waste utilization in construction. *J. Environ. Eng.* **1984**, *110*, 1117–1133.
12. Kamon, M.; Nontananandh, S. Combining industrial wastes with lime for soil stabilization. *J. Geotech. Eng.* **1991**, *117*, 1–17.
13. Chindaprasirt, P.; Jaturapitakkul, C.; Sinsiri, T. Effect of fly ash fineness on microstructure of blended cement paste.*Constr. Build. Mater.* **2007**, *21*, 1534–1541.
14. Chappex, T.; Scrivener, K. Alkali fixation of C–S–H in blended cement pastes and its relation to alkali silica reaction.*Cem. Concr. Res.* **2012**, *42*, 1049–1054.
15. Lim, D.I.; Park, Y.A. Late Quaternary stratigraphy and evolution of a Korean tidal flat, Haenam Bay, southeastern Yellow Sea, Korea. *Mar. Geol.* **2003**, *193*, 177–194.
16. Chung, S.G.; Kim, G.J.; Kim, M.S.; Ryu, C.K. Undrained shear strength from field vane test on Busan clay. *Mar. Georesour. Geotechnol.* **2007**, *25*, 167–179.
17. Taylor, H.F.W. *Cement Chemistry*, 2nd ed.; Thomas Telford Publishing Ltd.: White Plains, NY, USA, 1997.
18. Hewlett, P.C. *Lea's Chemistry of Cement and Concrete*, 4th ed.; Butterworth-Heinemann: Burlington, MA, USA, 2004.
19. Kuo, W.T.; Wang, H.Y.; Shu, C.Y.; Su, D.S. Engineering properties of controlled low-strength materials containing waste oyster shells. *Constr. Build. Mater.* **2013**, *46*, 128–133.
20. Ministry of Environment of South Korea. *Standard Methods for Wastes*; Ministry of Environment of South Korea: Seoul, Korea, 2012.
21. Japan Cement Association. *Manual of Soft Ground Improvement using Solidifying Agents of Cement based System*, 2nd ed.; Japan Cement Association: Tokyo, Japan, 1994.
22. American Society for Testing Materials. *Standard Test Method for Compressive Strength of Cylindrical Concrete Specimens*; ASTM C 39-96; ASTM International: West Conshohocken, PA, USA, 1998.
23. Eatone, A.D.; Closceri, L.S.; Greenberg, A.E. *Standard Methods for the Examination of Water and Wastewater*, 19th ed.; American Public Health Association (APHA): Washington, DC, USA, 1995.
24. Li, G.; Xu, X.; Chen, E.; Fan, J.; Xiong, G. Properties of cement-based bricks with oyster-shells ash. *J. Clean. Prod.* **2015**,*91*, 279–287.
25. Ding, Y.; Fang, Y.; Fang, H.; Zhang, Q.; Zhang, F.; Oh, W.C. Study on the retarding mechanism and strength loss of gypsum from hydrolyzed wheat protein retarder. *J. Korean Ceram. Soc.* **2015**, *52*, 28–32.
26. Peng, J.; Zhang, J.; Chen, M.; Wu, L.; Qu, J. Influence of macromolecule-type retarders on the hydration process of building gypsum and its retarding mechanism. *J. Chin. Ceram. Soc.* **2008**, *36*, 896–900. (In Chinese).
27. Li, X.G.; Xu, R.Q.; Rong, X.N. Assessment of strength development in cement-admixed artificial organic soil with GX07. *J. Cent. South Univ.* **2012**, *19*, 2999–3005.
28. Park, S.B.; Tia, M. An experimental study on the water-purification properties of porous concrete. *Cem. Concr. Res.***2004**, *34*, 177–184.

CHAPTER 7

MATERIAL AND STRUCTURAL PERFORMANCE EVALUATIONS OF HWANGTOH ADMIXTURES AND RECYCLED PET FIBER-ADDED ECO-FRIENDLY CONCRETE FOR CO_2 EMISSION REDUCTION

Bon-Min Koo, Jang-Ho Jay Kim, Sung-Bae Kim and Sungho Mun

Keywords: Eco-friendly concrete, Hwangtoh concrete, admixtures, short plastic fibers, recycling PET plastic, compressive strength, crack control, structural ductility

INTRODUCTION

Concrete is the most widely used construction material in the world due to its low cost, high availability, and simple constructability. However, the use of cement is a main contributor to high-energy usage, CO_2 and dust emissions, natural resource depletion, air pollution, ozone layer destruction, global warming, and continuous environmental deterioration. In addition, practices related to concrete construction exploit natural resources due to a lack of careful planning, which has triggered an urgent need to find alternative construction materials to mitigate environmental damage [2,3]. In terms of health issues, there have been fervent public outcries over illnesses associated with concrete structures, such as sick building syndrome, dermatitis, allergic diseases, and atopic skin diseases. Therefore, many technologically advanced countries such as the United States and Japan are trying to reduce the use of cement and concrete while searching for alternative structural materials for housing construction [3,4,5,6,7]. Due to the high availability, consistency, reliability, and affordability of concrete, however, concrete will likely remain a primary construction material in the future. As a result of the Kyoto Protocol for regulating CO_2 emissions and energy consumption in the international community, construction industries all over the world are trying to reduce cement usage, find alternative materials, and use recycled materials for future construction projects [4,5,6,7,8,9,10,11].

In order to develop more optimal eco-friendly concrete and reduce cement usage, research on mineral admixtures made from industrial by-products, such as fly ash, blast furnace slag, and silica fume, is being widely conducted [2,5,12,13,14]. For example, one recent alternative construction material development study explores the use of concrete mixed with crushed oyster shells for reducing coarse and fine aggregate usage [15,16]. There is a great deal of active research on the properties of concrete mixed with fly ash due to the cost advantages of fly ash over other admixtures. However, because of quality control difficulties associated with concrete mixed with fly ash, alternative studies on other types of admixture-added concrete have also been pursued. Numerous works on concrete admixtures show that the major obstacle in using admixtures in ordinary Portland cement (OPC) concrete for actual construction is the 30% maximum limit in the cement substitution ratio. According to most studies, when the maximum substitution percentage is violated, reductions in properties of strength, crack resistance, and overall material performance are observed [14,17,18,19,20]. Therefore, a solution to the problem of how to increase the admixture substitution ratio in a concrete mix without decreasing the

Citation: Bon-Min Koo, Jang-Ho Jay Kim, Sung-Bae Kim and Sungho Mun, Material and Structural Performance Evaluations of Hwangtoh Admixtures and Recycled PET Fiber-Added Eco-Friendly Concrete for CO2 Emission Reduction, doi:10.3390/ma7085959.

final performance of the material is needed in order to develop useable eco-friendly concrete. In addition, a simple method of recycling industrial waste products for admixture applications in concrete mixtures must be established in order to develop innovative eco-friendly concrete that is not harmful to human health.

Hwangtoh (*i.e.*, a type of red clay) is receiving substantial interest as a possible eco-friendly construction material. Accordingly, Hwangtoh is the subject of much industry research [4,21,22,23,24]. Hwangtoh is a Kaolin material with abundant reserves all over the world, and has traditionally been considered an eco-friendly material. According to recent research, Hwangtoh can be used as a mineral admixture for eco-friendly concrete due to its pozzolanic characteristics [4,22,25,26]. However, any Hwangtoh admixture replacement ratio greater than 15%–20% of the total cement weight sharply degrades the material properties of Hwangtoh concrete. Therefore, a Hwangtoh admixture replacement ratio of 15%–20% of the total cement weight in a concrete mix has generally been accepted as the maximum replacement ratio [27,28]. Recent experimentation with Hwangtoh powder blended with mineral admixtures has been pursued in order to improve the pozzolanic properties of Hwangtoh, and to increase its replacement ratio in a concrete mix [28,29]. The findings of these studies show that blending Hwangtoh with slag powder cement provides an optimal proportion ratio to maximize the strength capacity of Hwangtoh concrete.

In order to use Hwangtoh as a structural material, the inherent shrinkage-cracking problem associated with Hwangtoh must be resolved. Based on the findings of research on fiber reinforced concrete and its ability to control cracking in the cement matrix, studies pertaining to the application of short plastic fibers in Hwangtoh concrete have been conducted [9,30,31,32,33,34]. In conjunction with efforts by the plastic fiber industry to recycle plastic waste so that waste materials can be utilized for manufacturing secondary products, several studies have been carried out in recent years to optimize-recycling of plastic waste into plastic fibers for making eco-friendly concrete [1,10,35,36,37,38,39].

Previous research on Hwangtoh has mostly consisted of studies that investigate the basic material and mechanical properties of Hwangtoh, together with mixing proportions and procedures, to evaluate its constructability [4,21,22,23,24,27,28,29]. The aim of this study is to develop a type of Hwangtoh concrete with sufficient compressive strength to meet the standards of modern construction by adding Hwangtoh-slag admixtures, and to examine the shrinkage crack-controlling capability of Hwangtoh concrete when short recycled PET fibers are added to the mix. With the goal of developing an alternative construction material with healthy and eco-friendly qualities, the focus of this work is evaluation of the newly developed Hwangtoh concrete material in terms of its durability and structural performance capacities. The material and durability properties (*i.e.*, shrinkage strain, compressive strength, elastic modulus, pH, carbonation depth, and creep strain) of plain Hwangtoh concrete, as well as the structural properties (*i.e.*, load-deflection relationships, ultimate failure load, ductility index, and cracking patterns) of RC beams cast with Hwangtoh concrete and recycled PET fibers are examined.

MATERIAL PROPERTIES AND MIX PROPORTIONS

Hwangtoh

Hwangtoh has traditionally been considered an eco-friendly construction material for use in plastering walls and ondol (traditional Korean underfloor heating systems) in Korea. The main advantages of Hwangtoh are its high heat absorption capacity, auto-purification properties, antibiotic properties, and infrared ray emission capability [23]. However, the development of Hwangtoh has not been actively pursued in modern constructions, and the material has not been utilized in modern buildings due to its low compressive strength and low dry shrinkage crack resistant capacity. According to the results of recent research, however, Hwangtoh can be used as a substitute material for natural pozzolanic admixtures, such as flyash or pozzolan [4,22,24]. The potential of Hwangtoh as a construction material, especially in the construction of housing, is extremely promising because its chemical and mineralogical properties are similar to those of Metakaolin and Kaolinite. Silicon dioxide (SiO_2) forms stable pozzolan products in Hwangtoh by combining with cement hydrate $Ca(OH)_2$at room temperature, thus acting as a pozzolan material, such as fly ash or silica fume [40]. The chemical equation for the pozzolanic reaction in Hwangtoh is written as follows:

$$3Ca(OH)_2 + 2SiO_2 \rightarrow 3CaO{\cdot}2SiO_2{\cdot}3H_2O \quad (1)$$

Recycled PET Fibers

Polyethylene terephthalate (PET) is the most widely used plastic material in the world for manufacturing secondary products such as beverage bottles. However, PET bottles are commonly thrown away after a single use, which contributes to significant environmental problems. Therefore, methods to recycle PET bottle waste have been examined in order to improve PET recycling applications. The use of PET plastic fibers recycled from PET bottle waste in construction projects has been actively studied. Short PET fibers can be utilized to effectively reduce shrinkage strain and to enhance the ductility of concrete, especially to mitigate the inherent shrinkage problem of Hwangtoh concrete [35,36,41,42].

Optimal Hwangtoh Replacement Ratio Selection

Our experiments aimed to reduce the use of ordinary Portland cement (OPC) in concrete mixes as much as possible by replacing OPC with Hwangtoh and blast furnace slag powders. Before the actual casting of the specimens, trial tests were performed to find the optimal replacement ratio of Hwangtoh and blast furnace slag powders. In a previous study on the use of Hwangtoh powder as an admixture in a concrete mix, serious degradations in concrete performance were observed when the Hwangtoh to cement substitution ratio exceeded 20%. Therefore, a substitution ratio exceeding 20% is generally not recommended in concrete mix proportions [27,28]. Significant workability degradation has also been reported in many Hwangtoh concrete studies [4,21,27,43]. Hence, it is important to develop a concrete mix proportion using admixtures of Hwangtoh and blast furnace slag powder without decreasing the overall properties of strength, crack resistance, and workability in the final material. While a reduction in workability can be controlled by adding a water-absorbing chemical admixture to satisfy the target slump, this study adheres to the principle of adding the maximum amount of blast furnace slag, Hwangtoh, and other eco-friendly admixtures in the mix proportion.

Trial mix proportion tests were carried out by setting the water-to-cement ratio (W/C) as a constant, while varying the Hwangtoh and slag powder replacement ratios. The results from the trial tests are shown in Figure 1. The trial test results reveal that when the Hwangtoh replacement ratio was 20%, little variation in strength (independent of the slag replacement ratio) was observed. The workability decreased as the Hwangtoh replacement ratio increased due to the high porosity of Hwangtoh, which causes rapid water absorption during the mixing process. However, as shown in Figure 1b, when OPC, Hwangtoh, and blast furnace slag powders were mixed, the slump of the mix was higher than the slump of the OPC mix. This result is likely caused by the relatively smooth surface texture of blast furnace slag powder particles, which reduces both water molecule attachments on particle surfaces and friction in the cement paste-aggregate interface.

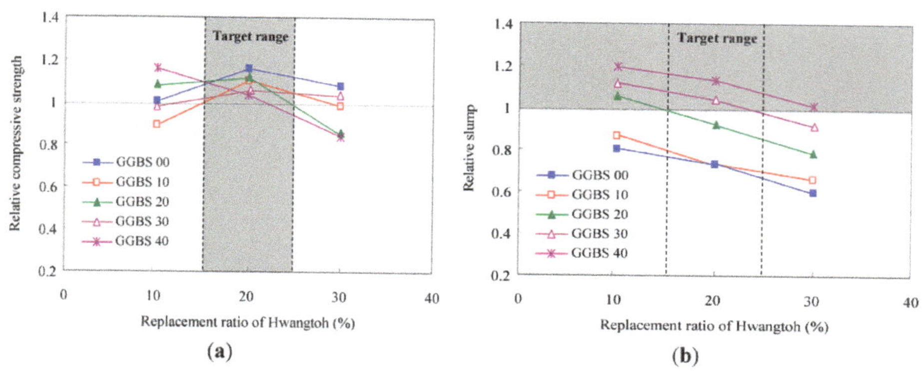

Figure 1. Compressive strength and slump of Hwangtoh concrete according to pre-tests (**a**) Compressive strength and (**b**) Slump.

In this study, replacement ratios of 20% and 30% were used for the Hwangtoh and slag powders, respectively, for a 50% total cement replacement ratio in the concrete mix. In order to improve the high rate of shrinkage in Hwangtoh concrete, a 0.5% volume of short plastic fibers made from recycled PET plastic bottles was also added to the mix. From the trial tests, it was found that the optimal PET fiber volume percentage to control crack formation in the concrete was approximately 0.5%.

The material properties of the Hwangtoh used in this study are shown in Table 1. Two types of Hwangtoh were employed in the study, namely K-type and C-type Hwangtoh produced in Korea and China, respectively. Recycled PET fiber used in the study was an embossed type with dimensions of 0.2 mm × 1.3 mm × 50 mm and density of 1.38 g/cm^3. Elastic modulus, tensile strength, and ultimate elongation percentage of the fiber were 1.2 × 10^4 MPa, 420.7 MPa, and 11.2%, respectively. The final mixture proportions selected for the experiments are shown in Table 2.

Table 1. Chemical composition of Hwangtoh.

Composition type	SiO_2	Al_2O_3	Fe_2O_3	MgO	CaO	K_2O	TiO_2	Na_2O
KH	42.5	36.6	4.05	0.69	0.57	0.41	0.23	0.18
CH	57.0	18.0	6.15	0.32	8.91	0.2	0.25	0.20
GGBS (Slag)	41.2	34.2	11.7	8.81	-	0.31	-	0.29

Table 2. Concrete mixture proportions.

Specimens	W/B (%)	S/A (%)	Replacement ratio of Hwangtoh (%)	Replacement ratio of GGBS (%)	Unit weight (kg/m^3)						RPET fiber (%)	SP (%)
					W	C	HT	GGBS	S	G		
Control			-	-		339.4	-	-			-	
SC			-	30		237.6	-	101.8			-	
KH			20	-		271.5	67.9	-			-	
CH			20	-		271.5	67.9	-			-	
KHS	55	44.8	20	30	186.7	169.7	67.9	101.8	746.7	958.1	-	0.7
CHS			20	30		169.7	67.9	101.8			-	
KHSP			20	30		169.7	67.9	101.8			0.5	
CHSP			20	30		169.7	67.9	101.8			0.5	
RPET			-	-		339.4	-	-			0.5	

EXPERIMENT METHODS

Compressive Strength and Elastic Modulus Tests

A total of 27 specimens were used for compression and elastic modulus tests. Hwangtoh concrete cylinders with dimensions of 100 mm × 200 mm were compression tested at 28 days from initial casting. These cylindrical specimens were tested with a universal testing machine (UTM) with a maximum load capacity of 2000 kN. The compressive strength of each specimen type was determined by averaging three specimen strength values. After casting, the cylindrical specimens were covered with plastic wrap for 24 h, after which the forms were removed. The specimens were then stored in a humidity chamber at a temperature of 20 ± 1 °C and a humidity of 60% ± 5% until testing. Compressive strength and elastic modulus experiments were performed according to Korean Standards 2405 and 2438, respectively [44,45]. A photo of the cylindrical specimens mixed with Hwangtoh and slag cement is shown in Figure 2.

Method for pH Tests

In order to measure the pH of Hwangtoh cement paste, Hwangtoh specimens with W/C of 55% were prepared. For comparative purposes, additional specimens made exclusively from Korean and Chinese Hwangtoh (called KH100 and CH100, respectively) were prepared. The formwork was removed after 24 hours of casting. In order to prevent evaporation of moisture from the specimens, the specimens were wrapped with plastic wrap before being put into a curing chamber with a temperature of 20 ± 2 °C and a relative humidity of 64% for 200 days. The loss of alkali residuals in the powder (sieve size 300 μm) was minimized by drying the power in a drying chamber for 24 h prior to testing. A volume of 3.0 g of the dried power was subsequently mixed with 40 mL of distilled water. After stirring the wet powder for 10 min, the pH was measured using a pH-measuring device. The pH correction factor was used to set the initial pH values at 4.01, 7.00, and 10.01 prior to measurement [46,47].

Figure 2. 100 mm × 200 mm Hwangtoh and slag cement concrete cylindrical specimens.

Carbonation Depth Test

In order to evaluate the effect of Hwangtoh on the carbonation properties of Hwangtoh concrete, carbonation depths at various curing ages were measured. The accelerated carbonation test was performed following the test guidelines of KS F 2584 [48]. The specimens were wet-cured for four weeks, followed by dry curing for another four weeks. Before the test, the specimens were coated with epoxy at both ends. The accelerated carbonation test was performed in a carbonation test chamber shown in Figure 3 under temperature, relative humidity, and CO_2 conditions of 20 ± 2 °C, 60% ± 5%, and 5% ± 0.2%, respectively. Carbon depth was measured based on the location of phenolphthalein color change at testing times of one, four, eight, 12, and 25 weeks, following the guidelines of KS F 2596 [49].

Creep Test Method

Currently, there are no guidelines available for the creep testing of recycled PET fiber-added Hwangtoh concrete. Accordingly, the concrete creep test guidelines of ASTM C 512 were used for this study [50]. A cylindrical specimen with a diameter of 150 mm and a height of 300 mm, containing air cured for seven days, was used for the test herein. The creep test was performed inside a humidity chamber with a temperature and relative humidity of 22 °C and 60%, respectively. The specimens of slag cement concrete, Hwangtoh concrete, and recycled PET fiber-added Hwangtoh concrete were creep tested by applying a load equivalent to 40% of the load used for the 28-day compressive strength tests, using a creep test device, as shown Figure 4. The creep load was applied to the specimens by compressing the springs attached to the bottoms of the specimens using an actuator. When the target load was applied, the spring length was fixed by a bolt that secured the steel plate to the bottom of the specimen. After application of the load, dial gauges were installed on the left and right sides of the specimen to measure deflection at an interval of seven days.

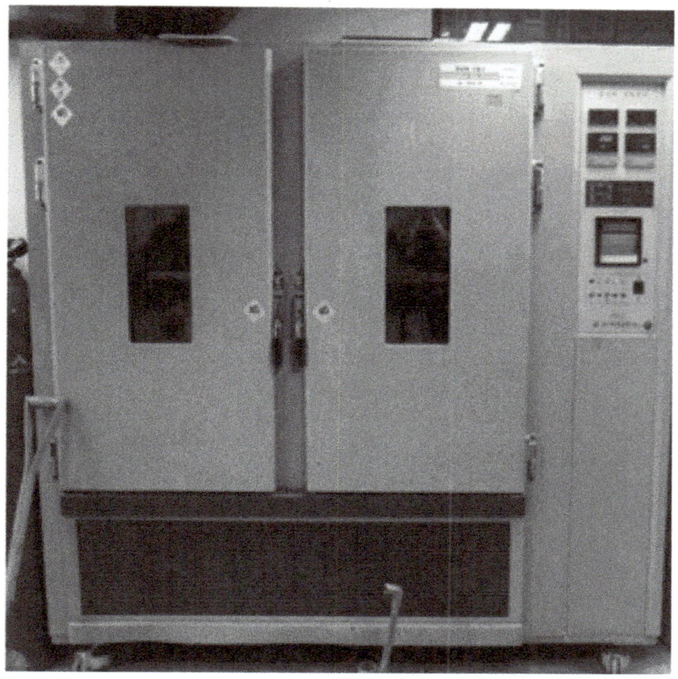

Figure 3. Carbonation test chamber.

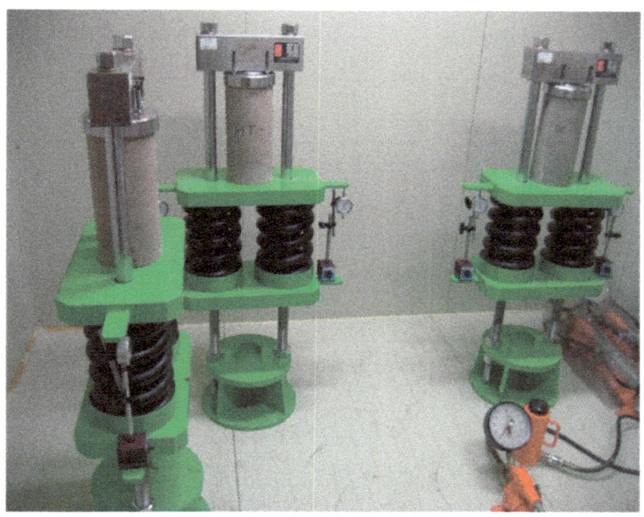

Figure 4. Creep test setup.

Drying Shrinkage Cracking Test Methods

To evaluate the drying shrinkage cracking capacity of Hwangtoh concrete, free and restrained drying shrinkage strain experiments were performed according to Korean Standard 2595, which details the drying shrinkage cracking test method for restrained concrete [51]. The specimens were cast and placed in a humidity chamber at a curing temperature of 20 ± 1 °C and a relative humidity of 60% ± 5% until testing. Removal of the form was performed seven days after casting. In order to measure the restrained drying shrinkage strain after removal of the form, strain gauges were attached to the central part of the upper and lower surfaces of the specimen and to the central part of the restraining plate in order to measure the concrete and restraining plate strains, respectively. The specimens were checked daily and the location and time of cracking were recorded. The dimensions and shape of a drying shrinkage cracking specimen are shown in Figure 5.

Reinforced Concrete (RC) Flexural Test Methods

All specimens were tested for flexural strength 28 days after casting. The specimens were reinforced with three D13 rebars (with a diameter of 13 mm) as tension reinforcement and D10 rebars (with a diameter of 10 mm) with 150 mm spacing as shear reinforcement. The dimensions and details of an RC flexural test specimen are shown in Figure 6. The specimens were tested using a four-point loading setup with hinge-roller supports, as shown in Figure 7. Stroke control loading with a vertical displacement rate at the mid-span of 0.025 mm/sec was applied with a UTM with a capacity of 2000 kN. A strain gauge and a linear variable differential transducer (LVDT) were attached to the bottom of the concrete surface of the specimen at the mid-span point, in order to measure the tensile strain and vertical deflection, respectively. Crack initiation and propagation were visually monitored and recorded throughout the flexural test.

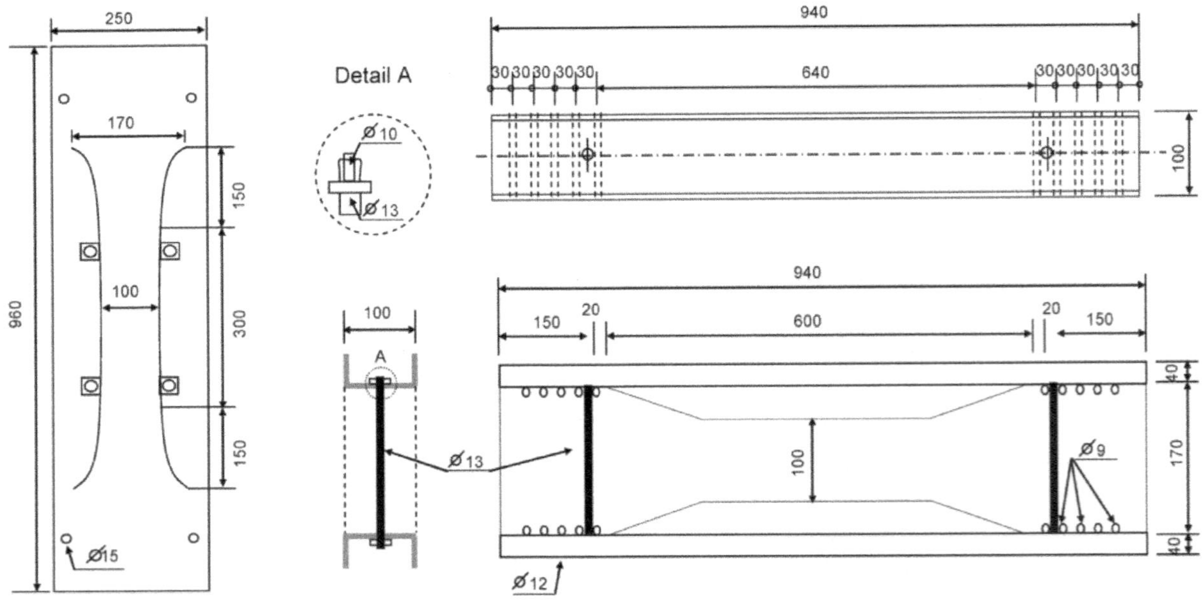

Figure 5. Restrained frame and concrete mold dimensions and details (units: mm).

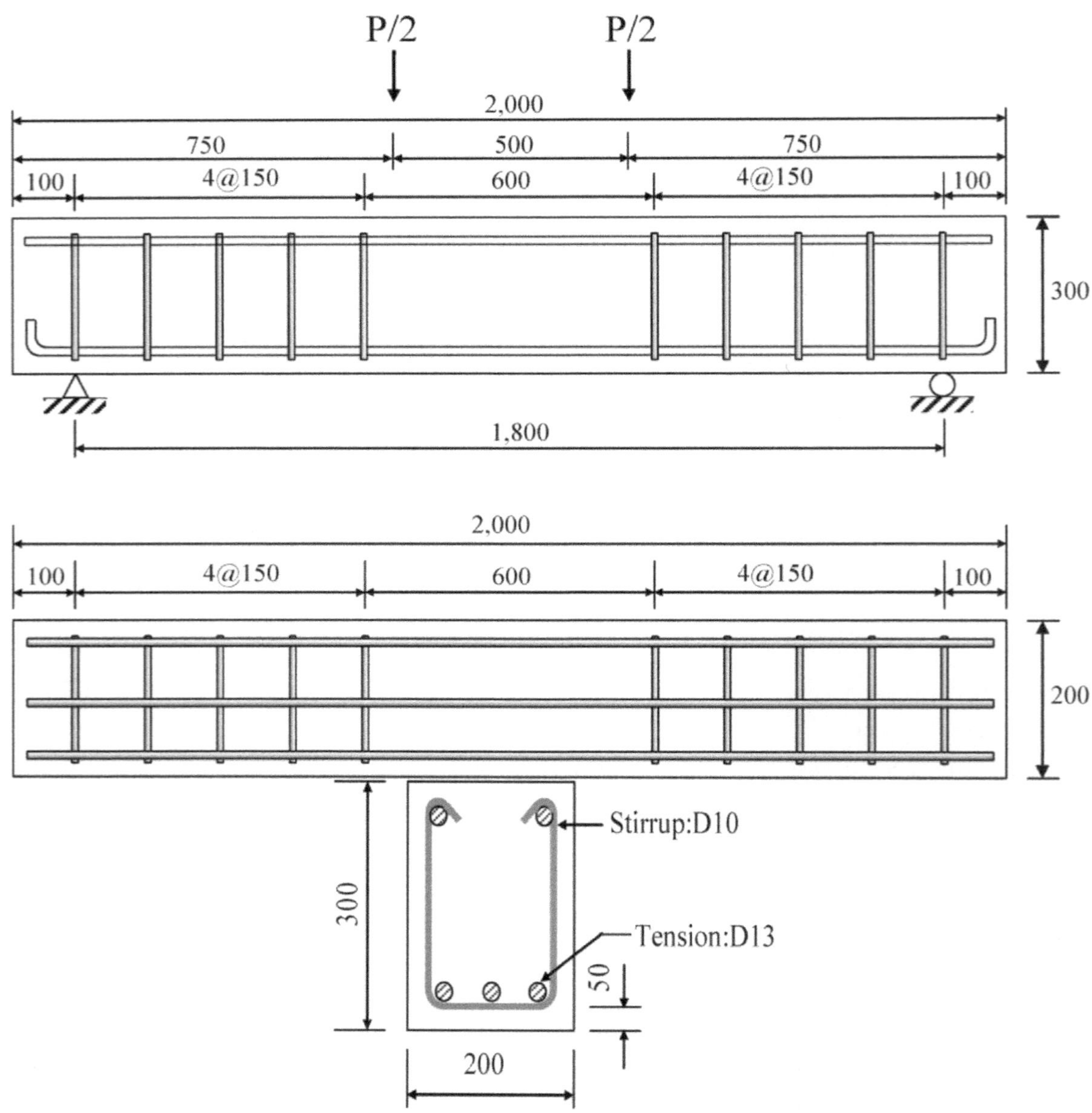

Figure 6. Dimensions and details of RC beams (units: mm).

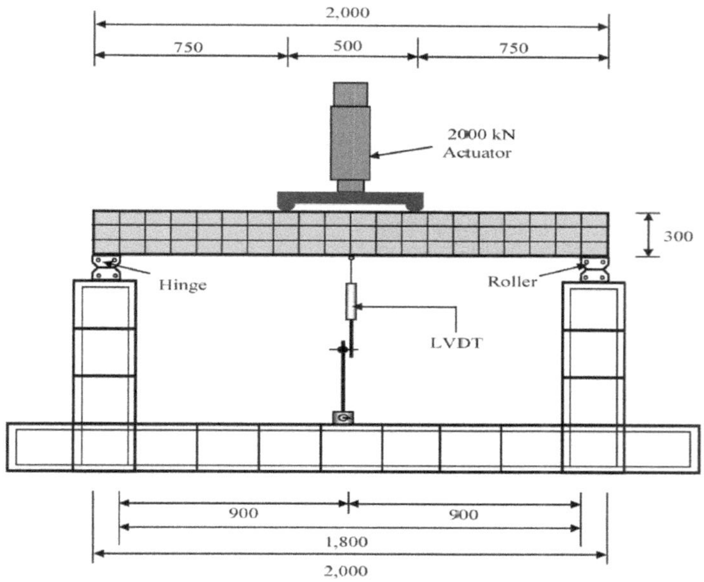

Figure 7. Test setup and data acquisition locations (units: mm).

EXPERIMENT RESULTS AND DISCUSSION

Compressive Strength and Elastic Modulus

The compressive strength test results for the Hwangtoh concrete are shown in Figure 8a. Control and slag concrete (SC) specimens were cast with plain OPC and slag cement concrete, respectively. To define our working acronyms, KH and CH specimens were cast with concrete containing Korean and Chinese Hwangtoh powder, respectively, while KHS and CHS specimens were cast with Korean and Chinese Hwangtoh with slag powder concrete, respectively. KHSP and CHSP specimens were cast with KHS and CHS concrete containing PET fibers, respectively, and an RPET specimen was cast with plain concrete containing PET fibers. As shown in Figure 8a, the control specimen demonstrated the highest compressive strength of 30 MPa. In contrast, an SC specimen, the KH and CH specimens mixed with Hwangtoh, and the KHS and CHS specimens mixed with Hwangtoh and slag exhibited lower capacities of compressive strength. These results demonstrate that the pozzolanic reactions of Hwangtoh and slag powders are not as active as the pozzolanic reaction of cement. This in turn causes less hydration in the mix, and produces an internal structure that is less dense. Recycled PET fiber-reinforced KHSP and CHSP specimens exhibited the lowest compressive strength out of all the specimens. In particular, the CHSP specimen displayed the lowest compressive strength (27 MPa) due to the addition of PET fibers. The behavior of the fibers within the matrix can be explained by drawing an analogy between fibers and internal voids, which decrease the effective cross-section and strength of a specimen. However, a compressive strength comparison between the different Hwangtoh concrete specimens and the control specimen show that the strength reduction is less than 2%–6% and is thus relatively insignificant for structural member construction.

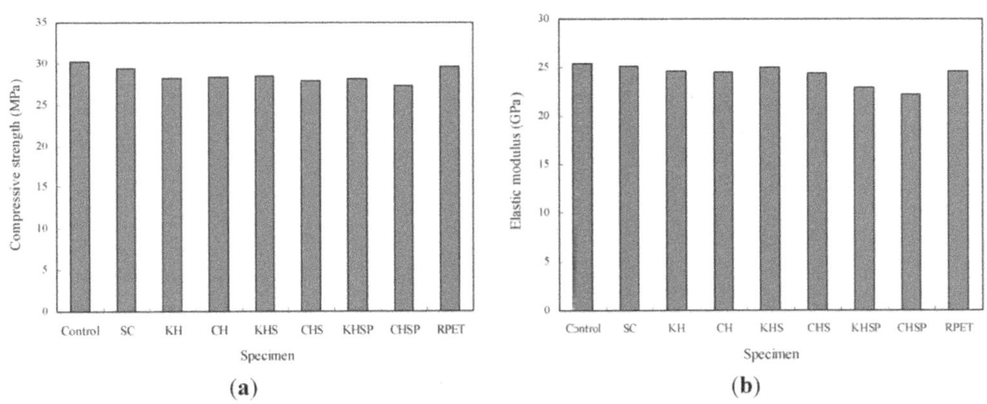

Figure 8. Compressive strength and elastic modulus of Hwangtoh concrete from main tests (**a**) Compressive strength and (**b**) Elastic modulus.

The elastic modulus test results for the various Hwangtoh concrete specimens are shown in Figure 8b as a bar graph. Similar to the compressive strength test results, the control specimen exhibited the highest elastic modulus at 25 GPa, while the moduli of the other Hwangtoh concrete specimens were in the range of 23–25 GPa. The results of the compressive strength and elastic modulus test for the Hwangtoh concrete and plain concrete specimens reveal similar capacities among the types of material. Indeed, the differences between the two materials are within an acceptable margin, indicating that their capacities may be considered equivalent for structural usage.

Compression Failure Behavior

The compressive failure modes of the cylindrical specimens are shown in Figure 9. As illustrated in Figure 9a, the failure mode of the Hwangtoh concrete was in line with the cone failure type, which is a typical concrete failure mode observed in compressed specimens. However, the recycled PET fiber-reinforced Hwangtoh concrete specimens failed due to a large expansion strain in the center, along with multiple vertical cracks, as shown in Figure 9b. These failure modes indicate that the recycled PET fibers restrained concrete spalling and crack growth. The fibers also dispersed internal stresses uniformly over the cross-section, thereby preventing localized failure. These results suggest that synthetic fibers, such as recycled PET and polypropylene (PP) fibers, are beneficial in preventing premature concrete failure. In addition, recycled PET fiber is effective in controlling crack growth in Hwangtoh concrete.

Figure 9. Failure mode of cylindrical Hwangtoh concrete specimens (**a**) Hwangtoh specimen without fibers and (**b**) Hwangtoh specimen with RPET fibers.

Measurement of pH

Measurement data on pH values are shown in Figure 10. The figure shows that the pH of the control specimen (general OPC concrete) was approximately 12.45, while most of the other specimens (with the exception of the PET fiber-reinforced Hwangtoh specimen) had lower pH values. The results show that, as the amount of slag and Hwangtoh admixtures increased (thereby decreasing the amount of OPC in the mix), the pH values decreased. In addition, the results show that PET fibers do not have a significant effect on the pH of concrete. The measured pH values of KH 100 and CH 100 specimens were approximately 8.88 and 9.33, respectively, showing a reduced pH, which is characteristic of Hwangtoh. The results of a previously published study report that the pH of calcium oxide ($Ca(OH)_2$) is approximately 12.5 [52]. Thus, the pH value of the control specimen was controlled by calcium oxide, as expected. The pH values of the Hwangtoh specimens, however, were lower due to reduction in the amount of calcium oxide production caused by the partial replacement of OPC with slag and Hwangtoh in the Hwangtoh specimens. It is important to note that the amount of calcium oxide production during the hydration processes of slag and Hwangtoh concrete is significantly less than the amount of calcium oxide production in OPC, thereby resulting in lower pH values in slag and Hwangtoh concrete.

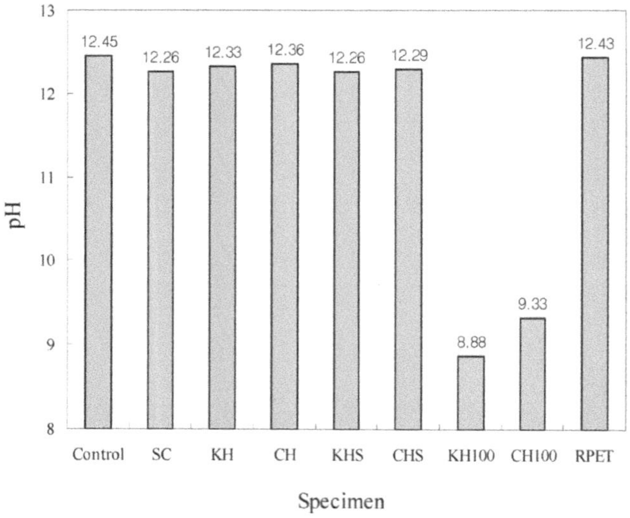

Figure 10. Results of pH measurement.

Carbonation Depth

Carbonation depth data are shown in Figure 11. The measurements were obtained at one, four, eight, 12, and 25 weeks from the start of the test. The results show that the control and PET fiber-added concrete specimens had the least carbonation depth. Carbonation depth values ascend in order from slag cement, Hwangtoh, Hwangtoh slag cement, and Hwangtoh slag cement-PET fiber specimens, with slag concrete having the lowest carbonation depth. The basic cause of carbonation in concrete is penetration of CO_2 gas into the concrete. In this way, the interior microstructure compactness of concrete dictates its carbonation resistance. In addition, interior defects or interfaces decrease the carbonation resistance of concrete. Based on these fundamental mechanics, it is logical that slag and Hwangtoh-added concrete would have less carbonation resistance, because lesser amount of calcium oxide are produced in these specimens than in OPC specimens. Additionally, the addition of PET fibers increases the damage-like interfaces in the engineered specimens, resulting in lower carbonation resistance.

Creep Strain

The creep strain test results for slag concrete, Hwangtoh concrete, and Hwangtoh-slag-PET fiber concrete measured up to 120 days are shown in Figure 12. The strain drastically increased in the early testing period within the initial 20 days, and then the rate of strain increase became less drastic. The creep strain measurements at 120 days ascend in order from Hwangtoh-slag (610×10^{-6}), slag (780×10^{-6}), to Hwangtoh-slag-PET fiber (871×10^{-6}) concrete, with Hwangtoh-slag concrete having the lowest creep strain at 120 days. The creep strain test results demonstrate that Hwangtoh is less susceptible to creep strain than slag cement, because the proportion of cement is reduced by the addition of the Hwangtoh admixture in the Hwangtoh-slag concrete mix relative to the slag concrete mix. In addition, the addition of recycled PET fibers increased the interfaces in the concrete, resulting in significant creep strain from the closing of interfaces as the duration of loading increased.

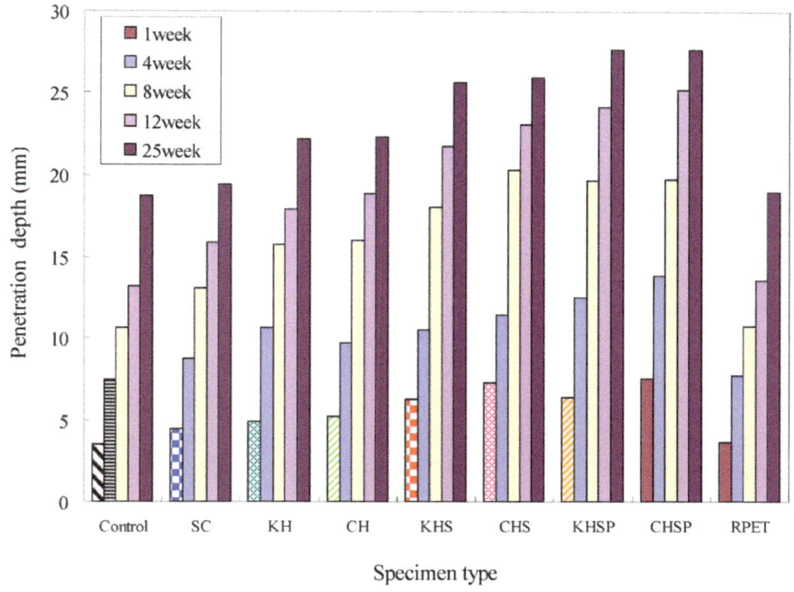

Figure 11. Carbonation depth results.

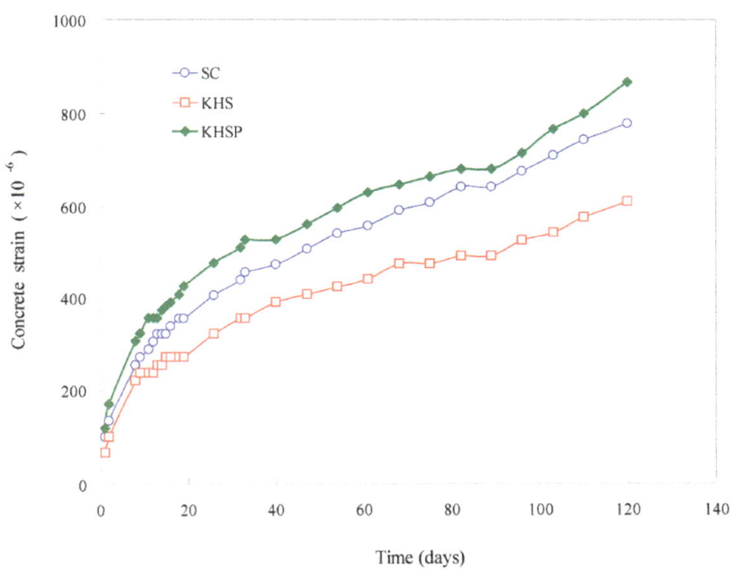

Figure 12. Creep strain test results.

Free and Restrained Drying Shrinkage Cracking Behaviors

The measured free drying shrinkage strain in the test specimens is shown in Figure 13a. The control specimens exhibited the lowest free drying shrinkage strain, while the PET fiber-reinforced Hwangtoh concrete specimens showed 150%–200% more shrinkage strain than the control specimen at an age of 59 days. This result is indirectly explained by the results of the compressive strength test for the Hwangtoh concrete reinforced with recycled PET fibers. The reduced compressive strength of concrete reinforced with Hwangtoh admixture and recycled PET fibers is due to the lower density of its internal matrix microstructure and the increased amount of voids from fiber interfaces, which cause the material to be more susceptible to strain and cracking. The increased free drying shrinkage strain of the Hwangtoh concrete is the direct consequence of the high porosity and water absorption characteristics of the material due to the addition of the Hwangtoh admixture [24]. The introduction of recycled PET fibers into the cement matrix increases the pore volume and reduces the rigidity, which also increases the strain [35,53].

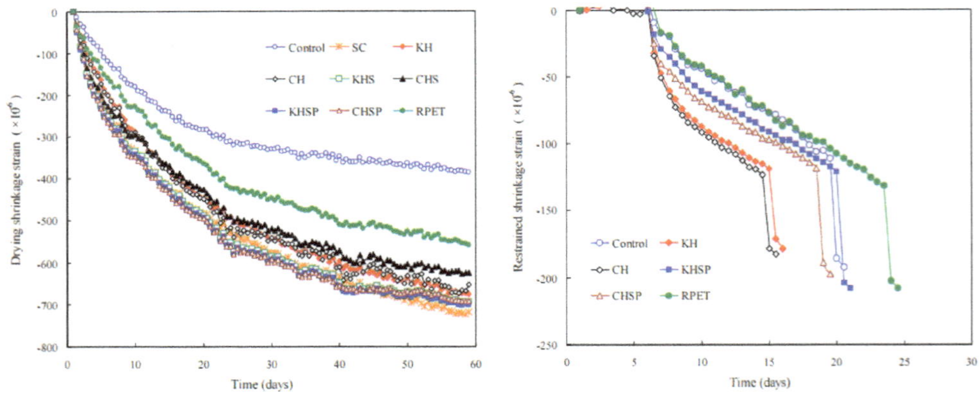

Figure 13. Free and restrained drying shrinkage strain test results (**a**) Free drying shrinkage strain results and (**b**) Restrained drying shrinkage strain and cracking results.

For the restrained drying shrinkage strain tests, the cracking times from the initial casting for all specimens are listed inTable 3. As measured, the curves of the restrained drying shrinkage strain as a function of curing time are shown in Figure 13b. The earliest full-through crack occurrence times were observed for the CH and KH specimens at 15 and 16 days, respectively, while the latest occurrence of cracking was observed for the PET fiber-reinforced RPET specimen at 25 days. The control specimen had full-through cracks at 20 days, and the CHSP and KHSP specimens had a full-through crack at 19 and 21 days, respectively. As observed from the test results, the full-through cracking time of the Hwangtoh concrete specimen was about five days earlier than the full-through cracking time of the control specimen. The CHSP and KHSP specimens exhibited times that were four to five days later than the time of the CH and KH specimens due to the crack-controlling characteristics of the PET fibers. Indeed, the delay in full-through crack formation in PET fiber-reinforced material comes from the control of shrinkage micro-cracks due to the bridging effects of PET fibers in the matrix.

Table 3. Restrained drying shrinkage cracking test results.

Specimens	Full-through cracking time (Days)
Control	20
KH	16
CH	15
KHSP	21
CHSP	19
RPET	24

Load-Deflection Relation

The load-deflection test results for different admixture-added RC specimens are shown in Figure 14 and summarized inTable 4. As seen in Figure 14, all of the specimens with different admixtures exhibited similar elastic behaviors prior to the occurrence of until cracking. Before the yielding of the tension rebars, crack initiation occurred earlier in the control specimen than in the Hwangtoh concrete specimens. However, once the rebars yielded, the Hwangtoh concrete and the control specimens showed similar overall behavior with slight deviations depending on the admixture type. The flexural capacities of the specimens reinforced with recycled PET fibers are shown in Figure 15. The recycled PET fiber-reinforced specimens exhibited better flexural capacity than the control specimen. From the results, it is safe to assume that recycled PET fibers can control shrinkage cracks and improve the structural ductility of reinforced Hwangtoh concrete members. However, the CHSP and KHSP specimens showed a lower flexural capacity than the RPET specimen. This behavior can be attributed to the lower bond strength between Hwangtoh concrete and recycled PET fibers. Accordingly, this issue needs to be addressed in future studies on the bond capacity of Hwangtoh concrete.

Table 4. Flexural strength test results for Hwangtoh concrete beams.

Specimens	P_{cr} (kN)	Δ_{cr} (mm)	P_y (kN)	Δ_y (mm)	P_u (kN)	Δ_u (mm)	Ductility Index (Δ_u/Δ_y)	Relative Ductility Index
Control	42.5	0.84	112.1	4.09	129.56	29.6	7.24	1
SC	50.8	0.94	121.1	4.03	129.95	26.1	6.48	0.90
KH	45.5	0.82	120.2	4.29	130.54	26.4	6.15	0.85
CH	46.8	0.77	117.6	4.16	129.16	38.6	9.28	1.28
KHS	43.9	1.02	118.2	4.35	131.52	27.1	6.23	0.86
CHS	45.3	1.18	112.5	4.17	126.62	31.5	7.55	1.04
KHSP	42.9	0.76	117.2	4.29	140.00	59.8	13.94	1.93
CHSP	49.4	0.79	118.8	4.27	146.60	63.4	14.85	2.05
RPET	49.0	0.74	117.0	4.11	152.39	59.8	14.55	2.01

Notes: *P* and Δ are the load and deflection of the member, respectively; *cr* is the initial crack condition; *y* and *u* are the yield and ultimate conditions, respectively.

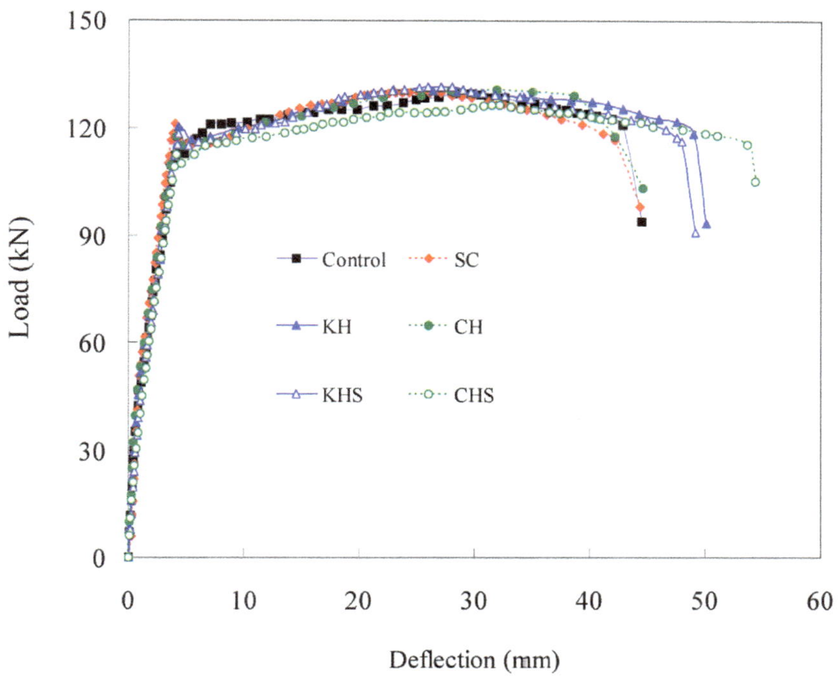

Figure 14. Load-deflection curves for Hwangtoh and slag cement concrete specimens.

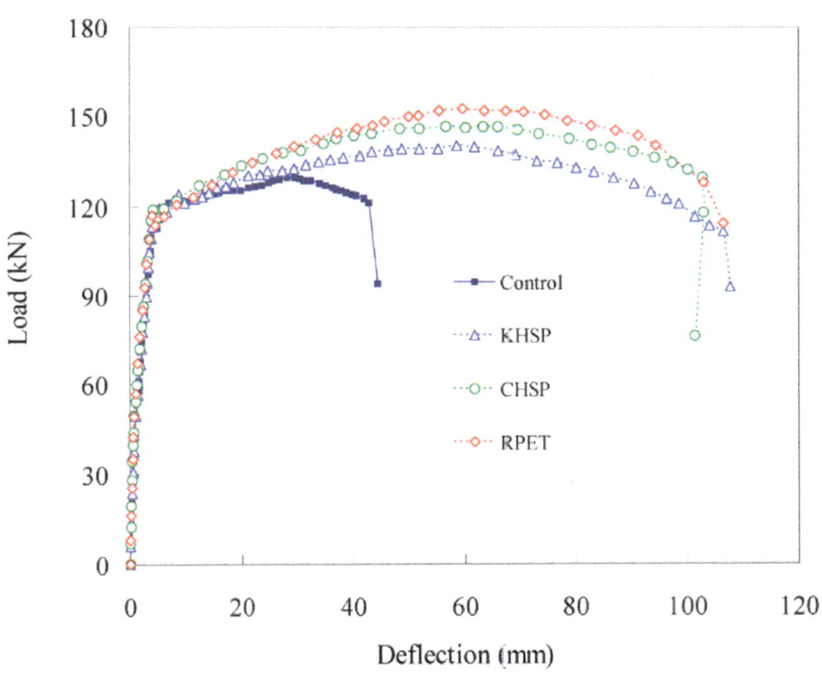

Figure 15. Load-deflection curves for recycled PET fiber RC beams.

Ductility Index

The ductility index is used to evaluate the energy absorbing capacity of reinforced concrete members, which, along with the strength, is one of the most important factors in evaluating the safety of a structural member. In this study, the ductility index (Δu/Δy), which is defined as the ratio of deflections at the ultimate load, Δu, to the deflections at the time of tension rebar yielding, Δy, is used to evaluate the safety of a member (see Table 4). As calculated, the results show that the specimens with and without recycled PET fibers had similar ductility indices, even though slight deviations were observed. The load-deflection relationships shown in Figure 14 and Figure 15 clearly validate the possibility of using Hwangtoh concrete as an alternative material to standard concrete.

The recycled PET fiber reinforced specimens exhibited better ductility indices than the non-reinforced specimens. As shown in Table 4, the recycled PET fiber-reinforced concrete specimens had relative ductility indices in the range of 1.93–2.05, which were approximately two fold greater than the ductility indices of the specimens without recycled PET fibers. The higher ductility index of PET fiber-reinforced concrete is due to the crack-bridging and controlling effects of recycled PET fibers in the concrete matrix. In addition, the fibers are beneficial in transferring tensile stresses in the member by delaying the fiber pull-out time and inhibiting crack propagation. The control of crack growth during loading slows the overall flexural failure of an RC specimen, thereby improving its strength and ductility (*i.e.*, the energy absorbing capacity). These results validate the results of a previous study by Lin *et al.* [54], wherein fiber-reinforced concrete is shown to be superior in resisting tensile stresses due to its crack-bridging capability and fiber pull-out resistance.

Cracking Modes

The failure modes of the RC specimens are shown in Figure 16. In order to understand the failure behavior of specimens mixed with a Hwangtoh admixture, the stress distribution and the crack growth for each load step were recorded until failure. All specimens were designed with a span-to-depth ratio (A/D) of 2.6, and were reinforced with shear stirrups to enforce flexure failure. The control specimen exhibited typical flexural tensile cracking failure, simultaneous to yielding of tension rebar. When the ultimate failure load was reached, catastrophic flexural failure occurred. All the specimens with the Hwangtoh admixture demonstrated failure behavior similar to the failure behavior that of the control specimen. The AH and AHS specimens mixed with Korean Hwangtoh powder, however, exhibited a slightly higher flexural capacity than the control specimen. The KHSP, CHSP, and RPET specimens showed partial crushing failures in the upper compressive region followed by rebar yielding. In contrast to the typical rebar yielding flexural failure behavior observed in the specimens without recycled PET fibers, compressive crushing failure behavior was observed in the specimens with PET fibers. This observation is due to the improved tensile stress transferring and crack-controlling capabilities of the PET fibers in the bottom tensile region of the PET fiber-reinforced concrete. In addition, the Hwangtoh concrete specimens reinforced with recycled PET fibers (*i.e.*, KHSP and CHSP) exhibited significantly larger deflections than the deflections of the Hwangtoh concrete specimens without recycled PET fibers (*i.e.*, KH, CH, KHS, and CHS). These findings validate the addition of recycled PET fibers in Hwangtoh concrete to improve the crack-controlling capacity of RC members.

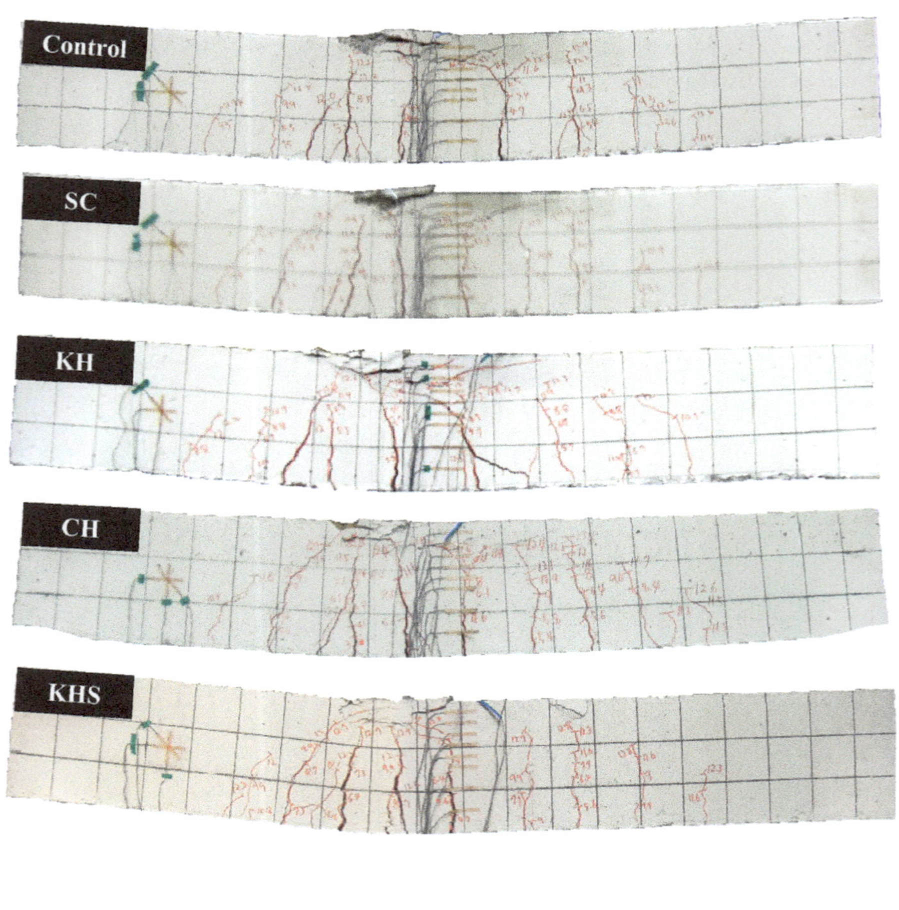

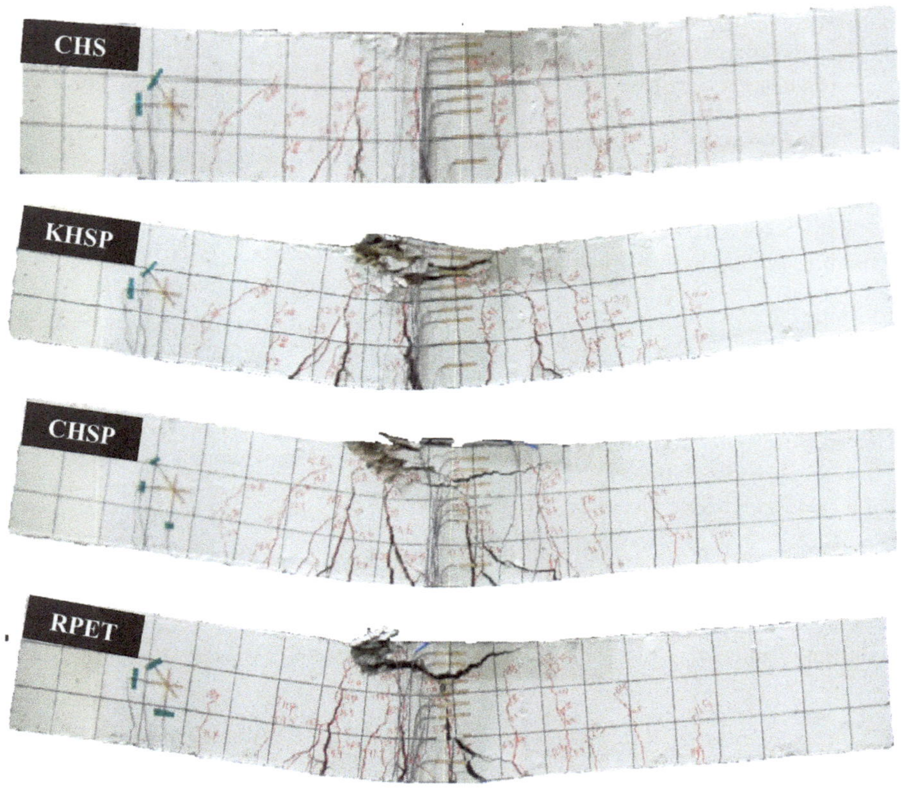

Figure 16. Failure mode and crack patterns of various RC specimens.

CONCLUSIONS

This study evaluates the material, durability, and structural capacities of Hwangtoh concrete. Plain and RC specimens cast with Korean and Chinese Hwangtoh admixtures and short recycled PET fibers are considered herein. The results of basic material property tests and RC structural member capacity tests are summarized as follows.

(1) From the trial tests, the optimum replacement ratio of Hwangtoh powder to cement is 20%. In addition, without using a chemical water-reducing admixture, desired workability can be achieved through the addition of blast furnace slag powder. By replacing cement with Hwangtoh and blast furnace slag powders in proportions of up to 50%, eco-friendly Hwangtoh concrete with material properties equivalent to the properties of OPC concrete can be developed.

(2) According to the results of the tests herein, Hwangtoh concrete exhibits a slightly lower compressive strength and elastic modulus than plain concrete. However, the differences (less than 2%–6%) are minute, thus, validating the possibility of using Hwangtoh concrete as a structural construction material.

(3) The free drying shrinkage strain of Hwangtoh concrete is approximately 1.5 to 2.0 times greater than the free drying shrinkage strain of plain concrete. The cracking time of Hwangtoh concrete is approximately five days earlier than the cracking time of plain concrete. However, the cracking time of the recycled PET fiber-reinforced Hwangtoh concrete is similar to the cracking time of plain concrete. Accordingly, these finding suggest that the crack-prone characteristics of Hwangtoh concrete can be mitigated by adding short plastic fibers, such as PP or PET fibers, to the material.

(4) Hwangtoh concrete measures slightly lower pH values than the pH values of ordinary cement concrete. The carbonation depth of Hwangtoh concrete is larger than the carbonation depth of ordinary cement concrete due to the decreased production of calcium oxide caused by the replacement of ordinary cement with Hwangtoh. Creep strain is highest in Hwangtoh-PET fiber concrete, followed by slag cement concrete and Hwangtoh concrete. These findings show that there is a tradeoff between drying shrinkage strain resistance and creep strain resistance, which comes with the addition of PET fibers.

(5) The flexural failure behavior of Hwangtoh concrete RC beams is very similar to the flexural failure behavior control specimens herein. The fiber-bridging capability of members reinforced with recycled PET fibers in the tensile region improves the flexural capacity of the RC members by delaying ultimate failure.

(6) This study's evaluation of the ductility index reveals almost no difference between Hwangtoh and plain concrete without recycled PET fibers. RC specimens with recycled PET fibers, however, have relative ductility indices that are approximately twice as great as the ductility indices of concrete without recycled PET fibers. Therefore, the ductility of RC members cast with Hwangtoh concrete is significantly improved by the addition of recycled PET fibers.

(7) Since all of the performance evaluations are conducted under normal environments in dry condition, additional durability tests of the material in a variety of aggressive environments, such as seawater, must be addressed in future works.

ACKNOWLEDGMENTS

This study was supported by the National Research Foundation of Korea (NRF) grant, which is funded by the government of Korea (MSIP) (No.2011-0030040).

AUTHOR CONTRIBUTIONS

Bon-Min Koo, the first author, was a main researcher of the study and the writer of the manuscript; Jang-Ho Jay Kim, the corresponding author, was the principal investigator of the study. Sung-Bae Kim was a co-researcher of the study. Sungho Mun was an advisor to the study.

Conflicts of Interest

The authors declare no conflict of interest.

REFERENCES

1. Fratermali, F.; Spadea, S.; Berardi, V.P. Effects of recycled PET fibers on the mechanical properties and seawater curing of Portland cement-based concretes. *Constr. Build. Mater.* 2014, *61*, 293–302.
2. Mehta, P.K. Greening of the concrete industry for sustainable development. *Concrete Int.* 2002, *24*, 23–28.
3. Miyake, J.; Matsushita, H. Evaluation method for consistencies of mortars with various mixture proportions. *J. Adv. Concrete Technol.* 2007, *5*, 87–97.
4. Yang, K.H.; Hwang, H.Z.; Kim, S.Y.; Song, J.K. Development of a cementless mortar using hwangtoh binder. *Build. Environ.* 2007, *42*, 3717–3725.
5. Malhota, V.M. A review: Fly ash, slag, silica fume, and rice husk ash in concrete. *Concrete Int.* 1993, *15*, 23–28.
6. Malhota, V.M. Introduction: Sustainable development and concrete technology. *Concrete Int.* 2002, *24*, 22.
7. Ishida, T.; Miyahara, S.; Maruya, T. Chloride binding capacity of mortars made with various portland cements and mineral admixtures. *J. Adv. Concrete Technol.* 2008, *6*, 287–301.
8. Dias, W.P.S.; Nanayakkara, S.M.A.; Ekneligoda, T.C. Performance of concrete mixes with OPC-PFA blends. *Mag. Concrete Res.* 2003, *55*, 161–170.
9. Alhozaimy, A.M.; Shannag, M.J. Performance of concretes reinforced with recycled plastic fibres. *Mag. Concrete Res.*2009, *61*, 293–298.
10. Jo, B.W.; Tae, G.H.; Kim, C.H. Uniaxial creep behavior and prediction of recycled-PET polymer concrete. *Constr. Build. Mater.* 2007, *21*, 1552–1559.
11. Sriravindrarajah, R.; Wang, N.D.H.; Wen Ervin, L.J. Mix design for pervious recycled aggregate concrete. *Int. J. Concrete Struct. Mater.* 2012, *6*, 239–246.
12. Jo, B.W.; Park, S.K.; Park, M.S. Strength and hardening characteristics of activated fly ash mortars. *Mag. Concrete Res.*2007, *59*, 121–129.
13. Plank, J.; Schroefl, C.; Gruber, M.; Lesti, M.; Sieber, R. Effectiveness of polycarboxylate superplasticizers in Ultra-High Strength Concrete: The importance of PCE compatibility with silica fume. *J. Adv. Concrete Technol.* 2009, *7*, 5–12.
14. Hossain, A.B.; Fonseka, A.; Bullock, H. Early age stress development, relaxation, and cracking in restrained low W/B ultrafine fly ash mortars. *J. Adv. Concrete Technol.* 2008, *6*, 261–271.
15. Yang, E.I.; Yi, S.T.; Leem, Y.M. Effect of oyster shell substituted for fine aggregate on concrete characteristics: Part I. Fundamental properties. *Cement Concrete Res.* 2005, *35*, 2175–2182.
16. Yang, E.I.; Kim, M.Y.; Park, H.G.; Yi, S.T. Effect of partial replacement of sand with dry oyster shell on the long-term performance of concrete. *Constr. Build. Mater.* 2010, *24*, 758–765.
17. Bougara, A.; Kadri, E.; Ezziane, K. Efficiency of granulated blast furnace slag replacement of cement according to the equivalent binder concept. *Cement Concrete Compos.* 2010, *32*, 226–231.
18. Saeki, T.; Sasaki, K.; Shinada, K. Estimation of chloride diffusion coefficient of concrete using mineral admixtures. *J. Adv. Concrete Technol.* 2006, *4*, 385–394.
19. Won, J.P.; Lee, Y.S.; Park, C.G.; Park, H.G. Durability characteristics of controlled low-strength materials containing recycled bottom ash. *Mag. Concrete Res.* 2004, *56*, 429–436.
20. Dhir, R.K.; McCarthy, M.J.; Caliskan, S.; Ashraf, M.K. Concrete pressure on formwork: Influence of cement combinations and superplasticising admixtures. *Mag. Concrete Res.* 2009, *61*, 407–417.
21. Go, S.S.; Lee, H.C.; Lee, J.Y.; Kim, J.H.; Chung, C.W. Experimental investigation of mortars using activated Hwangtoh.*Constr. Build. Mater.* 2009, *23*, 1438–1445.
22. Go, S.S.; Chung, C.W.; Struble, L.J.; Lee, H.C. Pozzolanic activity of Hwangtoh clay. *Constr. Build. Mater.* 2010, *24*, 2638–2645.
23. Kim, K.W.; Kim, S.M.; Park, M.J.; Kim, H.J.; Park, J.C. Characteristics of a reddish residual soil (Hwangtoh) finishing material with water-soluble adhesive for residential building. *Constr. Build. Mater.* 2010, *24*, 1542–1546.
24. Lee, K.C.; Her, J.H.; Kwon, S.K. Red clay composites reinforced with polymeric binders. *Constr. Build. Mater.* 2008, *22*, 2292–2298.
25. Dunster, A.M. The pozzolanic reaction of metakaolin and its effects on portland cement hydration. *J. Mater. Sci.* 2000,*28*, 1345–1350.
26. Bai, J.; Wild, S.; Sabir, B.B.; Kinuthia, J.M. Workability of concrete incorporating pulverized fuel ash and metakaolin.*Mag. Concrete Res.* 1999, *51*, 207–216.
27. Choi, H.Y.; Hwang, H.Z.; Kim, M.H.; Kim, M.H. A study on the development of Hwangtoh admixture for the application of cement mortar. *J. Archit. Inst. Korea* 2000, *16*, 95–102. (In Korean).
28. Choi, H.Y.; Kim, M.H.; Kim, M.H.; Hwang, H.Z.; Choi, S.W. Experimental study on the properties of concrete

by the kinds of admixture and the replacement ratios of activated Hwangtoh. *J. Korea Concrete Inst.* 2001, *13*, 123–129. (In Korean).

29. Jung, Y.B.; Yang, K.H.; Hwang, H.Z.; Chung, H.S. An evaluation of mechanical properties of fresh and hardened concrete mixed with Hwang-toh and ground granulated blast furnace slag. *J. Archit. Inst. Korea* 2006, *22*, 13–20. (In Korean).
30. Banthia, N.; Sheng, J. Fracture toughness of micro-fiber reinforced cement composites. *Cement Concrete Compos.* 1996,*18*, 251–269.
31. Dwarakanath, H.V.; Nagaraj, T.S. Deformational behavior of fiber-reinforced concrete beams in bending. *J. Struct. Eng. ASCE* 1992, *118*, 2691–2698.
32. Ochi, T.; Okubo, S.; Fukui, K. Development of recycled PET fiber and its application as concrete-reinforcing fiber.*Cement Concrete Compos.* 2007, *29*, 448–455.
33. Siddique, R.; Khatib, J.; Kaur, I. Use of recycled plastic in concrete: A review. *Waste Manag.* 2008, *28*, 1835–1852.
34. Foti, D. Use of recycled waste pet bottles fibers for the reinforcement of concrete. *Compos. Struct.* 2013, *96*, 396–404.
35. Kim, S.B.; Yi, N.H.; Kim, H.Y.; Kim, J.H.J.; Song, Y.C. Material and structural performance evaluation of recycled PET fiber reinforced concrete. *Cement Concrete Compos.* 2010, *32*, 232–240.
36. Kim, J.H.J; Park, C.G.; Lee, S.W.; Lee, S.W.; Won, J.P. Effects of the geometry of recycled PET fiber reinforcement on shrinkage cracking of cement-based composites. *Compos. Part B Eng.* 2008, *39*, 441–450.
37. Won, J.P.; Jang, C.I.; Lee, S.W.; Lee, S.L.; Kim, H.Y. Long-Term performance of recycled PET fiber-reinforced cement composites. *Constr. Build. Mater.* 2010, *24*, 660–665.
38. Fraternali, F.; Ciancia, V.; Chechile, R.; Rizzano, G.; Feo, L.; Incarnato, L. Experimental study of the thermo-mechanical properties of recycled PET fiber-reinforced concrete. *Compos. Struct.* 2011, *93*, 2368–2374.
39. Foti, D. Preliminary analysis of concrete reinforced with waste bottles PET fibers. *Constr. Build. Mater.* 2011, *25*, 1906–1915.
40. Hwang, H.Z. A Study on the Method of Activating Kaolin and Mortar and Concrete Mixed with Active Kaolin. Ph.D. Thesis, Seoul National University, Seoul, Korea, 1 March 1997.
41. Pereira de Oliveira, L.A.; Castro-Gomes, J.P. Physical and mechanical behavior of recycled PET fiber reinforced mortar. *Constr. Build. Mater.* 2011, *25*, 1712–1717.
42. Fraternali, F.; Farina, I.; Polzone, C.; Pagliuca, E.; Feo, L. On the use of R-PET strips for the reinforcement of cement mortars. *Compos. Part B Eng.* 2013, *46*, 207–210.
43. Hwang, H.Z.; Roh, T.H.; Yang, J.H. A study on the proper mixing design of concrete with Hwangtoh binder. *J. Archit. Inst. Korea* 2007, *23*, 73–80. (In Korean).
44. Korea Standards Association. *Testing Method for Static Modulus of Elasticity and Poisson's Ratio in Compression of Cylindrical Concrete Specimens*; KS F 2438; Korea Standards Association: Seoul, Korea, 2002. (In Korean)
45. Korea Standards Association. *Method of Test for Compressive Strength of Concrete*; KS F 2405; Korea Standards Association: Seoul, Korea, 2005. (In Korean)
46. Song, H.W.; Jung, M.S.; Ann, K.Y.; Lee, C.H. Chloride threshold value for steel corrosion considering chemical properties of concrete. *Korean Soc. Civil Eng.* 2009, *29*, 75–84.
47. Arya, C.; Xu, Y. Effect of cement type on chloride binding and corrosion of steel in concrete. *Cement Concrete Res.* 1995,*25*, 893–902.
48. Korea Standards Association. *Standard Test Method for Accelerated Carbonation of ConcreteStandard Test Method for Accelerated Carbonation of Concrete*; KS F 2584; Korea Standards Association: Seoul, Korea, 2010. (In Korean)
49. Korea Standards Association. *Method for Measuring Carbonation Depth of Concrete*; KS F 2596; Korea Standards Association: Seoul, Korea, 2004. (In Korean)
50. American Society for Testing and Materials (ASTM). *Standard Test Method for Creep of Concrete in Compression*; ASTM C 512; ASTM: West Conshohocken, PA, USA, 2010.
51. Korea Standards Association. *Method of Test for Drying Shrinkage Cracking of Restrained Concrete*; KS F 2595; Korea Standards Association: Seoul, Korea, 2004. (In Korean)
52. Gonzalez, J.A.; Ramirez, E.; Bautista, A. Protection of steel embedded in chloride-containing concrete by means of inhibitors. *Cement Concrete Res.* 1998, *28*, 577–589.
53. Rebeiz, K.S.; Fowler, D.W.; Paul, D.R. Recycling plastics in polymer concrete for construction applications. *J. Mater. Civil Eng. ASCE* 1993, *5*, 237–248.
54. Lin, Z.; Kanda, T.; Li, V.C. On interface property characterization and performance of fiber reinforced cementitious composites. *J. Concrete Sci. Eng. RILEM* 1999, *1*, 173–184.

CHAPTER 8

FABRICATION OF CEMENT, POLYMER, AND METAL-MATRIX FIBRE COMPOSITES BY GRAVITY-ASSISTED LIQUID-BINDER INFILTRATION

Panagiotis Frantzis

INTRODUCTION

The use of composite materials in the engineering industry has been rapidly increasing over the last fifty years or so. As a result, the science and technology of composite materials has progressed significantly. Developments in the construction industry include high volume fraction steel fibre reinforced composites fabricated by binder (cement slurry) gravity-assisted infiltration into preformed arrays of discontinuous or continuous fibres, known as SIFCON (slurry infiltrated fibre concrete) as reported elsewhere [1–5]. Effective infiltration was shown to be necessary to achieve optimum mechanical properties. It was concluded that understanding of the characteristics of infiltration was essential. Work into the rheological characteristics of magnesium phosphate and ordinary Portland cement binders has also been reported by Frantzis and Baggott [6, 7]. A relationship was derived, based on experimental observations, between viscosity and setting time that enabled the time available for effective gravity-assisted infiltration to be determined. Developments in the area of fabrication of metal fibre composites include pressureless liquid-binder infiltration techniques for the fabrication of composites with a magnesium-containing aluminium alloy binder reported by Aghajanian et al. [8, 9] and Dwivedi et al. [10]. The techniques developed employed preformed arrays of the fibres, contained within refractory vessels, infiltrated by the molten binder to identify, when possible, relationships between distance of infiltration and other process parameters. The results demonstrated the existence of linear relationships among infiltration distance, magnesium content, and process temperature.

The work reported here is a continuation of a large programme of experimental work on the fabrication of composites by gravity-assisted infiltration of (i) high volume fraction short steel fibre reinforced rapid hardening cements reported by Frantzis and Baggott [11], (ii) fabric (woven glass fibres) reinforced polymers with an epoxy resin binder, and (iii) short ceramic fibre reinforced metals with a modified (10%Mg + 1%Bi by weight) aluminium alloy binder. The characteristics of the interstices (that define the pore geometry) in arrays of discontinuous steel fibres (an array was regarded as set of fibres in random orientations) through which the binder (cement slurry) flows were considered to be a key parameter determining resistance to flow and therefore controlling gravity-assisted infiltration. However, the cost of commercially available short stainless steel fibres is too high and the preparation of arrays time consuming. To overcome these problems, a flow test was developed that was hoped to be able to simulate the flow of the binder through the arrays. The flow test was based on columns of woven stainless steel fibres infiltrated by the magnesium phosphate binder (the plane of the fabric was normal to the direction of flow) to identify possible relationships between interstice characteristics and infiltration distance. The distance of infiltration, considered to depend

Citation: Panagiotis Frantzis, "Fabrication of Cement, Polymer, and Metal-Matrix Fibre Composites by Gravity-Assisted Liquid-Binder Infiltration," ISRN Materials Science, vol. 2011, Article ID 757218, 9 pages, 2011. doi:10.5402/2011/757218.

upon the characteristics of the binder (rheology that included viscosity and setting/curing time, particle size, adhesion to fibres), was thought to be another key parameter of the fabrication process. A semiempirical expression characterising woven steel fibre interstice was identified based on fibre spacing and volume fraction (a measure of porosity), concepts that have proved helpful in explaining composite mechanical behaviour as discussed by Romualdi and Mandel [12], Kar and Pal [13], and Ganeshalingam et al. [14].

Further, an experimental approach was adopted that included a method to provide a measure of the average horizontal and vertical fibre spacing in preformed arrays of short steel fibres (prepared using a steel mould). The approach also included measurement of the infiltration distance, at the level where uniform infiltration had occurred, of the magnesium phosphate into the preformed arrays that was considered to simulate the actual infiltration process involved. The interstice characterisation was extended to include steel fibre arrays. Analysis of the data obtained has established that there was a common relationship between the expression characterising interstice and the distance of infiltration for both types of steel fibres. This indicated that the flow through woven fibres was similar to the flow through arrays, implying that a common mechanism of infiltration was in operation, despite the difference between the fibres. The columns of woven steel fibres were next replaced by columns of woven glass fibres. The columns were infiltrated by a liquid epoxy resin binder. The aim was to assess whether the flow test could provide with an inexpensive technique to assist in the fabrication of fabric- (woven glass fibres) based composites as discussed elsewhere [15–20]. This material is also increasingly used for civil engineering applications as discussed elsewhere [21–25].

This part of the experimental work reported here falls into the category of flow through porous media and there is a large number of work reported in the literature on liquid composite moulding/resin transfer moulding/resin infusion related to polymer composites as discussed elsewhere [26–36]. The experimental data obtained here might be correlated to Darcy's [37] equation applied to impregnation of fibre assemblies; the characteristics of the interstices might be quantified in a more consistent manner using tessellation or fractal dimensions as discussed by Summerscales et al. [38]; the interpretation of flow characteristics through clustered porous media might be modelled using the Kozeny-Carman [39] relationships, all well-established relationships and terminology that have been applied over long periods for both civil engineering and materials processing problems and these were not considered here. Details on processing as well as properties characterization of the fabricated composites were also not considered here. The aim was to produce a material in the laboratory with proven reliability and low cost. The flow test aims to provide with a universal hand lay-up fabrication technique capable of producing fabric-based composites with virtually no areas starved of binder thus eliminating the need to improve moulding conditions, a reliable product at very low cost but no scientific shortcomings. The interstice characterisation was therefore extended to include woven glass fibres. Analysis of the data obtained has established that there was a common relationship between the expression characterising interstice and the distance of infiltration for both binders and type of fibres. This indicated that the flow through woven fibres was similar to the flow through arrays and hence similar to the actual infiltration process involved, implying that a common mechanism of infiltration was in operation (that could be modelled by Darcy's equation but not considered here) despite the difference in the nature between the binders and the much more complex nature of the fibres. It was therefore suggested that the flow test could be used to assist in the fabrication of polymer fibre composites and a technique was subsequently proposed for use in research laboratories. No similar work has been reported in the literature before. Finally, an attempt was made to fabricate small-scale samples of short ceramic fibre reinforced metals. A modified test to the one used in the industry reported by Aghajanian et al. [8] and Dwivedi et al. [10] has therefore been developed based on gravity and employing a magnesium-containing aluminium alloy with the addition of bismuth to increase the fluidity of the binder.

MATERIALS

All materials used were arbitrarily selected and commercially available. Five (square) apertures (openings) of woven (wave styled) stainless steel fibres were selected, 0.57, 0.81, 1.21, 1.56, and 2.01 mm^2 with fibre diameters 0.26, 0.46, 0.34, 0.55, and 0.80 mm, respectively (an aperture or size of 0.57 mm^2 corresponds to horizontal spacing of 0.57 mm × 0.57 mm). Stainless steel fibres, 20, 25, and 35 mm long, kidney-shaped cross-sections of nominal diameters 0.12, 0.42, and 0.50 mm, respectively, were selected. Fabrics (woven E-glass fibres) with apertures 0.65 × 0.85 mm^2, 0.65 × 1.83 mm^2, and 0.65 × 2.60 mm^2 (the material most commonly used in practice and available at that time) were also selected. The glass fibre dimensions were 0.1 mm thickness and 1.0 mm width. The epoxy resin used was a two-part cold cured Araldite range adhesive (used for experimental purposes only) that when mixed with the hardener at a concentration of 5 parts per hundred of resin by weight the curing time at room temperature was 2 hours and 40 minutes approximately. The rapid strengthening cementitious binder (water/cement = 0.17) was a magnesium phosphate (ASR-1) with 2% (by weight) addition of a retarder (sodium tetraborate) to extend the setting time (from 4 minutes to 25 minutes) of the fresh binder. The short ceramic fibres used were in the form of a preform (20% Saffil in a SiO_2 binder) and the composition of the aluminium alloy (9%-10% Si, 0.14% Fe, 0.02% Cu, 0.05% Mo, 0.4% Mg, 0.02% Ni, 0.04% Zn, 0.02% Pb+Sn, 0.1% Tin) included the addition of 10% Mg by weight.

METHODS

Flow of Magnesium Phosphate Binder through Columns

A flow test was developed consisting of a column of square layers of woven steel fibres of dimensions 50 mm by 50 mm placed approximately at 45° to each other to simulate more closely the infiltration position, Figure 1. The column was placed on top of a Perspex hollow cylinder and was kept rigid by the application of moderate (hand) pressure via a horizontal beam as shown. The beam was supported between two vertical bars. The binder was poured through a cylindrical opening (of diameter 14.5 mm) at the centre of the horizontal beam to control the amount of binder flowing (a constant rate of fluid) through the column by maintaining a constant head of fluid. The number of layers in the column was varied to give different infiltration distances. The test was stopped shortly after setting time when the binder was trapped between the layers. The layers were carefully removed at the end of each test one by one starting from the bottom and visually examined. In a typical binder flow path, from the first layer to the terminating, the trapped binder could be seen on the upper surface of the terminating layer but there was no binder on the lower surface. The height of the column between the first layer and the terminating, measured by replacing the appropriate number of layers to form a column and using a travelling microscope, provided an accurate measure of infiltration distance. It should be noted that similar flow paths were observed with all apertures used, giving defined measurable infiltration distances, except for the coarsest (2.01 mm^2) layers where the binder infiltrated right through the longest practical column that could be set up, detected immediately through the walls of the Perspex cylinder. To estimate the infiltration distance, a further series of tests was carried out with increasing column height noting the time to infiltrate in each case.

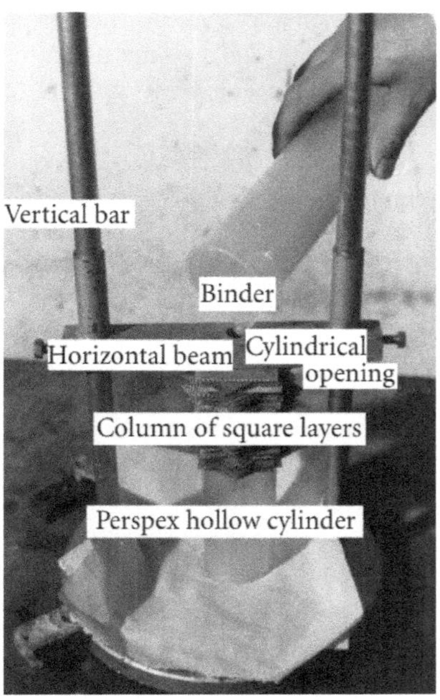

Figure 1: Flow test method apparatus.

Epoxy Resin

Layers of woven glass fibres replaced the layers of woven steel fibres and a similar procedure was followed. All flow tests gave defined measurable infiltration distances. The tests were stopped shortly after curing time when the flow had been terminated and the binder was trapped between the layers.

Flow of Magnesium Phosphate Binder through Arrays

An experimental approach was adopted to provide a measure of the average horizontal and vertical fibre spacing in preformed arrays of various steel fibre volume fractions, and including measurement of the infiltration distance of the binder into the arrays. Arrays of fibre volume fractions of 2%, 3%, and 5% were made using the 20 mm long fibres (it was not possible to prepare arrays of greater fibre volume fraction) and 14%, 15%, and 16% and 8%, 13%, and 16% using the 25 mm and 35 mm long fibres, respectively. The various arrays of randomly oriented steel fibres were prepared using a steel mould 118 mm long, 56 mm deep, and 56 mm wide. The fibres (were first weighed to give the appropriate fibre volume fraction and then) were dispersed into the mould by hand as evenly as possible. Pressure, applied using a small hydraulic press, was necessary to obtain arrays with higher fibre volume fraction, from 5 up to 16%. Estimates of horizontal spacing were made from photographs of known magnification of the top surface of the arrays. The estimation consisted of characterising horizontal spacing as the side of the equivalent square, formed between fibres, corresponding to the perimeter of the interstice, and subsequent measurement of the side using the travelling microscope. The vertical values of fibre spacing were estimated by observing and thus assuming that fibre orientation in the arrays made using the 20 mm and 25 mm long fibres varied between 0 to 45° to the horizontal. These observations identified vertical spacing to be given by as reported by Romualdi and Mandel [12] and Ganeshalingam et al. [14]:

$$s_v = \frac{s_h}{4} + d, \quad (1)$$

where s_v is the vertical fibre spacing, s_h the horizontal fibre spacing, and d the fibre diameter. In the case of the arrays made using the 35 mm long fibres, experimental observations showed that the fibres were almost aligned and thus vertical spacing was assumed to be equal to fibre diameter. Note that it was not possible to obtain clear, sharp, and with good contrast photographs of vertical surfaces. The arrays were infiltrated by the binder, removed from the mould after the binder had hardened, and cut across the centre to enable measurements of the infiltration distance to be made using the travelling microscope. The measurements of the infiltration distance were taken at the level where uniform infiltration had occurred, ignoring sporadic infiltration below this level, attributing this to localised regions of lower fibre volume fraction, as Figure 2 shows.

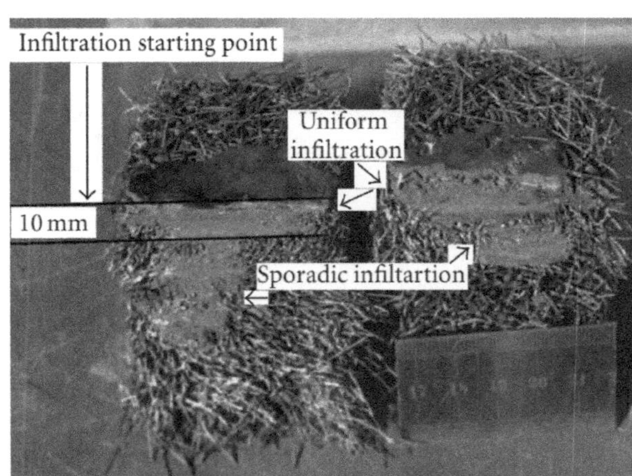

Figure 2: Typical measurements of the infiltration distance in arrays of fibres (sample shown is cut in half)—big arrow indicates starting point and direction of infiltration and small arrows indicate uniform and sporadic infiltration, respectively.

Aluminium Alloy Binder

Pieces of Al 10% Mg alloy (about 50 g) were placed into a cone-shaped ceramic pot (7 cm height, 4.6 cm large diameter, 2.3 cm base diameter, and 0.5 cm thickness) and the preform was placed either between the metal or on top. The pot was placed in a furnace and a supply of a nitrogen-containing gas (10% hydrogen-90% nitrogen) through a ceramic seal at a rate of 500 cm^3/min was maintained, as Figure 3 shows. 1% (by weight) of Bi was also added in some cases either as a solid directly into the pot or as an alloying element in the melt. The furnace, initially at room temperature, was heated to about 870°C for 24 hours and then it was switched off and allowed to cool for a further 24 hours. The gas supply was maintained for the 48 hour heating cycle thus ensuring the presence of the gas for the entire fabrication process. The samples were next removed and prepared for energy-dispersive X-ray spectroscopy (EDX) and scanning electron microscopy (SEM) examination. This was carried out using a Cambridge 100 Stereo Scan scanning electron microscope at an accelerating voltage of 25 kV.

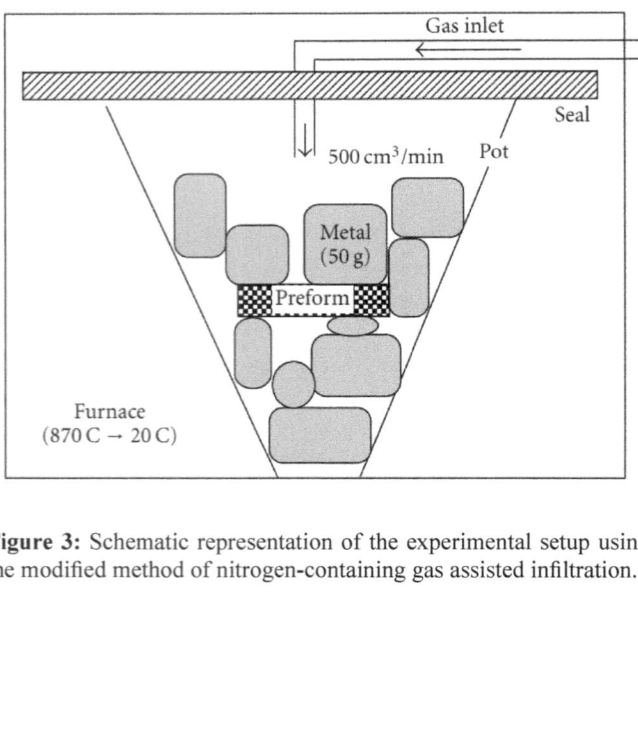

Figure 3: Schematic representation of the experimental setup using the modified method of nitrogen-containing gas assisted infiltration.

RESULTS AND DISCUSSION

Flow of Magnesium Phosphate through Columns

Flow of the binder through the columns of woven fibres was assisted by gravity. The 45° offset in consecutive layers was considered to simulate more closely the infiltration position during composite fabrication. Flow paths were clearly distinguished; they were vertical with no significant lateral divergence producing a cylindrical column of the binder (uniform infiltration), with diameter equal to that of the cylindrical opening at the centre of the horizontal beam. The results of the flow tests using the coarsest (2.01 mm^2) layers are presented in graphical form in Figure 4 and show a linear relationship between distance of infiltration (height of column) and time of flow. Extrapolation of the data was necessary and assuming that the infiltration distance could be determined by the setting time of the binder (25 minutes) a value of 1122 mm was identified (Table 1). It was assumed that the distance of infiltration could be related to interstice characteristics; possibly it could be proportional to some function of an interstice expression. To develop an expression it was necessary to establish interstice characteristics. These were considered to be the size, shape, and total volume of the interstices that involved equivalent fibre spacing and volume fraction of the fibres. A semiempirical expression characterising woven fibre interstice was therefore identified based on experience and as reported by Romualdi and Mandel [12] and Ganeshalingam et al. [14] as

$$\frac{s_h^2 s_v}{V_f}, \qquad (2)$$

where s_h^2 represented the spacing of the square aperture, s_v was the spacing between layers (measured using the travelling microscope), and V_f the volume fraction of the fibres. This indicated that larger apertures would lead to increasing distance of infiltration, increased vertical spacing would lead to increasing distance, and reduced total volume of space would lead to decreasing infiltration distance. The volume fraction was calculated by considering the volume of the column at the infiltration distance, the density of steel (7.85×103 kg/m^3), and the weight of the layers. Table 1 shows the results obtained.

Table 1: Flow test results for magnesium phosphate/woven steel fibres.

Volume fraction V_f (%)	Aperture s_h^2 (mm^2)	Spacing between layers s_v (mm)	Fibre diameter d (mm)	Infiltration distance, h (mm)	$s_h^2 s_v/V_f$ (mm^3 × 10^2)
26.3	0.57 × 0.57	0.47	0.26	3.2	0.58
26.5	0.81 × 0.81	0.95	0.46	7.4	2.35
20.4	1.21 × 1.21	0.54	0.34	6.1	3.88
23.7	1.56 × 1.56	0.92	0.55	103.3	9.44
21.4	2.01 × 2.01	1.76	0.80	1122	33.2

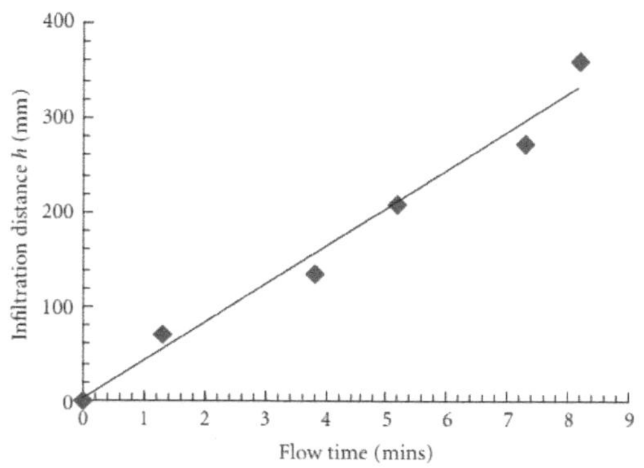

Figure 4: Linear relationship between infiltration distance h and flow time.

Flow of Magnesium Phosphate through Arrays

Flow of the binder through the arrays was assisted by gravity. The semiempirical expression characterising woven steel fibre interstice, $s_h^2 s_v/V_f$, was extended to include steel fibre arrays. The volume fraction was calculated by considering the volume of the mould, the density of steel, and the weight of the fibres. The measurements obtained are shown in Table 2. As expected, the infiltration distance increased with increasing spacing. The apparent discrepancies among the infiltration depths observed with the three different lengths of discontinuous fibres shown in Table 2 highlight the importance of characterising the whole interstice. The qualitative explanation of the smaller infiltration depths observed with the 35 mm long fibres (fibre diameter = 0.5 mm), compared with the 25 mm long fibres (fibre diameter = 0.42 mm) at volume fractions less than 14%, being the increased resistance to slurry flow provided by the greater horizontal orientation of the fibres. In the case of the relatively low infiltration with the 20 mm long fibres (fibre diameter = 0.12 mm) the qualitative explanation is that they are significantly finer than the other two fibres and therefore have much reduced vertical spacing.

Table 2: Flow test results for magnesium phosphate/steel fibre arrays.

Volume fraction V_f (%)	Fibre spacing (horizontal) s_h (mm)	Fibre spacing (vertical) s_v (mm)	Fibre diameter d (mm)	Infiltration distance h (mm)	$s_h^2 s_v/V_f$ (mm^3 × 10^2)
5	0.40	0.22	0.12	3.04	0.70
3	0.55	0.26	0.12	10.1	2.62
2	0.59	0.27	0.12	12.1	4.70
16	0.55	0.56	0.42	10.3	1.06
15	0.62	0.58	0.42	14.4	1.49
14	0.71	0.60	0.42	43.2	2.16
16	0.59	0.50	0.50	12.2	1.09
13	0.64	0.50	0.50	17.1	1.58
8	0.68	0.50	0.50	21.1	2.90

Flow of Epoxy Resin through Columns

Flow paths produced a trapezoidal column (truncated cone) of the binder, most probably due to much higher adhesion to glass fibres compared to magnesium phosphate and steel fibres, the big base being on the terminating layer with diameter a maximum four times that of the cylindrical opening at the centre of the horizontal beam. Table 3 shows the results obtained (density of glass = 2.58 × 10^3 kg/m^3). The semiempirical expression characterising woven steel fibre interstice, $s_h^2 s_v/V_f$, was extended to include woven glass fibres. Adopting the assumptions made in this work, the interstice expression was plotted against infiltration distance in logarithmic form. Data obtained using the magnesium phosphate (Tables 1 and 2) were superimposed on the data obtained using the epoxy resin. The plot is shown in Figure 5. A significant number of replicate tests were carried out and the average values are shown (typical scatter between replicates less than 5%). The sizes of the three glass fabrics (woven E-glass fibres) used in this work are the sizes of the material commonly used in practice. The best curve was fitted through the experimental data points and a common relationship could be identified as

$$\frac{s_h^2 s_v}{V_f} = 0.3h^{0.65} \tag{3}$$

with R-squared value of 0.53, a low but reasonable value bearing in mind the amount and difficulty of the experimental work. This indicated that the flow through columns was similar to the flow through arrays and hence similar to the actual infiltration processes involved. It was concluded that the flow through columns could be used to study the processes involved during composite fabrication, thereby providing a convenient test for the evaluation of likely infiltration characteristics. Hence, having established the applicability and validity of the flow test, Figure 6 was finally plotted that shows the data obtained using the epoxy resin and the glass fibres as a function of the interstice expression $s_h^2 s_v/V_f$ against infiltration distance h. The best curve was fitted through the experimental data points and a linear relationship could be identified as

$$\frac{s_h^2 s_v}{V_f} = 0.0744h - 0.5854 \tag{4}$$

with R-squared value of 0.9801 indicating a very good fit of the data obtained.

Table 3: Flow test results for epoxy resin/woven glass fibres.

Volume fraction V_f (%)	Aperture s_h^2 (mm²)	Spacing between layers s_v (mm)	Fibre dimension-sthickness, width (mm)	Infiltration distance h (mm)	$s_h^2 s_v/V_f$ (mm³ × 10²)
32.1	0.65 × 0.85	0.10	0.1, 1.0	10.3	0.17
32.1	0.65 × 0.85	0.12	0.1, 1.0	10.8	0.20
32.1	0.65 × 0.85	0.11	0.1, 1.0	10.5	0.19
32.0	0.65 × 1.83	0.11	0.1, 1.0	13.1	0.41
32.0	0.65 × 1.83	0.10	0.1, 1.0	12.8	0.37
32.0	0.65 × 1.83	0.12	0.1, 1.0	13.3	0.45
31.0	0.65 × 2.60	0.12	0.1, 1.0	16.4	0.65
31.0	0.65 × 2.60	0.10	0.1, 1.0	15.7	0.54
31.0	0.65 × 2.60	0.11	0.1, 1.0	16.1	0.60

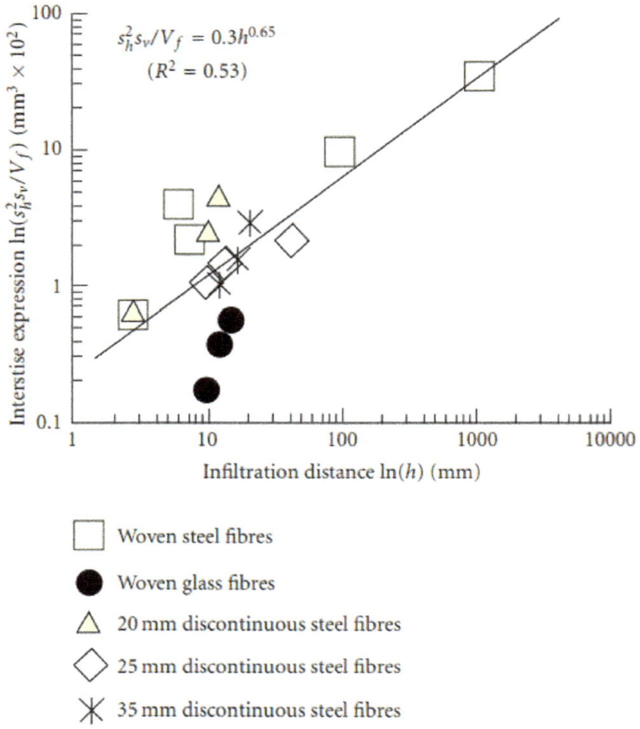

Figure 5: Logarithmic plot of $\ln(s_h^2 s_v/V_f)$ against infiltration distance $\ln(h)$ using columns of woven steel/glass fibres and arrays of steel fibres.

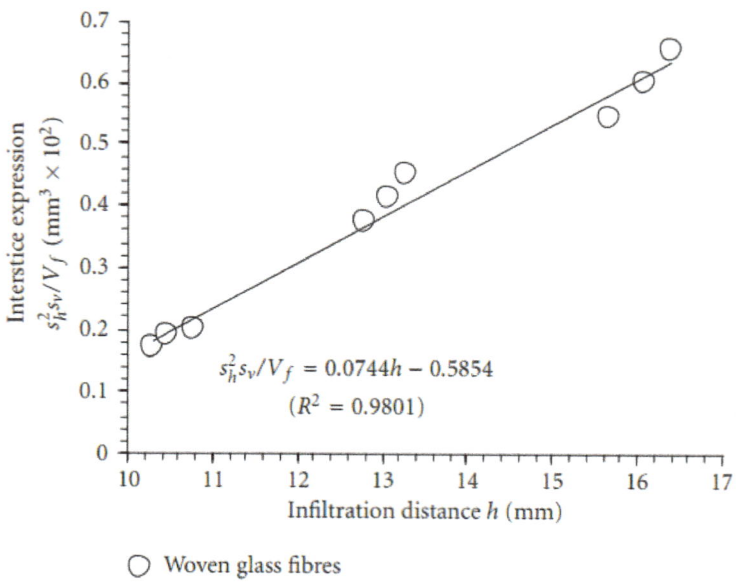

Figure 6: Plot of $s_h^2 s_v/V_f$ against infiltration distance h for epoxy resin/woven fibres.

Fabrication Technique

For the case of polymer composites, a hand lay-up fabrication technique is proposed.(i)A type of fabric reinforcement (at least three apertures) is acquired. An epoxy resin binder (rapid setting, cold or hot cured) is selected.(ii)Using the flow test (4) is established and hence the corresponding number of layers for effective infiltration is determined.(iii)The layers are laid up onto a mould and infiltrated by the binder. After a time interval sufficient for complete infiltration to occur, the product can be fully cured under the simultaneous application of heat and pressure or in air.(iv)The final product is a polymer laminate (thicknesses between 10 and 16 mm were achieved here at a fibre volume fraction of 31-32%).

The advantages of the fabrication technique were considered to be as follows.(i)Any type of fabric material like carbon fibre cloth used in the automotive industry as reported by Cunningham [40] or epoxy resin can be used.(ii)Less binder waste; virtually no areas starved of binder as reported by Cunningham [41]; only minimal final machining of the edges required to eliminate end effects; minimal technical skills and inexpensive tooling required; no initial investment required.

The disadvantages of the fabrication technique were considered to be(i)labour intensive; suitable for components of simple or complex shape with fibre volume fraction between 30–40% (like hulls of boats, roofs of cars, and small unmanned planes).

Aluminium Alloy Composites

Visual examination (using the travelling microscope) of the samples where the Bi was not used and the preform was placed on top of the metal showed that the preform was completely intact. Visual examination of the samples where the Bi was not used and the preform was placed between the metal showed that partial infiltration had occurred (the appearance of each sample was partly white, the colour of the preform, and partly with the colour of the aluminium alloy). Visual examination of the samples where the Bi was dissolved into the alloy and the preform was placed either on top or between the metal showed that the preform was completely intact and therefore fabrication was not successful. Visual examination of the samples where the Bi was added as a solid directly into the pot and the preform was placed between the metal showed that complete infiltration had occurred (the appearance of each sample resembled that of the aluminium alloy). The samples in the latter case were next prepared for microscopic examinations as follows: they were cut using a saw, placed in conducting mouldings, ground and polished using colloidal silica polishing suspension and finally washed and dried. Typical SEM images (Figure 7) taken on three points on the cut surfaces showed a fibrous microstructure (black/dark regions) within the aluminium alloy matrix (grey/white regions) at magnifications 50 μm and 125 μm, respectively. Similar results have been reported elsewhere [8, 9]. Typical EDX spectrum obtained (Figure 8) taken on three points on the cut surfaces showed strong peaks characteristic of Mg, Al, and Si but Bi peak was absent. To provide with a reasonable explanation, either the amount of Bi was not enough to be detected or the presence of unwanted gases in the furnace chamber might have initiated to reactions that resulted to the failure in detecting Bi which otherwise did not influence the success of the method developed. The results indicated that the addition of Bi can improve fabrication performance and that successful infiltration was obtained only when the Bi was not dissolved into the alloy. The results also indicated that a minimum amount of pressure was still necessary since the weight of the preforms was not enough for the infiltration process to start by the sinking of the preforms into the melt alloy, as reported in the industry by Aghajanian et al. [8] and Dwivedi et al. [10]. The pressure was a consequence of the weight of the metal above the preforms and therefore this method was thought to be a combination of gas-assisted and gravity-assisted infiltration. Further work is envisaged.

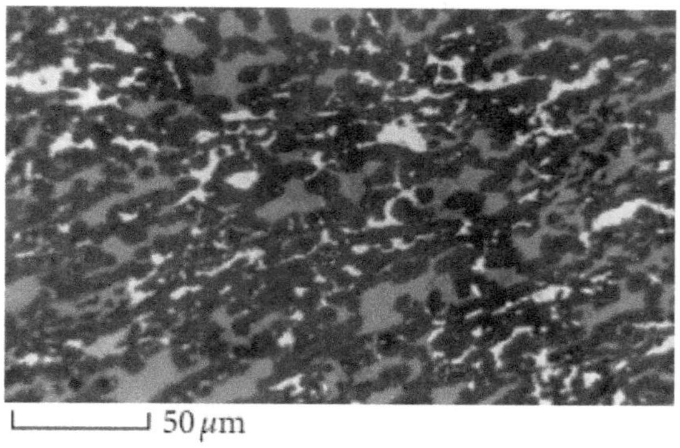

(a)

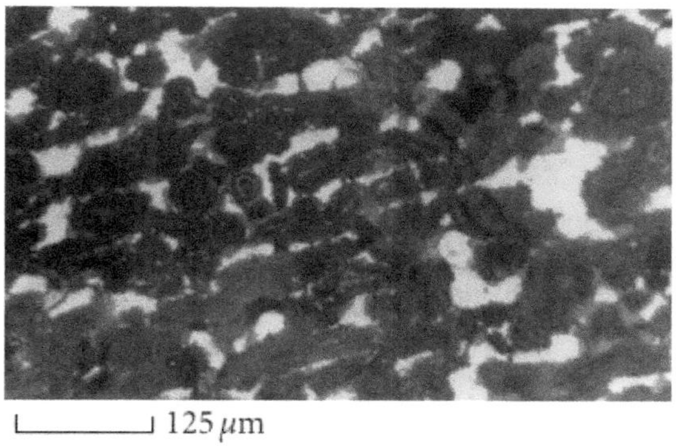

(b)

Figure 7: Typical SEM images of microstructure (preform placed between the metal) for Al 10%Mg 1%Bi (Bi not dissolved into the alloy) at magnifications 50 μm (a) and 125 μm (b), respectively, and showing a homogeneous sample.

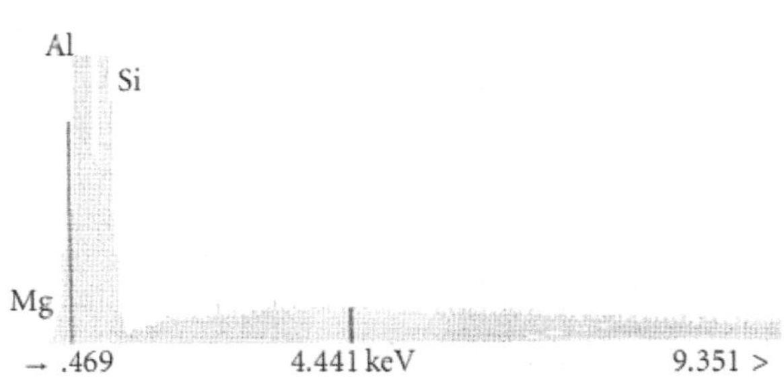

Figure 8: Typical X-ray results for Al 10% Mg 1% Bi alloy (Bi not dissolved into the alloy and preform placed between the metal).

CONCLUSIONS

This paper reports study to fabricate fibre composites by gravity-assisted liquid-binder infiltration into woven fibres and preformed arrays of short fibres. A flow test was developed that was shown to be able to simulate the actual infiltration processes involved during fabrication of cement/polymer fibre composites. A test was also developed that made use of a nitrogen-containing gas and the addition of Bi for the fabrication of metal fibre composites. The main findings of this work were therefore as follows.(1)Flow was identified to depend upon the interstice and the cement/polymer binder characteristics. The characteristics of infiltration were therefore simplified to two key parameters controlling flow: an expression of the interstice through which the binder flows, involving equivalent fibre spacing and volume fraction, and the distance of infiltration. (2)A common relationship was identified between the key parameters, implying that a common mechanism of infiltration was in operation, despite the difference in the nature between the cement/polymer binders and the much more complex nature of the interstices.(3)Although the experiments carried out here may at first appear to be basic, the flow test developed has never been reported in the literature before and a hand lay-up technique (of minimal cost and yet sophisticated enough) for the fabrication of polymer composites was subsequently proposed.(4)The test developed for the fabrication of metal fibre composites may be more attractive than those currently used in the industry because it is relatively simpler and with low cost, at least when intended for use for research purposes.

ACKNOWLEDGMENTS

The financial support of the UK Engineering and Physical Science Research Council is gratefully acknowledged. The writer would also like to thank John Staniar & Co, Manchester, UK (supplier of woven stainless steel fibres), Fibre Tech Ltd, Nottingham, UK (supplier of stainless steel fibres), Ciba Geigy plc (supplier of epoxy resin), and FEB International, Manchester, UK (supplier of magnesium phosphate ASR-1). The author would also like to thank W. J. Cantwell, P. Fox, and G. J. Tatlock for their useful discussions, interest, and support, and D. Whitehurst and M. Hughes for their valuable assistance (University of Liverpool, Liverpool, UK), R. Baggott for his many useful discussions and guidance, and P. B. Unsworth and I. Hanbridge for their valuable assistance (University of Salford, Manchester, UK).

REFERENCES

1. D. R. Lankard and C. Josifek, "SIFCON: slurry infiltrated fibrous concrete," in Proceedings of the Symposium on Fibre Reinforced Concrete, pp. 7.15–7.23, RILEM, Cacham, Cedex, France, 1987.
2. A. Bentur and S. Mindess, Fibre Reinforced Cementitious Composites, Elsevier Science, London, UK, 1990.
3. R. Baggott and P. Frantzis, "Pseudo metallic repair materials," in Proceedings of the International Tokyo Symposium on Strategies and Technologies for Maintenance and Modernisation of Building (CIB W70) Part 5: Building Defects and Materials Technology, pp. 527–534, Tokyo, Japan, 1994.
4. A. M. Brandt, Cement-Based Composites, Materials Mechanical Properties and Performance, E&FN Spon/Chapman & Hall, London, UK, 1995.
5. International Union of Testing and Research Laboratories for Materials and Structures (RILEM), American Concrete Institute (ACI), Centre for Advanced Cement Based Materials (ACBM), the University of Michigan, and the University of Stuttgart, "High performance fiber reinforced cement composites 2 (HPFRCC2)," in Proceedings of the 2nd International RILEM Workshop, RILEM Proceedings 31, A. E. Naaman and H. W. Reinhardt, Eds., E&FN Spon/Chapman & Hall, Ann Arbor, Mich, USA, 1996.
6. P. Frantzis and R. Baggott, "Effect of vibration on the rheological characteristics of magnesia phosphate and ordinary Portland cement slurries," Cement and Concrete Research, vol. 26, no. 3, pp. 387–395, 1996. ·
7. P. Frantzis and R. Baggott, "Rheological characteristics of retarded magnesia phosphate cement," Cement and Concrete Research, vol. 27, no. 8, pp. 1155–1166, 1997. ·
8. M. K. Aghajanian, M. A. Rocazella, J. T. Burke, and S. D. Keck, "The fabrication of metal matrix composites by a pressureless infiltration technique," Journal of Materials Science, vol. 26, no. 2, pp. 447–454, 1991.
9. M. K. Aghajanian, R. A. Langensiepen, M. A. Rocazella, J. T. Leighton, and C. A. Andersson, "The effect of particulate loading on the mechanical behaviour of Al_2O_3/Al metal-matrix composites," Journal of Materials Science, vol. 28, no. 24, pp. 6683–6690, 1993. ·
10. R. K. Dwivedi, J. T. Burke, G. H. Shirnky, M. K. Aghajanian, and S. D. Keck, "Flotation process for the formation of metal matrix composite bodies," Patent Number US5529109, 1996.
11. P. Frantzis and R. Baggott, "Factors influencing magnesia phosphate slurry infiltration into steel fibre arrays," in Proceedings of the International Workshop on Fibre Reinforced Cement and Concrete, R. N. Swamy and V. Ramakrishan, Eds., pp. 46–61, National Science Foundation (USA), Centre for Cement and Concrete, University of Sheffield, Sheffield, UK, 1994.
12. J. P. Romualdi and J. A. Mandel, "Tensile strength of concrete affected by uniformly distributed and closely spaced short lengths of wire reinforcement," Journal of the American Concrete Institute, vol. 61, no. 6, pp. 657–671, 1964.
13. J. N. Kar and A. K. Pal, "Strength of fibre reinforced concrete," American Society of Civil Engineers [ASCE], Journal of the Structural Division, vol. 98, no. ST5, pp. 1053–1068, 1972.
14. R. Ganeshalingam, P. Paramasivam, and G. K. Nathan, "An evaluation of theories and a design method of fibre cement composites," International Journal of Cement Composites and Lightweight Concrete, vol. 3, no. 2, pp. 103–114, 1981. ·
15. F. L. Matthews and R. D. Rawlings, Composite Materials: Engineering and Science, Chapman & Hall, London, UK, 1995.
16. B. T. Åström, Manufacturing of Polymer Composites, Chapman & Hall, London, UK, 1997.
17. W. D. Callister Jr., Materials Science and Engineering: An Introduction, John Wiley & Sons, New York, NY, USA, 4th edition, 1997.

18. G. Akovali, Handbook of Composite Materials, Rapra Technologies Limited, Shropshire, UK, 2001.

19. National Physical Laboratory (NPL), SQP Good Practice Guide, Teddington, Middlesex, UK, 2003.

20. C. Vasile and A. K. Kulshreshtha, Handbook of Polymer Blends and Composites, vol. 1&2, Rapra Technologies Limited, Shropshire, UK, 2003.

21. J. M. Illston, Construction Materials, E&FN Spon, London, UK, 1994.

22. Fibreforce Replacing Metals Where It Matters Brochure, Fibreforce Composites Limited, Runcorn, Cheshire, UK, 1997.

23. S. Wallace, "Design of falls creek trail bridge: a fiber reinforced polymer composite bridge,"Transportation Research Record, vol. 1, no. 1652, pp. 133–142, 1999. ·

24. D. Greek, "Naval gazing keeps tube ship-shape: a composite material used to patch warships could help keep London's underground train fleet moving too," Professional Engineering Magazine, vol. 13, no. 6, pp. 39–40, 2000.

25. "Bridge system undergoes tests," Professional Engineering Magazine, vol. 14, no. 20, p. 12, 2001.

26. V. K. Harten, "Production by resin transfer moulding," in Composite Materials in Maritime Structures, R. A. Shenoi and J. F. Wellicome, Eds., pp. 86–126, Cambridge University Press, Cambridge, UK, 1993.

27. S. G. Advani and M. V. Bruschke, "Governing equations for flow and heat transfer in stationary fibre beds," in Flow and Rheology in Polymer Composites Manufacturing, S. G. Advani, Ed., pp. 465–515, Elsevier Science, Amsterdam, The Netherlands, 1994.

28. C. L. Tucker and R. B. Dessenberger, "Resin transfer moulding phenomena in polymeric composites," inFlow and Rheology in Polymer Composites Manufacturing, S. G. Advani, Ed., pp. 257–323, Elsevier Science, Amsterdam, The Netherlands, 1994.

29. C. Williams, J. Summerscales, and S. Grove, "Resin infusion under flexible tooling (RIFT): a review,"Composites Part A, vol. 27, no. 7, pp. 517–524, 1996.

30. K. Potter, Resin Transfer Moulding, Chapman & Hall, London, UK, 1997.

31. C. D. Rudd, A. C. Long, K. N. Kendall, and C. G. E. Mangin, Liquid Moulding Technologies, Woodhead Publishing, Cambridge, UK, 1997.

32. T. M. Kruckenberg and R. Paton, Resin Transfer Moulding for Aerospace Structures, Kluwer Academic Publishers, Dordrecht, The Netherlands, 1998.

33. D. Cripps, T. J. Searle, and J. Summerscales, "Open mould techniques for thermoset composites," inComprehensive Composite Materials Encyclopaedia, R. Talreja and J. A. Manson, Eds., vol. 2 of Polymer Matrix Composites, pp. 737–761, Elsevier Science, Oxford, UK, 2000.

34. R. S. Parnas, Liquid Composite Moulding, Hanser Gardner Publications, 2000.

35. G. Polaski, J. Means, B. Stull, et al., Bonding Elastomers: A Review of Adhesives and Processes, vol. 15, Report 177, Rapra Technologies Limited, Shropshire, UK, 2005.

36. J. Summerscales and T. J. Searle, "Review: low pressure (vacuum infusion) techniques for moulding large composite structures," in Proceedings of the Institution of Mechanical Engineers (IMechE), Part L, pp. 45–58, Journal of Materials-Design and Applications, 2005.

37. H. P. G. Darcy, Les Fontaines Publiques de la Ville de Dijon, Dalmont, Paris, 1856.

38. J. Summerscales, F. J. Guild, N. R. L. Pearce, and P. M. Russell, "Voronoi cells, fractal dimensions and fibre composites," Journal of Microscopy, vol. 201, no. 2, pp. 153–162, 2001.

39. P. C. Carman, "Fluid flow through granular beds," Transactions of the London Institute of Chemical Engineers, vol. 15, pp. 150–166, 1937.

40. J. Cunningham, "Composites have proved themselves in many ways in specialised industries-so what is holding them back from wider use?" Professional Engineering Magazine, vol. 20, no. 10, pp. 39–40, 2007.

41. J. Cunningham, "Sum of its parts: how manufacturers are racing to overcome production with composites as demand from the aviation industry begins to soar," Professional Engineering Magazine, vol. 21, no. 10, pp. 28–29, 2008.

CHAPTER 9

MICROSTRUCTURAL ANALYSIS AND RHEOLOGICAL MODELING OF ASPHALT MIXTURES CONTAINING RECYCLED ASPHALT MATERIALS

Augusto Cannone Falchetto, Ki Hoon Moon and Michael P. Wistuba

Keywords: Recycling, reclaimed asphalt pavement, manufacturer waste scrap shingles, tear-off scrap shingles, microstructure, creep stiffness, back-calculation

INTRODUCTION

Using recycled materials is a top priority in the pavement industry not only for reducing construction costs but also for minimizing the environmental impact of road construction. Over the past decades, the use of different recyclable materials has been investigated by international, national, and local agencies; the list includes Reclaimed Asphalt Pavement (RAP), reclaimed Portland cement concrete, iron blast-furnace slag, fly ash, waste tire rubber, waste glass, and roofing shingles. Two sources in particular—RAP and Recycled Asphalt Shingles (RAS)—have been increasingly used for asphalt pavement applications.

RAP has been used in the U.S. for more than three decades [1,2,3]. According to Federal Highway Administration (FHWA), almost 73 million tons of RAP are reclaimed and 84% of that reclaimed (62 million tons) is used annually in asphalt pavement construction, making RAP the most recycled material in the U.S. [4]. A number of specifications that regulate the amount of RAP have been already in place for many years. Most of these limits were developed from observations of field performance of asphalt pavements built with recycled mixture.

Citation: Augusto Cannone Falchetto, Ki Hoon Moon and Michael P. Wistuba, Microstructural Analysis and Rheological Modeling of Asphalt Mixtures Containing Recycled Asphalt Materials, doi:10.3390/ma7096254.

In Europe, the use of RAP is common practice with some countries having more than 30 years of experience, such in the Netherlands [5], Denmark and Germany [6]. More than 4000 production sites and over 10,000 companies are involved in RAP recycling and in the production and placing of asphalt materials. Roughly 50 million tons of RAP are produced in Europe each year with over 70% of it being reused for pavement surfaces [7].

Roofing shingles represent another material that is available in large quantity for recycling. According to one estimate, almost 11 million tons of waste bituminous roofing materials are generated each year in the U.S. and most of them are discarded into landfills [4,8,9]. The pavement recycling of roofing shingles finds limited application in Europe where this material is generally used for energy recovery as fuel supplement [10]. The use of RAS in hot mix asphalt has seen increased acceptance from government agencies and construction contractors in the U.S. only in recent years. The main obstacle is the potentially detrimental effect of the aged oxidized binder present in old roofing shingles.

LITERATURE REVIEW

A vast amount of literature in the past addressed the effect of RAP on recycled asphalt mixtures and, in most cases, it was concluded that RAP has a significant influence on materials' properties [11,12,13,14,15]. Little et al. [11] and McDaniel et al.[16] observed that higher RAP content are associated with increased viscosity and stiffness at high and intermediate temperatures, positively affecting rutting resistance;

nevertheless, limited relaxation properties and reduced fracture resistance were also experienced due to increased material brittleness. Other research efforts focused on design procedures, forensic evaluation and modeling [17,18]; specifications were also developed with recommendations for selecting RAP content based on traffic level. For example, the Minnesota Department of Transportation (MnDOT) Specification 2350/2360 [19] allows up to 40% (by weight) RAP based on traffic level and binder grade.

From a European perspective, two large research efforts were undertaken in the past under the umbrella of the European Union [14,15]; the main focus of these studies was on test methods and safe use of alternative pavement materials to obtain satisfactory environmental and functional performance. Two recent research projects [20,21] have addressed recycling through a holistic approach with the goal of raising the level of re-use of asphalt mixture with minimum material downgrading. The maximization of RAP recycling, together with the use of rejuvenator, was also addressed in the past [22] and in two ongoing research projects [23,24] in Germany, with the objectives of evaluating the field response of the material and to implement material selection recommendations for practitioners.

More recent studies centered their efforts in understanding and modeling the effective blending between virgin and aged binder in the recycled mixtures [25,26]. In particular, Kriz et al. [27] investigated the diffusion processes occurring at the interface between virgin and recycled binders concluding that the total thickness of the new-aged binder film plays a fundamental role in the process.

Regarding the re-use of roofing asphalt shingles, two distinct categories are available in the market: Manufacturer Waste Scrap Shingles (MWSS) and Tear-off Scrap Shingles (TOSS), from old roofs that have been exposed to solar radiation and high temperatures for extended periods of time. Both TOSS and MWSS, which are generally identified as RAS, contain a high amount of small aggregates and a much harder asphalt binder compared to that used in asphalt pavement mixtures: at 25 °C, the penetration values for asphalt binder in shingles ranges from 20 to 70 dmm, while traditional paving binders range from 50 to 300 dmm [28].

Newcomb et al. [28] were among the first to investigate the properties asphalt mixture prepared with different amounts of RAP, MWSS and TOSS. It was observed that adding MWSS did not significantly change the moisture susceptibility, however, TOSS did. Lower tensile strength was measured with increasing MWSS and TOSS contents. It was concluded that up to 5% of MWSS could be used in asphalt mixtures with a minimum impact on the mixture properties. However, addition of TOSS resulted in embrittled or stiffer mixture which is not desirable for cracking resistance at low temperatures. The 5% limit for MWSS was also proposed by other authors [29,30] that measured proportional decrease in tensile strength, creep stiffness and freeze-thaw resistance with increasing amount of roofing shingles. Burak and Ali [31] investigated several mechanical properties of asphalt mixtures containing roofing waste shingles from 1% to 5%. They observed that shingles could be used to improve Marshall-stability and rutting resistance. Other recent research efforts [32,33] investigated the cracking properties of mixture containing RAP and RAS, indicating that the use of rejuvenators and of a mix design which balances rutting and moisture damage, together with cracking requirements, can significantly improve the performance of the recycled mixture. McGraw et al. [34] evaluated the combined use of RAP and RAS showing the negative effect of TOSS on mixture strength and binder's critical cracking temperature. In a recent study [35] on the use of fractionated recycled asphalt pavement (FRAP) and RAS, it was found that the fibers contained in RAS may be beneficial to low temperature cracking resistance. Recent European studies have also addressed the use of shingles [36,37]. In one of this research works [37], both stiffness and fatigue properties of the recycled mixture were investigated showing an increase in mixture stiffness due to the inclusion of RAP and shingles (up to 5%).

Reusing asphalt shingles in asphalt mixture poses, therefore, significant challenges especially in cold climates, where good fracture resistance is critical for good pavement performance. This is particularly true for TOSS, which contain highly oxidized binders that are more prone to brittle failure. Unlike MWSS, only recently a provisional specification (MnDOT tear-off scrap asphalt shingles specification 2360, 2010) on the use of TOSS was released which limits the content to 5% (by weight) with the provision that at least 70% (by weight) of the required binder is new binder [38].

OBJECTIVE AND RESEARCH APPROACH

The present study investigates the effect of adding different amounts of RAP, MWSS, and TOSS to asphalt mixtures used for pavement applications based on changes in asphalt mixtures microstructure, and on modeling of mixtures and binders' low temperature properties, respectively.

The internal microstructure of asphalt mixture is analyzed using Digital Image Processing (DIP) of scanned Red Green Blue (RGB) scale images of small beam specimens having the same size of those used for creep tests in the Bending Beam Rheometer (BBR) [39]. Numerical evaluation of 2- and 3-point correlation functions, together with the estimation of the auto-correlation length (ACL), [40] of the aggregate, asphalt mastic, and air voids phases is performed to obtain detailed information on the microstructural spatial distribution.

The semi-empirical Hirsch [41] model and Huet [42] analogical model [43] coupled with transformation proposed by the research team of the Ecole Nationale des Travaux Publics de l'Etat (ENTPE) [44,45], are used to investigate the effect of adding recycled material on both asphalt binder and asphalt mixture creep stiffness. Back-calculation of the asphalt binder creep stiffness is performed using mixture creep stiffness data obtained with the Bending Beam Rheometer (BBR) [39,46].

The objective is to determine if changes in mixture behavior are due to variations in microstructure, from addition of recycled material, or to the effective blending of new and aged binder, or to a combination of both phenomena.

MATERIALS AND TESTING

Eight different asphalt mixtures (Table 1) provided from the Minnesota Department of Transportation (MnDOT) to the Department of Civil Engineering at University of Minnesota were used in this study. The entire set of mixtures was prepared with a virgin plain binder having Performance Grade (PG) PG58-28 [47] and target air voids content of 4%. Three types of virgin aggregates were blended to prepare the mixtures: pit-run-sand, quarried ¾ inch (19 mm) dolostone, and quarried dolostone manufactured sand. The recycled material consisted of different amounts of RAP, TOSS and MWSS: Table 2 presents the gradation of each recycled material used. Additional details on these materials can be found elsewhere [48].

Table 1. Mix design for the eight asphalt mixtures investigated.

Mix		Recycled Material			*VMA*	*VFA*	$V_{Air\ Voids}$	V_{Mastic}
ID	Description	RAP (% weight)	TOSS (% weight)	MWSS (% weight)	(%)	(%)	(%)	(%)
1	PG 58-28 Control	0	0	0	15.9	76.6	3.7	12.2
2	15% RAP	15	0	0	15.2	72.9	4.1	11.1
3	25% RAP	25	0	0	15.3	73.0	4.1	11.2
4	30% RAP	30	0	0	15.0	75.4	3.7	11.3
5	15% RAP 5% MWSS	15	0	5	15.6	75.0	3.9	11.7
6	15% RAP 5% TOSS	15	5	0	15.9	77.2	3.6	12.3
7	25% RAP 5% TOSS	25	5	0	15.4	73.9	4.0	11.4
8	25% RAP 5% MWSS	25	0	5	14.8	72.5	4.1	10.7

Table 2. Gradation of RAP MWSS and TOSS.

Sieve size (mm)	RAP Passing (%)	MWSS Passing (%)	TOSS Passing (%)
19.000	100.0	100.0	100.0
12.500	94.0	100.0	100.0
9.500	87.0	100.0	100.0
4.750	69.0	100.0	98.0
2.360	55.0	99.0	97.0
1.180	44.0	85.0	81.0
0.600	32.0	65.0	61.0
0.300	18.0	49.0	53.0
0.150	10.0	35.0	40.0
0.075	6.6	24.1	30.9

The eight asphalt mixtures were compacted into cylindrical specimens with a Superpave gyratory compactor (SGC). Five slices of asphalt mixture and, from these, specimens having the same size of BBR beams were cut and used for DIP analysis, first, and for BBR creep tests at low temperature in a second stage. Specimens were prepared and tested according to the procedure proposed by Marasteanu et al. [46]; a scheme of the different preparation steps is shown in Figure 1.

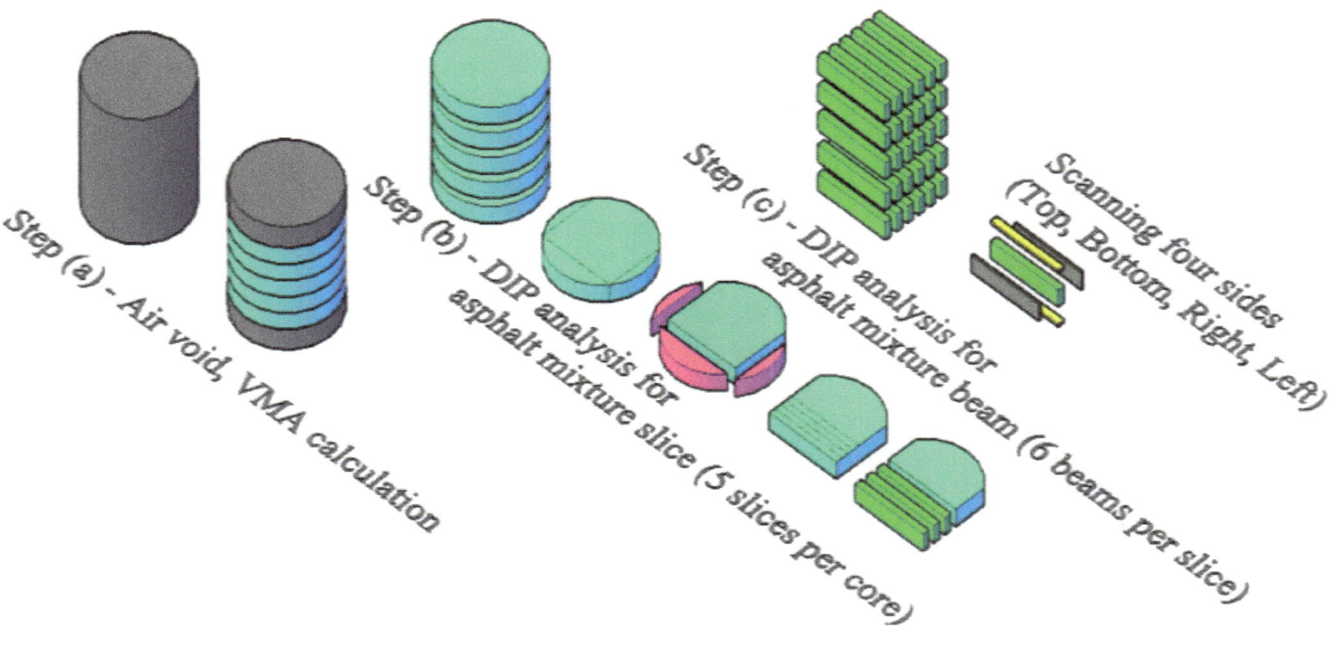

Figure 1. Specimens preparation for (digital image processing) DIP analysis and Bending Beam Rheometer (BBR) testing.

Asphalt mixtures beams (six replicates per mixture obtained from the central slice) were tested using the Bending Beam Rheometer (Figure 2a), and creep stiffness, S(t), was then calculated as:

$$S(t)=\frac{\sigma}{\varepsilon(t)}=\frac{P\cdot l^3}{4\cdot b\cdot h^3\cdot\delta(t)} \quad (1)$$

where σ is maximum bending stress in the beam; $\varepsilon(t)$ is the bending strain; P is the constant load applied at the mid-span of the specimen; l is the beam length (span) (102.0 ± 5 mm); b is the width of the specimen (12.7 ± 0.5 mm); h is the height of the specimen (6.25 ± 0.5 mm); $\delta(t)$ is the time dependent deflection at the mid-span of the beam; and t is the time variable. Due to the higher stiffness of asphalt mixtures, the load was increased to 4000 mN and test duration was set to 1000 s.

BBR tests [39] were also performed on the asphalt binders extracted from mixtures 2, 3, 4, 5, 6, 7 and 8 and, then, used for comparison purposes with the models' predictions; extraction and asphalt binder tests were conducted at MnDOT Office of Materials (Figure 2b). In addition, BBR creep stiffness was measured on the original PG58-28 binder after short term aging with the Rolling Thin-Film Oven Test (RTFOT) [49]; this value was assumed as control in the analysis. All binder tests were performed according to AASHTO T313-12-UL [39] and, therefore, limited to 240 s under a constant load of 1000 mN. To reduce errors associated with time-temperature superposition, binder and mixture creep stiffness were obtained at the same test temperature of −6 °C, which is equal to the binder PG low temperature limit +22 °C.

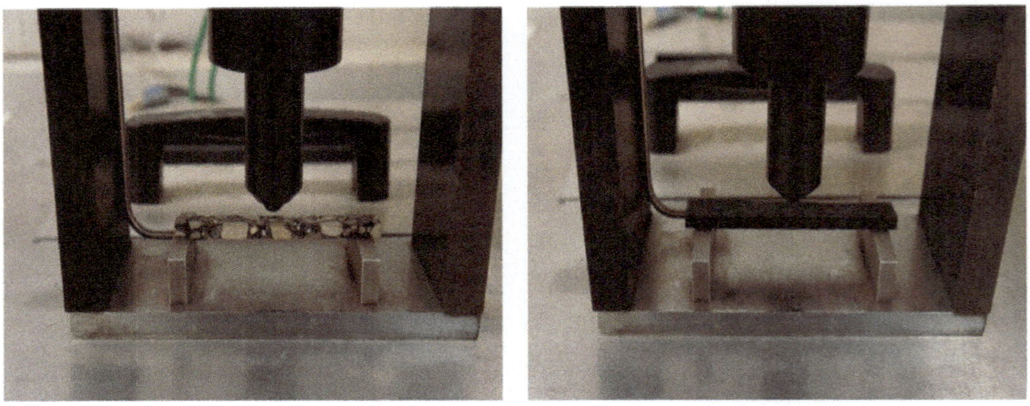

(a) (b)

Figure 2. Bending Beam Rheometer with thin (**a**) asphalt mixture and (**b**) asphalt binder beams.

MICROSTRUCTURAL ANALYSIS

Most research efforts have addressed asphalt mixture either as a two-phase material [50,51], or as a three-phase material [52,53]. Eriksen and Wegan [54] and Yue et al. [55] were among the first to investigate the microstructure of asphalt binders and mixtures, using DIP techniques, for obtaining material properties such as gradation, aggregate distribution, shape and orientation. Dai et al. [56] used DIP analysis to provide detailed information of asphalt mixture for Finite Element simulations of Indirect Tensile (IDT) strength test [57]. Kose et al. [58] and Partl et al. [59] and, more recently, other authors [60,61,62] have proposed advanced DIP techniques, and threshold algorithms for aggregate, asphalt mastic and air voids segmentation using RGB (Red-Green-Blue) scanned and gray scale converted asphalt mixture images. In some of these studies, the analysis of the materials microstructure was extended to a three-dimensional investigation based on 3D X-Ray Computer Tomography (CT) technology.

The microstructural information obtained with DIP techniques can be also used together with high-order microstructure functions, such as n-point correlation functions, to obtain improved upper and lower bounds of mechanical material properties [63]. Effective mechanical properties and failure characteristics of a heterogeneous material with the same volume fraction can be significantly influenced by the spatial distribution of its constituents [64]. In a recent study, Velasquez et al.[65] used DIP analysis and 2- and 3-point correlation functions to obtain specific input parameters for micromechanical models and detailed information on the Representative Volume Element (RVE) of asphalt mixture.

Two- and Three-Point Correlations: Theoretical Background

Correlation functions quantify the degree of influence among different parts of a material providing a numerical estimation of how they affect each other [62]. In the case of random heterogeneous materials, higher-order microstructural information is critical for describing the mechanical properties; several functions such as the n-point correlation functions, surface correlation functions, and lineal path functions, can be used for this purpose [63,65,66,67,68]. Among these set of functions, 2- and 3-point correlation functions present a simpler formulation compared to other complex statistical tools.

The n-point spatial correlation function measures the probability of finding all n points located on the space occupied by one of the phases of a heterogeneous material [40]. For instance, the 1-point correlation function, S_1, measures the probability that any point lies on phase 1 and this corresponds to the volumetric fraction of phase 1. The 2-point correlation function, S_2, calculates the probability that two points, separated by a certain distance, r, are located both in the same phase. Similarly, the 3-point correlation function, S_3, computes the probability that all the vertices of a triangle are located in the same phase.

The n-point correlation function of a two-phase random heterogeneous material in a d-dimensional Euclidian space, R^d, is defined according to Torquato [63] as:

$$S_n^{(i)}(x_1,x_2,x_3,\ldots,x_n) = \langle I^{(i)}(x_1)I^{(i)}(x_2)I^{(i)}(x_3)I^{(i)}(x_n)\rangle \quad (2)$$

where $\langle\ \rangle$ indicates the ensemble averaging. The indicator function $I^{(i)}(x)$ is defined as:

$$I^{(i)}(x) = \begin{cases} 1, & x \in V_i \\ 0, & x \in \overline{V_i} \end{cases} \quad (3)$$

where $V_i \in R^d$ is the volume occupied by the ith phase and $\overline{V_i} \in R^d$ is the volume occupied by the other phase.

The n-point correlation function is translationally invariant for a statistically homogeneous material. The function depends on the differences in the coordinate values of the x_i vectors, but not on their absolute position [63]. The n-point correlation function can also be expressed as:

$$S_n^{(i)}(x_1,x_2,x_3,\ldots,x_n) = S_n^{(i)}(x_{12},x_{13},x_{14},\ldots,x_{1n}) \text{ for } n \geq 1 \quad (4)$$

where $x_{ij} = x_j - x_i$ is the difference between the two vectors x_i and x_j, and x_1 is the reference vector.

The 1-point correlation function represents the volume fraction ϕ_i of the selected i^{th} phase; this is constant and it denotes the probability that a randomly selected point in the material belongs to specific phase:

$$S_1^{(i)} = \langle I^{(i)}(x)\rangle = \phi_i \quad (5)$$

The 2- and 3-point correlation function can be defined, respectively as [63]:

$$S_2^{(i)}(x_1,x_2) = \langle I^{(i)}(x_1)I^{(i)}(x_2)\rangle \quad (6)$$

$$S_3^{(i)}(x_1,x_2,x_3) = \langle I^{(i)}(x_1)I^{(i)}(x_2)I^{(i)}(x_3)\rangle \quad (7)$$

In the case of a translationally invariant isotropic material, the 3-point correlation function can be expressed as:

$$S_3^{(i)}(r_1,r_2) = S_3^{(i)}(|r_1|,|r_2|,u_{12}) \quad (8)$$

where $r_1 = x_2 - x_1$ and $r_2 = x_3 - x_1$ are vectors, and u_{12} is the cosine of the angle θ_{12} between vectors r_1 and r_2:

$$u_{12} = \cos\theta_{12} = \frac{r_1 \cdot r_2}{|r_1||r_2|} \quad (9)$$

In addition, if the microstructure of the material does not present long range order, the initial value of 2- and 3-point correlation functions is ϕ_i ($r = 0$) and for very large r ($r \to \infty$), it reaches the asymptotic limits of ϕ_i^2 and ϕ_i^3, respectively.

$$\lim_{r\to\infty} S_2^{(i)}(r) = \phi_i^2 \quad (10)$$

$$\lim_{|r_1|,|r_2|\to\infty} S_3^{(i)}(|r_1|,|r_2|,u_{12}) = \phi_i^3 \quad (11)$$

The n-point correlation function can be used to describe the microstructure of random heterogeneous materials such as asphalt mixtures; however, the computation process is complicated when an analytical solution is sought. Therefore, alternative techniques are required: a combination of DIP analysis and numerical algorithms provides a simpler approach for obtaining the values of the function [62,63,65].

Digital Image Processing (DIP) Analysis Overview

The internal structure of random heterogeneous materials, such as asphalt mixtures, can be described and analyzed using DIP techniques. A digitally converted image can be considered as a two dimensional independent function, f(x,y), where f is the function of color intensity and f(x,y) returns the color intensity of the image at the specific point (x,y). The entire digital image is made of the contribution of each pixel, which is associated to a specific location and intensity value [69]. For example, the matrix representation of an R×C size binary image (i.e., consisted of black (0) and white (1) colors) is expressed as:

$$f(x,y) = \begin{bmatrix} f(0,0) & f(0,1) & \ldots & f(0,N-1) \\ f(1,0) & f(1,1) & \ldots & f(1,N-1) \\ \ldots & \ldots & \ldots & \ldots \\ f(M-1,0) & f(M-1,1) & \ldots & f(M-1,N-1) \end{bmatrix} \quad (12)$$

In this study, a flat-bed scanner was used to acquire 2D images of asphalt mixture in RGB (Red, Green and Blue) color format. Scanning resolution was limited to 720 dpi since this is sufficient to detect aggregate particles larger than 75 μm while reducing the time-consuming process of analyzing very large files. In previous studies [56,65], RGB scale asphalt mixture images were converted into binary images (black color for air voids + asphalt binder + aggregates smaller than 75 μm, and white color for aggregates larger than 75 μm) using noise filtering function. In the present research, RGB scale images of asphalt mixture were converted, through the MATLAB™ Image Processing Toolbox [70], into 8-bit gray scale images (intensity ranges from 0, black, to 255, white) to retain more information of the material phases.

The following steps schematically summarize the DIP analysis procedure used in this research effort; detailed information about the DIP analysis technique used in this study can be found elsewhere [62].

1. First, the volume fraction of the actual air voids content and of the Voids in the Mineral Aggregate (VMA, %) were obtained for each mixture using the experimental data as:

$$Air\ void(\%) = \left(1 - \frac{G_{mb}}{G_{mm}}\right) \times 100\ (\%) \quad (13)$$

$$VMA(\%) = 100 - G_{mb} \cdot \frac{p_s}{G_{sb}} = 100 - G_{mb} \cdot \frac{1-p_b}{G_{sb}} \quad (14)$$

where: G_{mb} is the bulk specific gravity; G_{mm} is the maximum specific gravity; p_s is the aggregate content (mass); p_b is the asphalt content (mass); and G_{sb} is the bulk specific gravity of aggregate. Then, from Equation (14), VMA, which includes air voids (V_{air}) and the effective volume fraction of asphalt binder (V_{be}), was expressed as:

$$VMA = V_{air} + V_{be} \quad (15)$$

Based on the scanning resolution (i.e., 720 dpi), the following expression was used for computing VMA [62]:

$$VMA = V_{air} + (V_{be} + V_{agg<75\mu m}) = V_{air} + V_{mastic} \quad (16)$$

Equations (13), (14) and (16) were next used to determine the volume fraction of mastic, V_{mastic}.

2. Five slices, with thickness of 12.7 ± 5.0 mm, were cut from each asphalt mixture core, (Figure 1, Steps a and b) and then scanned in RGB scale. The digital images

of the asphalt mixture slices (3540 pixels × 3540 pixels) were then converted to gray scale and noise filtering applied [62,70].

3. Based on the experimental data of Table 1, two threshold values, one for the air voids volume fraction, T1, and one for the mastic volume fraction, T2, were calculated from the converted gray scale images of asphalt mixture slices. The values of T1 and T2 were computed by assuming that the air voids phase takes the darkest pixel range in the gray scale images (Figure 1, Step b).

4. All the asphalt mixture slices were cut into 240 asphalt mixture BBR beams (8 mixtures × 5 slices × 6 replicates) [46] (Figure 1, Step b). The four largest sides of the BBR mixture beams were then scanned and converted into gray scale images [62]. Therefore, a total of 960 images of asphalt mixture beams were used for DIP analysis (Figure Step c). Finally, based on the threshold values, T1 and T2, previously obtained, three-phase images of asphalt mixture beams were generated; meanwhile the DIP values of VMAd (total VMA,%), Vagg_d (volume fraction of aggregate,%), Vma_d (volume fraction of asphalt mastic,%) and Vair_d (volume fraction of air voids,%) were computed.

The flowchart of the entire DIP analysis procedure used in the present study for generating three-phase (aggregate, asphalt mastic and air voids) images of asphalt mixture is shown in Figure 3.

Table 3 presents the average values of VMA_d, V_{ma_d} and V_{air_d}, obtained from DIP analysis. Figure 4 and Figure 5 provide examples of three-phase images obtained through DIP analysis, for two asphalt mixture slices and for two beams made of two different mixture types.

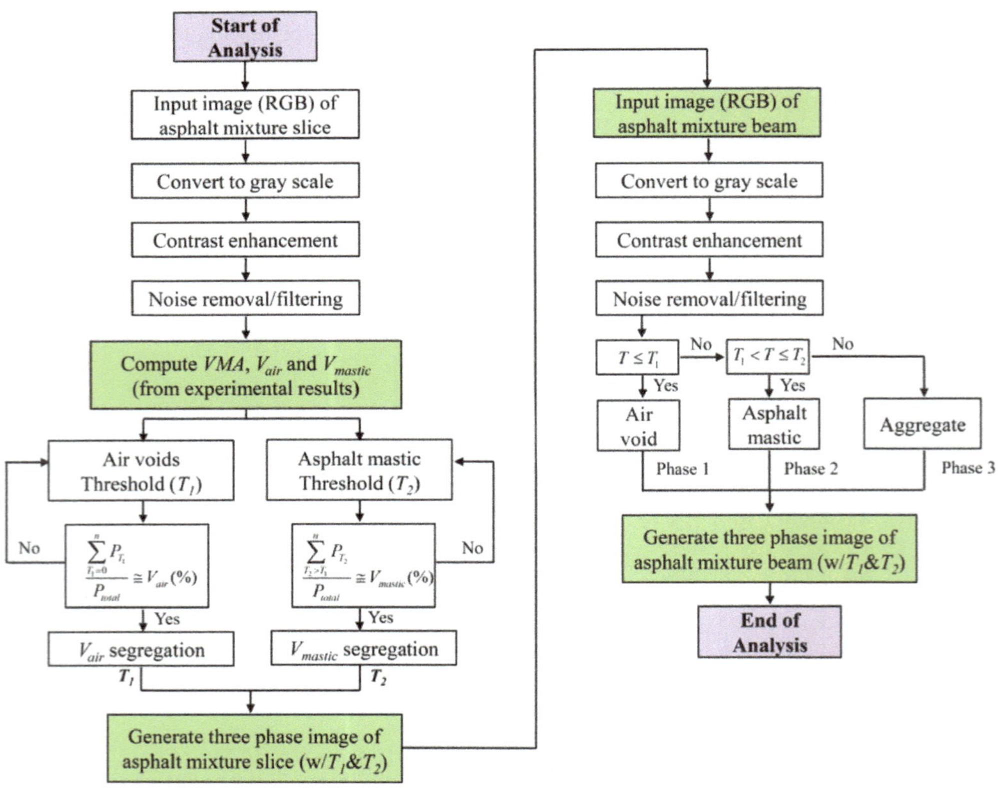

Figure 3. Flowchart of DIP analysis for generating three-phase images of asphalt mixtures. Notes: P: Number of pixels in the image; T: Threshold values (0:Black~255:White). In computing T_1 and T_2, error level was converged to 5% [$0 < T_1 < T_2 < 255$].

Table 3. Summary of the volumetric properties of asphalt mixture beams obtained from DIP analysis.

Mixture	VMA_d, %		V_{air_d}, %		V_{mastic_d}, %		VMA^*	V_{air}^*	V_{mastic}^*
	Ave	CoV	Ave	CoV	Ave	CoV	%	%	%
1	15.55	3.26	4.25	10.12	11.30	7.32	−0.35	0.55	−0.88
2	14.56	3.62	4.36	4.19	10.20	4.18	−0.64	0.26	−0.88
3	14.94	5.03	4.71	6.13	10.22	6.94	−0.36	0.61	−0.94
4	14.77	4.73	4.31	4.39	10.46	6.06	−0.23	0.61	3.65
5	14.25	6.77	4.31	5.87	9.94	10.42	−1.35	0.41	−1.76
6	15.08	3.70	4.11	4.46	10.97	5.55	−0.82	0.51	−1.31
7	14.42	3.44	4.64	4.17	9.79	5.77	−0.98	0.64	−1.59
8	14.08	2.91	4.67	6.53	9.40	6.46	−0.72	0.57	−1.33

Notes: $VMA^* = (VMA - VMA_d)$, $V_{air}^* = (V_{air} - V_{air\_d})$, $V_{mastic}^* = (V_{mastic} - V_{mastic\_d})$; Values of V_{air}, and V_{mastic}, and VMA can be referred from Table 1.

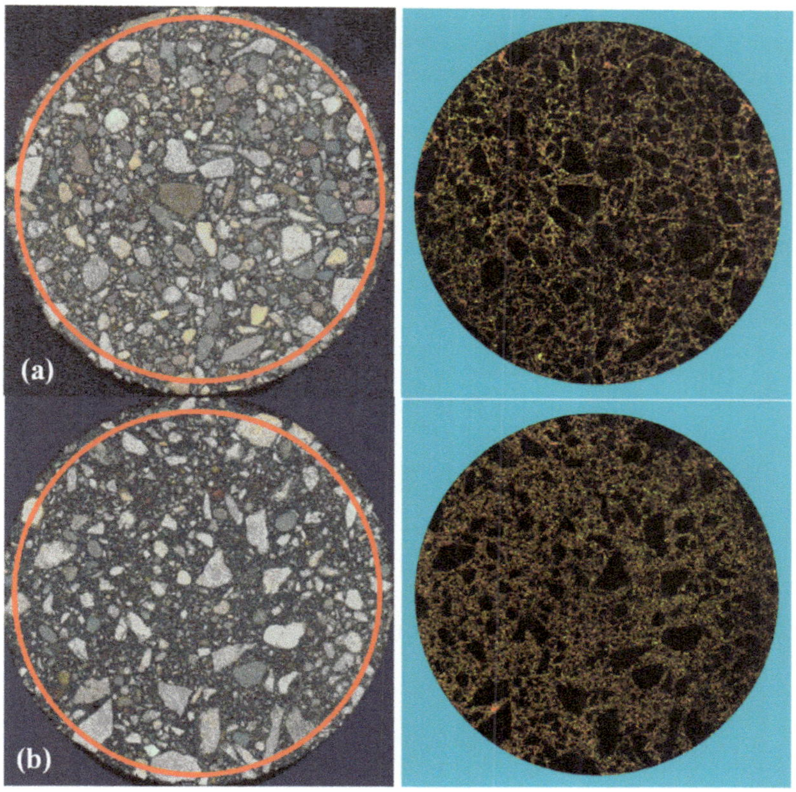

Figure 4. Images of asphalt mixture slices: original RGB and three-phase processed image. (**a**) Mixture 1; (**b**) Mixture 5.

Figure 4 and Figure 5 suggest good correspondence of the material phases between the original RGB-scale images and the three-phase image of asphalt mixture, indicating that the experimental volumetric parameters can be successfully used as input for DIP analysis. In addition, low values of Coefficient of Variation (CoV_{max} = 10.42%) were observed in all analyzed mixtures, including those containing RAP and RAS together (Table 2). This indicates that there is a consistently similar volume fractions distribution of the material phases among the different asphalt mixture beams cut from the same asphalt mixture cylinder.

Numerical Solutions of 2- and 3-Point Correlation Functions

The 2-point correlation function can be obtained using a discretized expression of Equation (6) accounting for the width and the height of the specific digital image. Nevertheless, when high resolution images containing a large number of pixels are used, a "brute force method" may be very time demanding and sometimes even prohibitive. Recently, an algorithm, based on Monte Carlo simulation and on binary images of the asphalt mixture BBR specimens, was proposed for calculating the correlation functions curves [65]; a simplified version of this method, applied to three-phase images, is used hereafter to determine the values of 2- and 3-point correlation functions of each asphalt mixture phase [62].

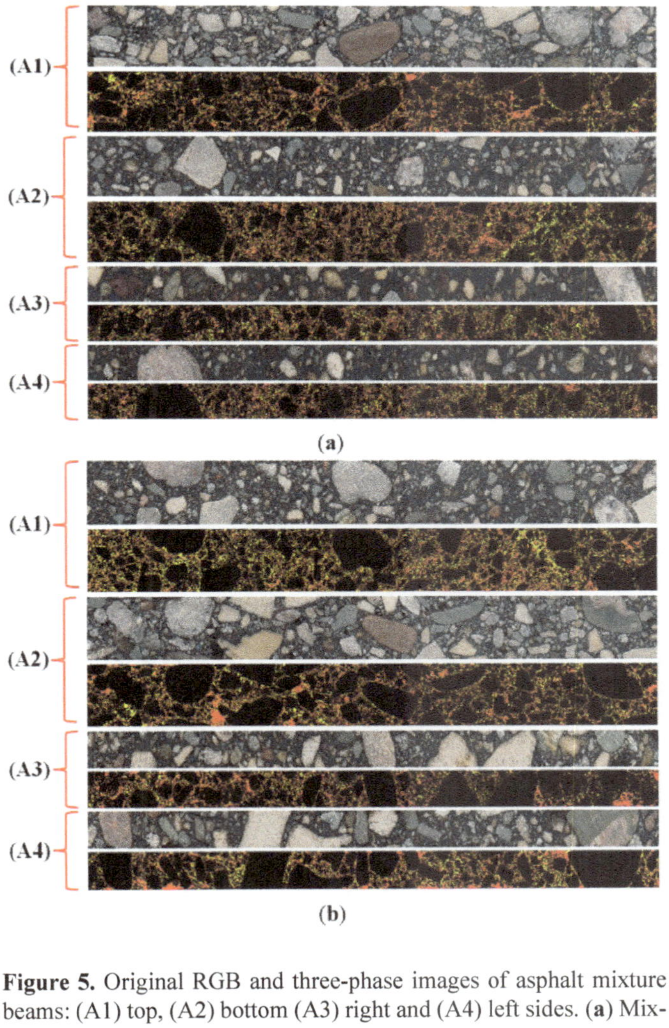

Figure 5. Original RGB and three-phase images of asphalt mixture beams: (A1) top, (A2) bottom (A3) right and (A4) left sides. (**a**) Mixture 1; (**b**) Mixture 5.

The procedure for computing 2-point correlation consists in dropping vectors of increasing length and random orientation (i.e., angle θ) in the digital image N times, while the number of times that both end points of the vectors are in the phase of interest is recorded (Figure 6). The process is iterated for vectors of lengths varying from 0 to half the size of the image for a number of drops N > 100,000 [62,65].

The three-phase images of the different beam specimens were used to compute 2-point correlation functions of aggregates, asphalt mastic and air voids. The single values of the 2-point correlation function of each specimen were calculated as the mean of the two larger sides of each beam image. Figure 7 presents the curves of the average 2-point correlation function for all the eight mixtures investigated. Small CoV were obtained: for example, in the case of the aggregate phase, a CoV_{max} = 3.8% was observed.

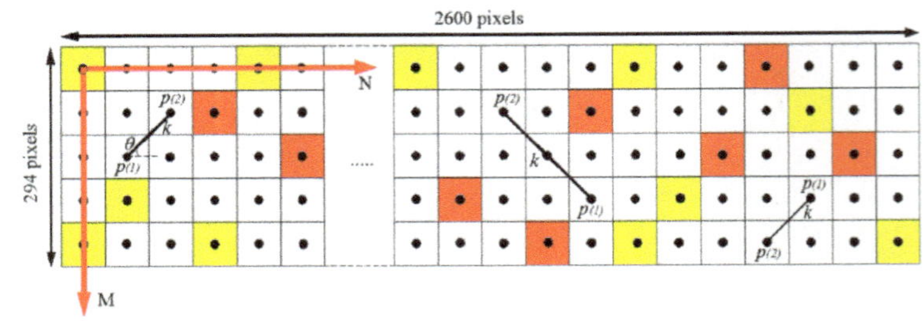

Figure 6. Schematic of the simplified algorithm for 2-point correlation function calculation.

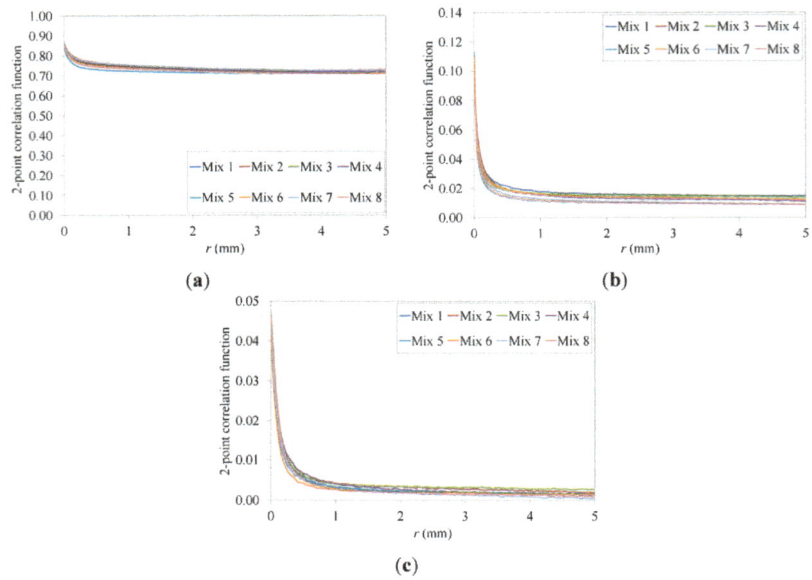

Figure 7. Two-point correlation functions for the entire set of eight asphalt mixtures used: (**a**) Aggregates; (**b**) Asphalt mastic; (**c**) Air voids.

The values of the 2-point correlation function do not fluctuate as the distance (r) increases for all eight mixtures investigated. The 2-point correlation function starts at approximately ϕ_i and smoothly drops to ϕ_i^2, corresponding to the asymptotic limit of Equation (10) as r approaches infinite [40,63]. All the curves are very close although not overlapping, suggesting a slightly different spatial arrangement of the three phases across the different mixtures most likely associated to the different amount of recycled material used in the mix design.

For the calculation of the 3-point correlation function, a set of lattice commensurate triangles [65] and the symmetry property of the function can be used to reduce the computational time. The set of triangles used for the estimation of the spatial correlation function are defined by three integers, l, m, and n (Figure 8) which are related by the following conditions:

$$m \leq l/2 \text{ and } m^2 + n^2 \leq 2ml \quad (17)$$

Each triangle defined by the integers (l, m, n) is randomly dropped N = 100,000 times in the three-phase image [62]. The number of times (N_{hits}) the three vertices of the triangle are in the phase of interest is counted (Figure 8).

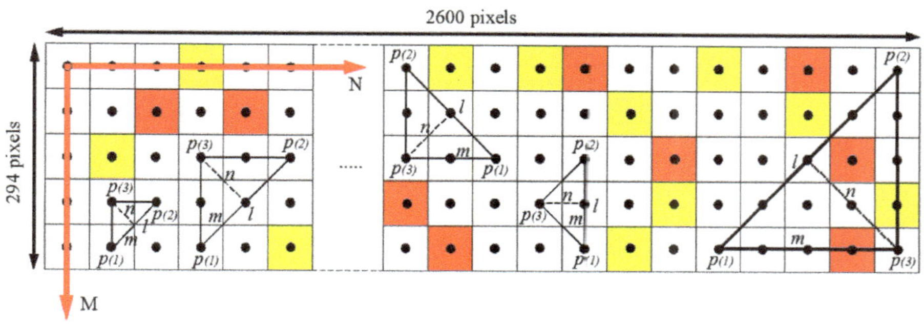

Figure 8. Schematic of the simplified algorithm for 3-point correlation function calculation.

In addition, to further reduce computation time and memory demand, only symmetrical triangles in the size range of 0.0 mm < l ≤ 5.2 mm were used. Although this restriction does not allow obtaining the all of the 3-point correlation functions, this method significantly reduces the computation time (from 4~5 h to 2~3 min for each simulation) while still providing important information on the global trend of the function [62]. The curves of the average 3-point correlation function for all eight mixtures investigated are shown in Figure 9.

Similarly to the 2-point correlation function, small CoV were observed with a CoV_{max} = 4.1% for mixture 1. The average 3-point correlation function has a similar pattern for all eight asphalt mixtures considered. No large fluctuations in the spatial distribution of the materials phases were observed as the size of the triangle increases. S_3 begins at ϕ_i and smoothly drops to ϕ_i^3, as described by Equation (11), showing no evidence of clustering or preferred paths in the microstructure [62,63]. Also, in this case, the curves of the function are close, but not overlapping, confirming the influence of the recycled materials on the distance required for the function to reach a stable value.

Autocorrelation Length

Based on the results of 2- and 3-point correlation functions, it was decided to investigate the auto correlation length (ACL) associated with the correlation function of each mixture. ACL measures how quickly a random event decays, or the distance over which two or three points can be treated as independent for the specific random process [40,62,71]. For example, Bazant and Pang [72] associated the ACL with the estimation of the dimensions of the Representative Volume Element (RVE) of the material. Some authors [40,71,73] used exponential expressions to analytically describe the autocorrelation functions (ACF) and identified the ACL as the distance over which the ACF decays to 1/e of its original value [73]; however, it was recently shown that this may not apply to asphalt mixture [62]. Therefore, in this study, ACL was estimated, for all the material phases, as the distance over which the function assumes its theoretical value (ϕ_i^2 and ϕ_i^3 for 2- and 3-point correlation functions, respectively) in a stable manner within a 5% error (95% confidence interval). The following ACL values were obtained (Table 4).

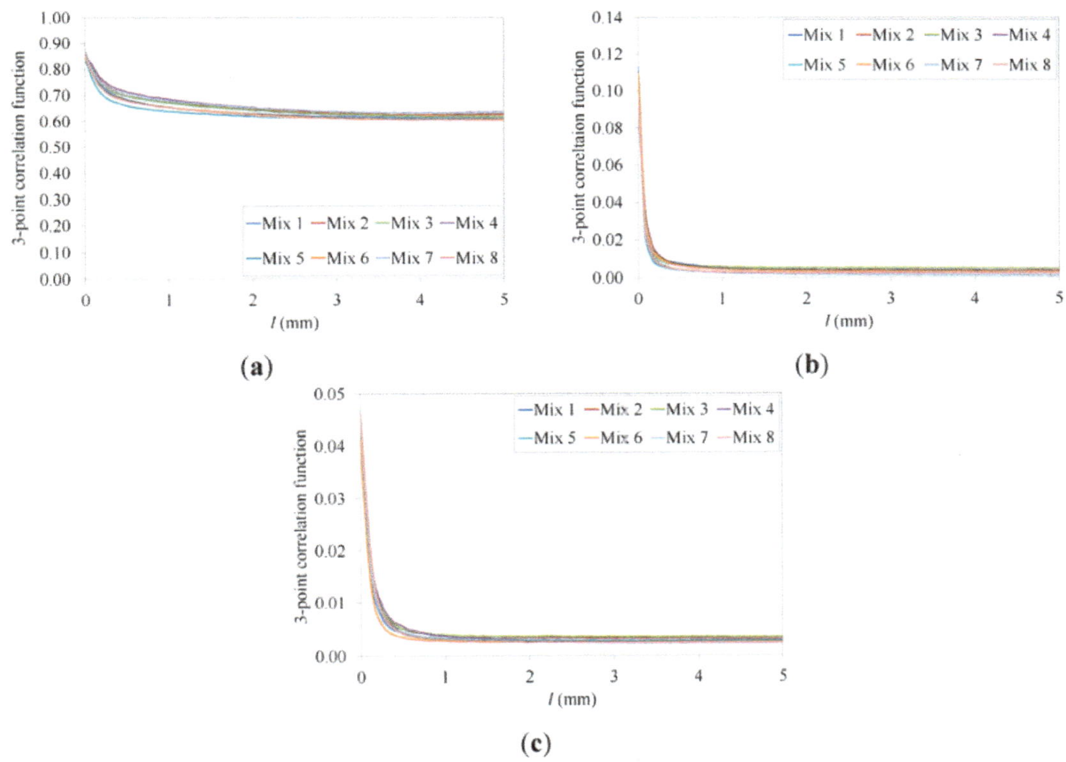

Figure 9. Three-point correlation functions for the entire set of eight asphalt mixtures used: (**a**) Aggregates; (**b**) Asphalt mastic; (**c**) Air voids.

Table 4. Auto Correlation Length (ACL) of 2- and 3-point correlation functions.

ID	Material	2-Point correlation function (mm)			3-Point correlation function (mm)		
		Aggregates	Mastic	Air voids	Aggregates	Mastic	Air voids
1	PG 58-28 Control	3.97	0.59	1.02	2.33	0.40	0.88
2	15% RAP	3.97	0.64	1.48	2.34	0.48	1.37
3	25% RAP	4.06	0.64	1.44	2.42	0.48	1.53
4	30% RAP	4.09	0.66	1.51	2.45	0.49	1.61
5	15% RAP 5% MWSS	1.57	0.25	0.72	1.37	0.22	0.73
6	15% RAP 5% TOSS	2.75	0.42	0.76	2.10	0.32	0.78
7	25% RAP 5% TOSS	2.84	0.47	1.14	2.18	0.40	0.97
8	25% RAP 5% MWSS	2.03	0.30	0.76	1.69	0.28	0.76

From Table 4, it can be first observed that the values of ACL increase with respect to the three phases: aggregates, air voids and asphalt mastic, respectively. This indicates that ACL reflects the influence of the spatial distribution of each material phase rather than the overall volume fraction. The addition of RAP results in larger ACLs for all phases, suggesting a longer randomness range and thus larger RVE dimensions.

A quite significant opposite trend is associated with the presence of RAS. It may be hypothesized that this is due to the higher amount of fine particles which are present in shingles. Nevertheless, the aged TOSS material seems to introduce a smaller randomness decay compared to the MWSS. It is interesting to note that the ACL of the 2-point correlation function is larger than that observed for the 3-point correlation function, suggesting a slightly faster decay of the microstructure randomness associated to the potential presence of clustering effect and/or connected paths. In addition, the maximum value of ACL was found to be equal to 4.09 mm (Mixture 4, Aggregate phase) which is smaller than the thickness (=6.35 mm) and width (=12.70 mm) of the BBR beam used for mixture testing. Therefore, BBR specimens can be assumed to be large enough for obtaining the low temperature creep properties of asphalt mixture.

RHEOLOGICAL ANALYSIS

Based on the results of the microstructural analysis of the eight asphalt mixtures evaluated in the present study, the change in rheological properties due to the addition of aged recycled materials is investigated in the next section.

Analogical and Semi Empirical Models

In previous research efforts [26,74], analogical models were used to obtain the creep stiffness of asphalt binders from the creep stiffness of the corresponding asphalt mixtures (inverse problem). It was found that Huet model [42] fitted very well the experimental data obtained from BBR tests at low temperatures for both asphalt binders and mixtures. This model is composed of two parabolic elements, $D_1(t) = a_1(t/\tau)^h$ and $D_2(t) = a_2(t/\tau)^k$, and a spring with stiffness E_∞, combined in series (Figure 10).

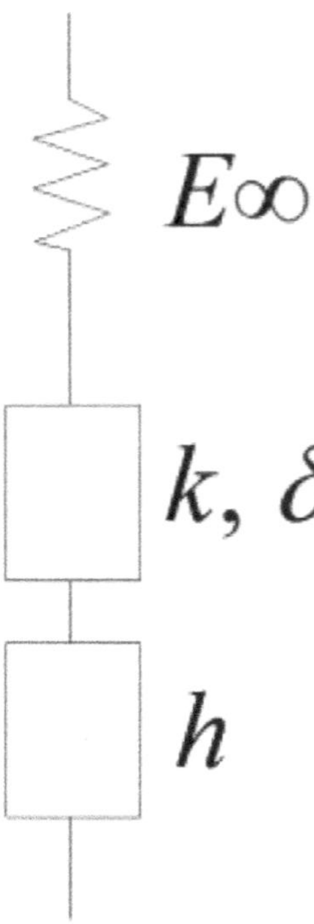

Figure 10. Huet model [42].

The analytical expression of the Huet model for creep compliance is:

$$D(t) = \frac{1}{E_\infty}\left(1 + \delta\frac{(t/\tau)^k}{\Gamma(k+1)} + \frac{(t/\tau)^h}{\Gamma(h+1)}\right) \tag{18}$$

where D(t) is the creep compliance; E_∞ is the glassy modulus; h, k are exponents such that $0 < k < h < 1$; δ is a dimensionless constant; t is the time, Γ is the gamma function, τ is the characteristic time varying with temperature accounting for the Time Temperature Superposition Principle (TTSP): $\tau = a_T(T)\tau_0(T_S)$; a_T is the shift factor at temperature T; τ_0 is the characteristic time determined at reference temperature T_S.

The authors [26,74] found the following relationship between the characteristic time of mixture, τ_{mix} and the characteristic time of the corresponding binder, τ_{binder}, at the reference temperature T:

$$\tau_{mix}(T) = 10^{\alpha}\tau_{binder}(T) \tag{19}$$

where α is a regression coefficient depending on mixture type and aging [45]. This expression corresponds to the Equation proposed by Di Benedetto [45] and obtained from 2S2P1D model and complex modulus data [44,45]. Based on Equations (18) and (19), a transformation that relates the creep stiffness of the asphalt binder $S_{binder}(t)$ to the creep stiffness of the corresponding asphalt mixture $S_{mix}(t)$ can be written:

$$S_{mix}(t) = S_{binder}(t10^{-\alpha}) \cdot (E_{\infty\_mix}/E_{\infty\_binder}) \quad (20)$$

where E_{∞_mix} is the mixture glassy modulus and E_{∞_binder} is the binder glassy modulus. Expression (19) does not depend on the original model used to obtain it and represents the low temperatures formulation of the ENTPE transformation [44,45,74].

A different back-calculation method based on the semi empirical Hirsch [41] formulation was used by several authors [17,65,74,75]. In this model, the effective response of the material is obtained assembling the mixture components (air voids, asphalt binder and aggregate) in parallel and in series. The general expression is:

$$E_{mix} = Pc\left[E_{agg}V_{agg} + E_{binder}V_{binder}\right] + (1-Pc)\left[\frac{V_{agg}}{E_{agg}} + \frac{(1-V_{agg})^2}{E_{binder}V_{binder}}\right]^{-1} \quad (21)$$

where E_{mix} is the effective modulus of the mixture, E_{agg} and V_{agg} are the modulus and volume fraction of the aggregate; E_{binder} and V_{binder} are the modulus and volume fraction of binder; $P_c = 0.1\ln(E_{binder}/a) + 0.609$ is the contact volume (an empirical factor); E_{binder} is the relaxation modulus of the binder in GPa; and a is a constant equal to 1 GPa.

Back-Calculation

In this section, asphalt binder creep stiffness is back-calculated from the creep stiffness experimental data of asphalt mixture with the aim of obtaining useful information on the blending process and the degree of blending between the new virgin binder and the aged binder contained in RAP and RAS. For such a purpose, Huet model was introduced into Equation (11) as:

$$S_{mix}(t, h, k, \delta, \tau_{mix}) = S_{binder}(t, h, k, \delta, 10^{\alpha}\tau_{mix}) \cdot (E_{\infty\_mix}/E_{\infty\_binder}) \quad (22)$$

where the five constants δ, k, h, E_∞, and τ can be estimated by minimizing the sum of the distances between asphalt mixture experimental data and the predicted values obtained from Equation (22) at n time points. In previous studies [44,45,74], it was found that Huet model parameters assume the same values for binder and corresponding mixtures. Therefore, the difference between asphalt binder and asphalt mixture is given by the characteristic times, τ_{binder} and τ_{mix}, which, based on Equation (19), are directly associated to parameter α. In this study, the value of α was obtained during the minimization process starting from initial values of δ, k, h, E_∞, τ found in literature [42,44,45,74]. In addition, the recent recommendation proposed by Moon et al. [76], which limits the value of k between 0.25 and 0.40 for plain binder having PG58-28, was used.Table 5 presents the Huet model parameters for the eight asphalt mixtures tested at −6 °C and for the corresponding back-calculated asphalt binders; the values of α are given in the last column of the table.

Table 5. Huet model parameters for mixtures and corresponding back-calculated binders.

ID	Material	δ	k	h	E_{∞_binder} (MPa)	E_{∞_mix} (MPa)	Log (τ_{binder})	Log (τ_{mix})	α
1	PG 58-28 Control	6.67	0.28	0.71	2979	29,895	−0.770	2.390	3.16
2	15% RAP	6.45	0.25	0.58	2982	29,921	−0.824	3.646	4.47
3	25% RAP	6.02	0.30	0.57	2985	29,926	−0.658	3.492	4.15
4	30% RAP	6.83	0.25	0.62	2980	29,899	−0.824	3.726	4.55
5	15% RAP 5% MWSS	6.84	0.27	0.64	2988	29,931	−0.509	3.591	4.10
6	15% RAP 5% TOSS	6.85	0.25	0.66	2986	29,932	−0.456	4.274	4.73
7	25% RAP 5% TOSS	6.04	0.28	0.59	2989	29,926	−0.409	4.171	4.58
8	25% RAP 5% MWSS	6.11	0.28	0.59	2991	29,962	−0.495	3.965	4.46

Table 5 presents values of parameters h and k which are in agreement with the values obtained in different research efforts [45,74,76], while δ is higher compared to what was found in literature. As expected, there is a significant difference in the characteristic times of binders and corresponding mixtures. Nevertheless, it is interesting to observe that the presence of recycled materials significantly influences this parameter [25], although there is not a clear common trend for binder and mixture. Higher values of the transformation parameter, α, were observed when recycled material is added to the mixture in comparison with the control mixture prepared with virgin aggregates and virgin binder. This is true both when only RAP is used as well as when MWSS and TOSS are included in the mix design. Nevertheless, no statistically significant differences could be detected among the values of α obtained for the recycled mixtures. Furthermore, no specific trend of the parameterα could be identified and associated with the amount of recycled material.

In addition to the ENTPE transformation, back-calculation of asphalt binder creep stiffness was performed with the Hirsch model for comparison purposes. The simplified procedure adopted includes three steps. First, based on the volumetric properties of the mixtures (Table 1) and on Equation (21), plots of binder creep stiffness versus predicted mixture stiffness were generated for binder stiffness values between 50 and 1000 MPa. Figure 11 shows the predicted curves for mixtures 1 and 5.

Then, the following expression, $E_{mix} = aH\ln(E_{binder}) + b$, was fitted to log of mixture stiffness vs. log of binder stiffness curves and parameters a and b were determined. Finally, the binder stiffness was simply calculated over the entire range of loading times.

Comparison with Experimental Data

The values of the asphalt binder creep stiffness predicted from Hirsch model and ENTPE transformation were compared to the creep stiffness values obtained experimentally on the RTFOT aged original binder and on the extracted asphalt binders (mixtures 2–8). As described in Section 4, asphalt mixtures were tested for 1000 s while the test duration of asphalt binders was limited to 240 s as prescribed in the standard

[39]. Therefore, in the plots of Figure 12, Figure 13, Figure 14 andFigure 15, the curves of creep stiffness obtained from asphalt binders are shorter than those predicted through back-calculation from mixture data.

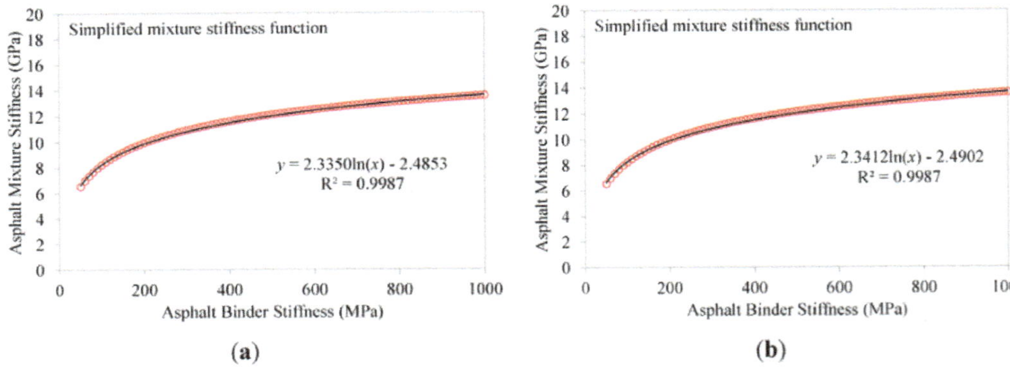

Figure 11. Simplified mixture stiffness function for mixtures 1 (**a**) and 5 (**b**), T = −6 °C.

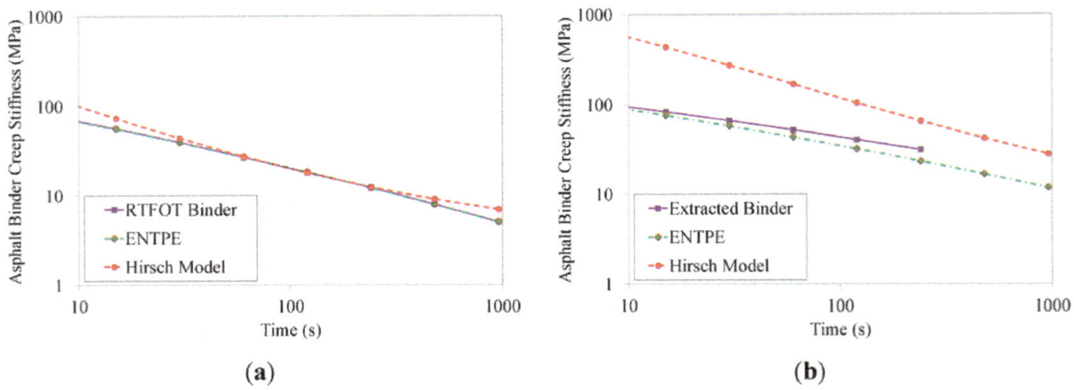

Figure 12. Back-calculated and extracted creep stiffness of asphalt binder: mix 1 (**a**) and 2 (**b**), T = −6 °C.

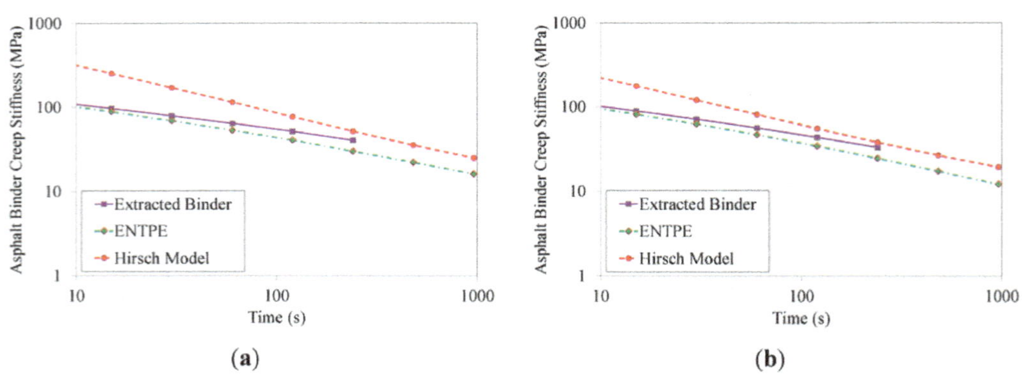

Figure 13. Back-calculated and extracted creep stiffness of asphalt binder: mix 3 (**a**) and 4 (**b**), T = −6 °C.

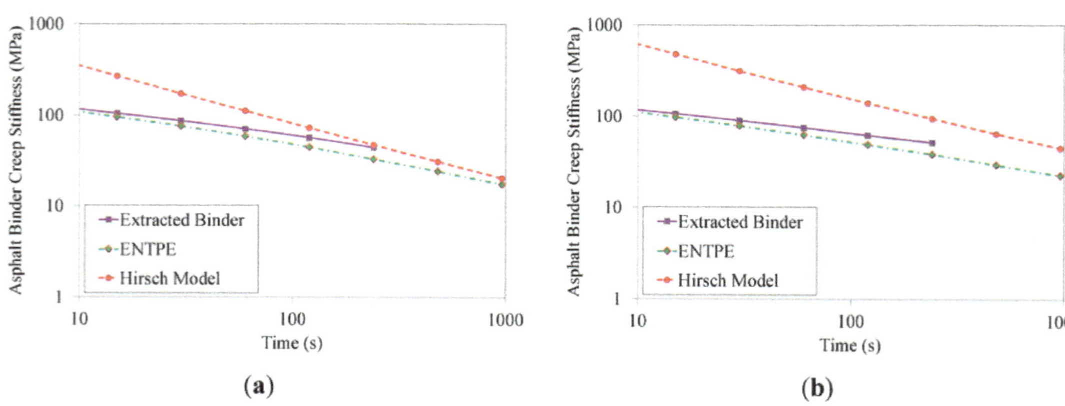

Figure 14. Back-calculated and extracted creep stiffness of asphalt binder: mix 5 (**a**) and 6 (**b**) , T = −6 °C.

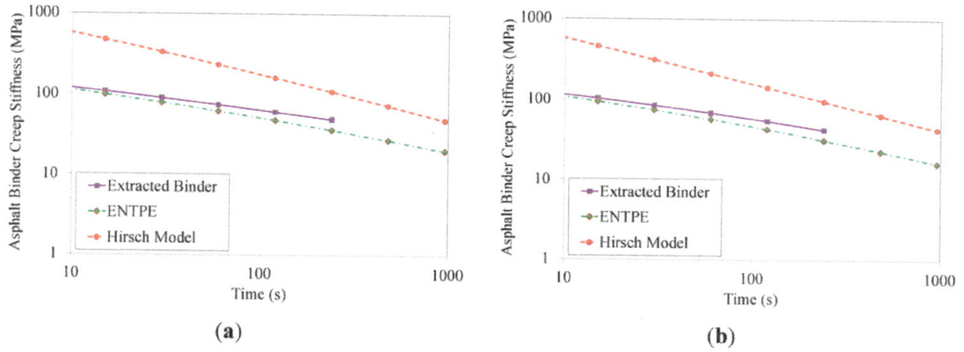

Figure 15. Back-calculated and extracted creep stiffness of asphalt binder: mix 7 (**a**) and 8 (**b**), T = −6 °C.

Hirsch model predictions returned much higher initial stiffness values with a significant decreasing trend compared to the experimental measurements, resulting in a significant over estimation of the creep curves obtained from the extracted binders. On the other hand, the ENTPE transformation provides a very accurate prediction of the creep stiffness of the RTFOT original binder, confirming the findings of previous studies on virgin materials [45,74,76]. A common trend is observed for stiffness curves of asphalt binders obtained from the recycled mixtures (2–8): they are initially asymptotically close to the creep stiffness curves of extracted binders, while they diverge toward lower stiffness values as time increases. This suggests that the relaxation properties obtained from ENTPE back-calculation are higher compared to the extracted binders. It can be hypothesized that one of the reasons for these observations is the presence of fine particles in the asphalt binder which could not be removed during the extraction process. However, this hypothesis would result in global higher stiffness values also for the part of the curve which is asymptotically approaching the model predictions, leading to an overall deviation of the experimental measurements from the back-calculated stiffness.

An alternative and more reasonable explanation for the difference between extracted binders and back-calculated creep stiffness curves is associated with the limited or partial blending occurring between new and recycled binder. Therefore, while the mixture creep behavior is affected by all new binders and only a part of the aged, oxidized binder, the extraction and recovery process resulted in a complete blending; this implies that the resulting blend has less relaxation capability since the properties are influenced by the entire aged binder component.

Bonaquist [77] has associated the insufficient heat transfer during mixing with partial binders blending in the mixture. An additional explanation of the limited binder blending can be provided on the basis of the work recently presented by Kriz et al. [27]; he identified two major processes which are involved in the binder blending: contact between the two binders, achieved mechanically, and binder blending after contact is made, achieved mainly by diffusion and strongly influenced by the total new-aged binder thickness. The limited or partial blending may be linked to the distribution of the binder film thickness within the asphalt mixtures microstructure and to the inability to obtain a proper binder contact when mixing. This can explain the discrepancy between the actual binder stiffness in the mixture and that obtained from extraction.

SUMMARY AND CONCLUSIONS

In this paper, the effect of adding different amounts of reclaimed asphalt pavement, manufactured waste scrap shingles and tear-off scrap shingles on the microstructural distribution and on the low temperature properties of asphalt mixtures was investigated. The microstructural analysis of the mixtures was coupled with the rheological investigation of both asphalt mixtures and corresponding asphalt binders. Digital Image processing and correlation functions were first used to evaluate the influence of the recycled material on the material microstructure. Then, the creep stiffness of the asphalt binder present in the mixtures was obtained with the transformation implemented by the Ecole Nationale des Travaux Publics de l'Etat (ENTPE) (coupled with Huet model) and the semi-empirical Hirsch model, respectively. Finally, the back-calculated predictions were compared to the experimental creep stiffness values measured on the short-term aged original binder and on the binders extracted from the other recycled mixtures.

The following conclusions can be drawn:

- Limited variations were observed in the 2- and 3-point correlation functions of the asphalt mixtures; however, the auto correlation length of aggregates, mastic and air voids is significantly influenced by the recycled material used. Nevertheless, while adding reclaimed asphalt pavement increases the auto correlation length, the use of recycled asphalt shingles shorten it, most likely due to the fine particles which are included in the recycled shingles.
- Smaller auto correlation length was observed for manufactured waste scrap shingles compared to tear-off scrap shingles; this indicates that larger representative volume elements are associated with the use of aged shingles.
- Hirsch predictions provide significant overestimations of the creep stiffness of original short-term aged binder and of the seven extracted asphalt binders.
- The creep stiffness of the original binder after short term aging is matched very closely by the ENTPE transformation, while the asphalt binders stiffness curves back-calculated from recycled mixtures data do not match the creep stiffness curves of extracted binders.
- It is hypothesized that blending between new and aged and oxidized binders occurred only partially due to

insufficient heat transfer and limited binder contact during mixing and/or to the distribution of the binder film thickness within asphalt mixtures. Therefore, the mixture creep properties were affected by all new binders and only a portion of the old binder. On the other hand, extraction and recovery process resulted in a complete blending which implies reduced relaxation capabilities since all aged binders contributed to the properties of the blend.

- The findings of the present research indicate that the low temperature stiffness properties of asphalt mixture are only partially influenced by the spatial distribution of its components, while the blending process of asphalt binder appears to play a fundamental role when recycled asphalt materials are added to the mixture.

ACKNOWLEDGMENTS

The authors would like to gratefully acknowledge the Minnesota Department of Transportation (MnDOT) for providing the asphalt mixtures used in this investigation and the asphalt binder creep stiffness data, and Mugurel I. Turos for the experimental support at the University of Minnesota.

AUTHOR CONTRIBUTIONS

The experimental work, analysis and modeling were performed by Augusto Cannone Falchetto and Ki Hoon Moon. Augusto Cannone Falchetto prepared, organized and wrote the paper with the contribution of Ki Hoon Moon. Michael Wistuba contributed to this scientific paper through revisions, suggestions, advising.

Conflicts of Interest

The authors declare no conflict of interest.

REFERENCES

1. National Cooperative Highway Research Program (NCHRP). Recommended Use of Reclaimed Asphalt Pavement in the Superpave Mix Design Method: Guidelines; Research Results Digest 253; National Academy Press: Washington, DC, USA, 2001.
2. National Cooperative Highway Research Program (NCHRP). Recommended Use of Reclaimed Asphalt Pavement in the Superpave Mix Design Method: Technician's Manual; Research Results Digest 252; National Academy Press: Washington, DC, USA, 2001.
3. McNichol, D. Paving the Way: Asphalt in America; National Asphalt Association: Lanham, MD, USA, 2005.
4. Hansen, K.; Newcomb, D.; Cervarich, M. Asphalt tops the charts for environmental stewardship-again. Asph. Pavement Mag. Natl. Asph. Pavement Assoc. **2011**, 16, 20–25.
5. Van den bergh, W.; van de Ven, M.F.C. Hot recycling of asphalt: Experiences in Flanders and the Netherlands. In Proceedings of the ENVIROAD 2009, Warsaw, Poland, 15–16 October 2009; 2009.
6. Schimmoller, V.E.; Holtz, K.; Eighmy, T.T.; Wiles, C.; Smith, M.; Malasheskie, G.; Rohrbach, G.J.; Schaftlein, S.; Helms, G.; Campbell, R.D.; et al. Recycled Materials in European Highway Environments: Uses, Technologies, and Policies; Final Report FHWA-PL-00-025; Federal Highway Administration, U.S. Department of Transportation: Washington, DC, USA, 2000.
7. Aggregates Business Europe, Route One Publishing Ltd. Available online: http://www.aggbusiness.com (accessed on 10 June 2014).
8. Dannhausen, W.O. Recycling alternatives studied and tried. Better Roads **1997**, 67, 25–26.
9. Marks, V.J.; Petermeier, G. Let me shingle your roadway. Transp. Res. Rec. **1997**, 1589, 54–57.
10. Shingle Recycling. Available online: http://www.shinglerecycling.org (accessed on 10 June 2014).
11. Little, D.H.; Holmgreen, R.J., Jr.; Epps, J.A. Effect of recycling agents on the structural performance of recycled asphalt concrete materials. J. Assoc. Asph. Paving Technol. **1981**, 50, 32–63.
12. Jung, D.; Vinson, T.S. Low temperature cracking resistance of asphalt concrete mixtures. J. Assoc. Asph. Paving Technol.**1993**, 62, 54–92.
13. Zubeck, H.K.; Zeng, H.-Y.; Vinson, T.S.; Janoo, V.C. Field validation of thermal stress restrained specimen test: Six case histories. Transp. Res. Rec. **1996**, 1545, 67–74.
14. Reid, M. Alternative Materials in Road Construction; Final Research Report as Part of the Transport RTD Programme of the European Union; European Commission: Brussels, Belgium, 1999.
15. Forum of European National Highway Research Laboratories (FEHRL). SAMARIS, Sustainable and Advanced Materials for Road Infrastructures; Final Research Report on the 5th Framework Programme of the European Union; European Commission: Brussels, Belgium, 2003.
16. McDaniel, R.S.; Soleymani, H.; Anderson, R.M.; Turner, P.; Peterson, R. Recommended Use of Reclaimed Asphalt Pavement in the Super-Pave Mix Design Method; NCHRP Web Document 30 (Project D9–12); Transportation Research Board (TRB); National Research Council: Washington, DC, USA, 2000.
17. Zofka, A.; Marasteanu, M.; Li, X.; Clyne, T.; McGraw, J. Simple method to obtain asphalt binders low temperature properties from asphalt mixtures properties. J. Assoc. Asph. Paving Technol. **2005**, 74, 255–282.
18. Buttlar, W.G.; Dave, E.V. A micromechanics-based approach for determining presence and amount of recycled asphalt pavement material in asphalt concrete. J. Assoc. Asph. Paving Technol. **2005**, 74, 829–883.
19. Minnesota Department of Transportation (MnDOT). Combined 2350/2360 Plant Mixed Asphalt Pavement, Standard Specifications for Construction; MnDOT: St. Paul, MN, USA, 2008.
20. DIRECT-MAT–Dismantling and recycling techniques for road materials–sharing knowledge and practices. In Bringing Together Best Practice across Europe on the Dismantling and Recycling of Road Materials–Introductory Article; Proceedings of 2th International

Conference Environmentally Friendly Roads, Warsaw, Poland; 2009.

21. Forum of European National Highway Research Laboratories (FEHRL). RE-ROAD–End of LIFE Strategies of Asphalt Pavements; European Commission: Brussels, Belgium, 2012.
22. Arand, W.; Renken, P. Verwendung Größtmöglicher Ausbauasphaltanteile im Straßenbau; Final Report, Research Project-Federal Ministry for Transport, FA Q 148669; Institut für Straßenwesen-Braunschweig Pavement Engineering Centre, Technische Universität Braunschweig: Braunschweig, Germany, 1994. (In German)
23. Wistuba, M.; Hauser, E.; Walther, A. Untersuchungen an mit Rejuvenatoren Modifizierten Asphalten und Ermittlung des Nutzungszeitraumes; F&E-Projekt (MaxRecycling) Ongoing Research Project, Federal State Authority of Lower Saxony for Road Building and Transportation; Institut für Straßenwesen-Braunschweig Pavement Engineering Centre, Technische Universität Braunschweig: Braunschweig, Germany, 2012. (In German)
24. Radenberg, M.; Wistuba, M.; Hauser, E. Einsatz von Rejuvenatoren bei der Wiederherstellung von Asphalt; Forschungsprojekt(Rejuvenatoren)Nr. FE 07.0250/2011/ LRB, Ongoing Research Project, Federal Ministry for Transport, Building and Urban Development, Ruhr-Universität Bochum (Lehrstuhl für Verkehrswegebau); Technische Universität Braunschweig (Institut für Straßenwesen) and Basalt-Actien-Gesellschaft (BAG): Braunschweig, Germany, 2012.
25. Mangiafico, S.; di Benedetto, H.; Sauzéat, C.; Olard, F. Influence of reclaimed asphalt pavement content on complex modulus of asphalt binder blends and corresponding mixes: Experimental results and modelling. Road Mater. Pavement Des. **2013**, 14 (Suppl. 1), 132–148.
26. Moon, K.H.; Falchetto, A.C.; Hu, J.W. Investigation of asphalt binder and asphalt mixture low temperature creep properties using semi mechanical and analogical models. Constr. Build. Mater. **2014**, 53, 568–583.
27. Kriz, P.; Grant, D.L.; Veloza, B.A.; Gale, M.J.; Blahey, A.G.; Brownie, J.H.; Shirts, R.D.; Maccarrone, S. Blending and diffusion of reclaimed asphalt pavement and virgin asphalt binders. Road Mater Pavement Des. **2014**. in press.
28. Newcomb, D.E.; Stroup-Gardiner, M.; Weikle, B.; Drescher, A. Influence of Roofing Shingles on Asphalt Concrete Mixture Properties; Final Report, No. MN/RC-93-09; Minnesota Department of Transportation (MnDOT): St. Paul, MN, USA, 1993.
29. Button, J.W.; Williams, D.; Scherocman, J.A. Roofing Shingles and Toner in Asphalt Pavements; Research Report 1344, Research Study No. 0-1344; Texas Transportation Institute: Austin, TX, USA, 1996.
30. Hanson, D.I.; Foo, K.Y.; Lynn, T.A. Evaluation of Roofing Shingles in HMA; Final Report; NCAT—National Center for Asphalt Technology, Auburn University: Auburn, AL, USA, 1997.
31. Burak, S.; Ali, T. Use of asphalt roofing shingle waste in HMA. Constr. Build. Mater. **2005**, 19, 337–346.
32. Mogawer, W.S.; Booshehrian, A.; Vahidi, S.; Austerman, A.J. Evaluating the effect of rejuvenators on the degree of blending and performance of high RAP, RAS, RAP/RAS. Mixtures Road Mater. Pavement Des. **2013**, 14 (Suppl. 2), 193–213.
33. Zhou, F.; Hua, S.; Scullion, T.; Lee, R. Balanced RAP/RAS mix design system for project-specific conditions. Road Mater. Pavement Des. **2014**. in press.
34. McGraw, J.; Zofka, A.; Krivit, D.; Schroer, J.; Olson, R.; Marasteanu, M. Recycled asphalt shingles in hot mix asphalt. J. Assoc. Asph. Paving Technol. **2007**, 76, 235–274.
35. Cascione, A.A.; Willimas, R.C.; Buttlar, W.G.; Ahmed, S.; Hill, B.; Haugen, D.S. Laboratory evaluation of field produced hot mix asphalt containing postconsumer recycled asphalt shingles and fractionated recycled asphalt pavement. J. Assoc. Asph. Paving Technol. **2011**, 80, 377–418.
36. Rasmussen, R. From Roof to Road-Innovative Recycling of Bitumen Felt Roofing Material; Final Report LIFE07 ENV/DK/000102; LIFE + Programme: Glostrup, Denmark, 2012.
37. Tapsoba, N.; Sauzéat, C.; di Benedetto, H.; Baaj, H.; Ech, M. Behaviour of asphalt mixtures containing reclaimed asphalt pavement and asphalt shingle. Road Mater. Pavement Des. **2014**, 15, 330–347.
38. Minnesota Department of Transportation. Tear-Off Scrap Asphalt Shingles Specification 2360; Minnesota Department of Transportation: St. Paul, MN, USA, 2010.
39. Determining the Flexural Creep Stiffness of Asphalt Binder Using the Bending Beam Rheometer (BBR); AASHTO T313–12-UL; American Association of State Highway and Transportation Officials (AASHTO): Washington, DC, USA, 2012.
40. Berryman, J.G. Measurement of spatial correlation functions using image processing techniques. J. Appl. Phys. **1985**,57, 2374–2384.
41. Hirsch, T.J. Modulus of elasticity of concrete affected by elastic moduli of cement paste matrix and aggregate. J. Am. Concrete Inst. **1962**, 59, 427–452.
42. Huet, C. Etude par une Méthode D'impédance du Comportement Viscoélastique des Matériaux Hydrocarbonés. Ph.D. Thesis, Faculté des Sciences de l'Université de Paris, Paris, France, 1963. (In French).
43. Sharpe, W.N. Springer Handbook of Experimental Solid Mechanics; Sharpe Editions: New York, NY, USA, 2008.
44. Olard, F.; di Benedetto, H.; Dony, A.; Vaniscote, J.C. Properties of bituminous mixtures at low temperatures and relations with binder characteristics. In Proceedings of the 6th RILEM Symposium on Performance Testing and Evaluation of Bituminous Materials, Zurich, Switzerland, 1416 April 2003; Partl, M., Ed.; RILEM Publications: Bagneux, France, 2003.
45. Di Benedetto, H.; Olard, F.; Sauzéat, C.; Delaporte, B. Linear viscoelastic behaviour of bituminous materials:

From binders to mixes. Road Mater. Pavement Des. **2004**, 5 (Suppl. 1), 163–202.

46. Marasteanu, M.; Velasquez, R.; Falchetto, A.C.; Zofka, A. Development of a Simple Test to Determine the Low Temperature Creep Compliance of Asphalt Mixture; IDEA Program Final Report NCHRP 133; Transportation Research Board of the National Academies: Washington, DC, USA, 2009.
47. American Association of State Highway and Transportation Officials (AASHTO). Standard Specification for Performance-Graded Asphalt Binder; AASHTO M320-10-UL; AASHTO: Washington, DC, USA, 2010.
48. McGraw, J.; Johnson, E.; Johnson, G.; Dai, S.; Linell, D.; Watson, M. Incorporation of Recycled Asphalt Shingles in Hot-Mixed Asphalt Pavement Mixtures; Final Report MN/RC 2010-08; Minnesota Department of Transportation: St. Paul, MN, USA, 2010.
49. American Association of State Highway and Transportation Officials (AASHTO). Test for Effect of Heat and Air on a Moving Film of Asphalt (Rolling Thin-Film Oven Test); AASHTO T240-13UL; AASHTO: Washington, DC, USA, 2013.
50. Masad, E.; Somadevan, N. Microstructural finite-element analysis of influence of localized strain distribution on asphalt mix properties. J. Eng. Mech. **2002**, 128, 1105–1014.
51. Abbas, A.R.; Papagiannakis, A.T.; Masad, E.A. Linear and nonlinear viscoelastic analysis of the microstructure of asphalt concretes. J. Mater. Civ. Eng. **2004**, 16, 133–139.
52. Wang, L.B.; Wang, X.; Mohammad, L.; Wang, Y. Application of mixture theory in the evaluation of mechanical properties of asphalt concrete. J. Mater. Civ. Eng. **2004**, 16, 167–174.
53. Li, Y.; Metcalf, J.B. Two-Step approach to prediction of asphalt concrete modulus from two-phase micromechanical models. J. Mater. Civ. Eng. **2005**, 17, 407–415.
54. Eriksen, K.; Wegan, V. Optical Methods for the Evaluation of Asphalt Concrete and Polymer-Modified Bituminous Binders; Note 244; Danish Road Institute: København, Denmark, 1993.
55. Yue, Z.; Bekking, W.; Morin, I. Application of digital image processing to quantitative study of asphalt concrete microstructure. Transp. Res. Rec. **1995**, 92, 53–60.
56. Dai, Q.; Sadd, M.; Parameswaran, V.; Shukla, A. Prediction of damage behaviors in asphalt materials using a micromechanical finite element model and image analysis. J. Eng. Mech. ASCE **2005**, 131, 668–677.
57. American Association of State Highway and Transportation Officials (AASHTO). Standard Method of Test for Determining the Creep Compliance and Strength of Hot Mix Asphalt (HMA) Using the Indirect Tensile Test Device; AASHTO T322–07-UL; AASHTO: Washington, DC, USA, 2007.
58. Kose, S.; Guler, M.; Bahia, U.H.; Masad, E. Distribution of strains within hot-mix asphalt binders–applying imaging and finite element techniques. Transp. Res. Rec. **2007**, 1728, 21–27.
59. Partl, M.N.; Flisch, A.; Jönsson, M. Comparison of laboratory compaction methods using X-ray computer tomography.Road Mater. Pavement Des. **2007**, 58, 139–164.
60. Zelelew, H.M.; Almuntashri, A.; Agaian, S.; Papagiannakis, A.T. An improved image processing technique for asphalt concrete X-ray CT images. Road Mater. Pavement Des. **2013**, 14, 341–359.
61. Sefidmazgia, N.R.; Teymourpoura, P.; Bahia, H.U. Effect of particle mobility on aggregate structure formation in asphalt mixtures. Road Mater. Pavement Des. **2013**, 14 (Suppl. 2), 16–34.
62. Moon, K.H.; Cannone Falchetto, A.; Jeong, J.H. Microstructural analysis of asphalt mixtures using digital image processing techniques. Can. J. Civ. Eng. **2014**, 41, 74–86.
63. Torquato, S. Random Heterogeneous Materials; Springer-Verlag: New York, NY, USA, 2002.
64. Torquato, S. Modelling of physical properties of composite materials. Int. J. Solids Struct. **2000**, 37, 411–422.
65. Velasquez, R.A.; Cannone Falchetto, A.; Marasteanu, M.O. From mixtures to binders: Can the inverse problem be solved? Road Mater. Pavement Des. **2010**, 11 (Suppl. 1), 225–249.
66. Beran, M. Statistical continuum theories. Trans. Soc. Rheol. **1965**, 9.
67. Corson, P. Correlation functions for predicting properties of heterogeneous materials I. Experimental measurements of spatial correlation functions in multiphase solids. J. Appl. Phys. **1974**, 45, 3159–3170.
68. Jiao, Y.; Stillinger, F.H.; Torquato, S. Modeling heterogeneous materials via two-point correlation functions: Basic principles. Phys. Rev. **2007**, E76.
69. Gonzalez, R.C. Digital Image Processing; Woods, R.E., Ed.; Prentice-Hall: Upper Saddle River, NJ, USA, 2002.
70. MATLAB R2012a Help Documentation; The MathWorks, Inc: Natick, MA, USA, 2012.
71. Zhang, Y.; Sandararajan, S. The effect of autocorrelation length on the real area of contact and friction behavior of rough surfaces. J. Appl. Phys. **2005**, 97.
72. Bažant, Z.P.; Pang, S.-D. Activation energy based extreme value statistics and size effect in brittle and quasibrittle fracture. J. Mech. Phys. Solids **2007**, 55, 91–134.
73. Zhang, Y.; Sandararajan, S. Generating random surfaces with desired autocorrelation length. Appl. Phys. Lett. **2006**, 88.
74. Falchetto, A.C.; Marasteanu, M.O.; di Benedetto, H. Analogical based approach to forward and inverse problems for asphalt materials characterization at low temperatures. J. Assoc. Asph. Paving Technol. **2011**, 80, 549–580.
75. Christensen, D.; Pellinen, T.; Bonaquist, R.F. Hirsch model for estimating the modulus of asphalt concrete. J. Assoc. Asph. Paving Technol. **2003**, 72, 97–121.

76. Moon, K.H.; Falchetto, A.C.; Marasteanu, M.O. Rheological modelling of asphalt materials properties at low temperatures: from time domain to frequency domain. Road Mater. Pavement Des. **2013**, 14, 810–830.

77. Bonaquist, R. Can I run more RAP? Hot Mix Asph. Technol. **2007**, 12, 11–13.

CHAPTER 10

CHANGES IN NANOSCALED MECHANICAL AND RHEOLOGICAL PROPERTIES OF ASPHALT BINDERS CAUSED BY AGING

Ben Liu, Junan Shen, and Xuyan Song

INTRODUCTION

Asphalt binder plays an important role in the performance of asphalt mixture, which accounted for 29% of rut depth at high temperature, according to the strategic highway research program (SHRP) research. Rheology is a very powerful tool for characterizing and quantifying materials properties. Since the rheological properties of asphalt change during production and continue to change subsequently in service, there is necessity to study the phenomenon of aging [1]. As a result of the global climate, air temperature of most areas in China in the summer is higher and higher and lasts for a long time, making the rutting and water damage particularly more prominent than ever before. The DSR is widely used for characterizing asphalt binders and is required for specifications in many countries. DSR can accurately measure the rheological responses of asphalt films adhesion to aggregate's surfaces [2]. As a special material, asphalt binder has viscoelastic property and is more complicated than other pavement materials. The testing involves using rotating thin film oven (RTFO) aged binders to mimic the state of binder aging in the mixture beams [3]. Because of its complexity in compositions, the rheological properties of asphalt binder are dependent on the aging, the temperature tested, and the test methods. A number of studies have recognized that binders of nearly identical SHRP performance grades can show differences in fracture properties that vary significantly [4–6]. In order to understand the changes in nanoscale caused by the aging of short and long terms, the frontier of atomic force microscope (AFM) was used to explore the nanomechanical forces. The adhesion, nanoforce curve, and rheological properties were tested on an asphalt binder before and after aging through rotating thin film oven (RTFO) aging.

MATERIALS AND METHODS

Materials

This research used Shell Pen 70# as asphalt. The properties of Pen asphalt were shown in Table 1.

Citation: Ben Liu, Junan Shen, and Xuyan Song, "Changes in Nanoscaled Mechanical and Rheological Properties of Asphalt Binders Caused by Aging," Journal of Nanomaterials, vol. 2015, Article ID 961924, 6 pages, 2015. doi:10.1155/2015/961924.

Table 1: Properties of Shell Pen 70# asphalt.

Test index	Unit	Pen 70	Specs	Test method
Penetration (25°C, 5 s, 100 g)	0.1 mm	68	60~80	T0604
Ductility (5 cm/min, 15°C)	cm	>100	≥100	T0605
Softening point (global method)	°C	58	≥46	T0606
Dynamic viscosity (60°C)	Pa·s	230	≥180	T0625
The residue after thin film oven				T0609
Quality loss	%	0.04	≤±0.8	T0609
Ductility (5 cm/min, 15°C)	cm	27	15	T0605
Penetration ratio (25°C, 5 s, 100 g)	%	69	≥58	T0604

It is indicated from Table 1 that the asphalt conformed to the requirements of the Chinese specifications.

In addition, the materials used for atomic force microscope test including acetone solution and carbon disulfide solution are all commercially available.

Methods

AFM analysis uses BRUKER company's icon type atomic force microscope. Its principle is that there is a small tip which is very sensitive to the weak force at the other end of micro-cantilever. Tip gently contacted the sample surface, due to the extremely weak repelling force between needle tip atoms and the sample surface atoms, the force will be kept constant in the scan control. A sharp tip probes the surface of a sample while collecting data. The cantilever tip is usually only several micros in length, and is located at the free end of a cantilever. Vertical displacement and lateral distortion in the cantilever is measured and mapped as the tip probes the surface of the specimen. The surface information will be measured by nanoscaled resolution. The operation mode of the atomic force microscope was classified in the form of acting force between tip and sample classification. There are mainly three kinds of operating modes: contact mode, noncontact mode, and tapping mode, with atomic force microscope structure model as shown in Figure 1.

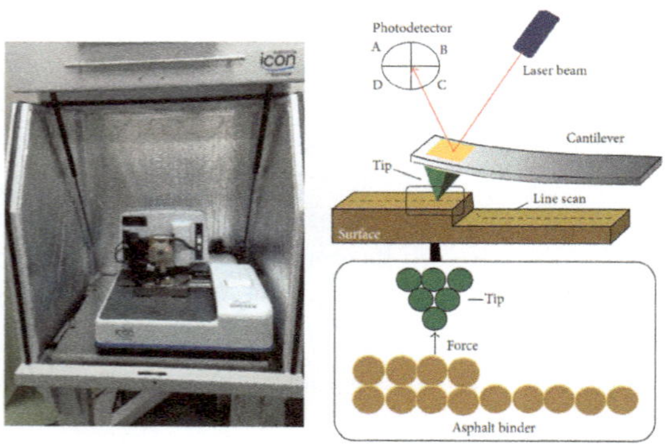

Figure 1: Atomic force microscope used in the study.

The nanomechanical properties of the asphalt before and after aging are conducted by contact model. The probe model of the cantilever used in the study is SCANASYST—AIR, and its elasticity coefficient is 0.4 N/m. After correction, deflection sensitivity value of the cantilever beam is 51.48 nN/V, and the K of the spring coefficient is 0.4840 N/m.

This study followed the steps below [7]:

(1) Weigh the asphalt before and after aging separately, with the weight ratio of carbon disulfide and asphalt 1:9.

(2) Wipe completely glass slide with acetone solution to remove surface dust and make sure the slide surface is clean.

(3) Absorb by glue rubber head a drop of the asphalt solution prepared and release it on a glass slide.(4)Let the drop of solution be a thin film.

Dynamic shear rheometer (DSR) is a machine that determines the rheological properties of asphalt binders for Superpave system: rutting resistance at high temperature and fatigue resistance at intermediate temperature. It belongs to the plate type rheometer. Asphalt samples were placed between parallel plates as shown in Figure 2, at a certain rate of rotating shearing, determining binder plural shear modulus G^* and phase angle of the δ, charactering its viscoelastic properties. G^* is material repeated shear deformation measurement of total resistance, composed of elastic (recoverable) part of G_1 and G_2 of part of the viscosity (irreversible). The δ is elastic and viscous deformation amount of relative index. The smaller the δ is, the closer the material is in the elastomer, complex shear modulus (G^*), and its two components: storage modulus (G_1) and loss modulus (G_2) [8], as shown in Figure 3.

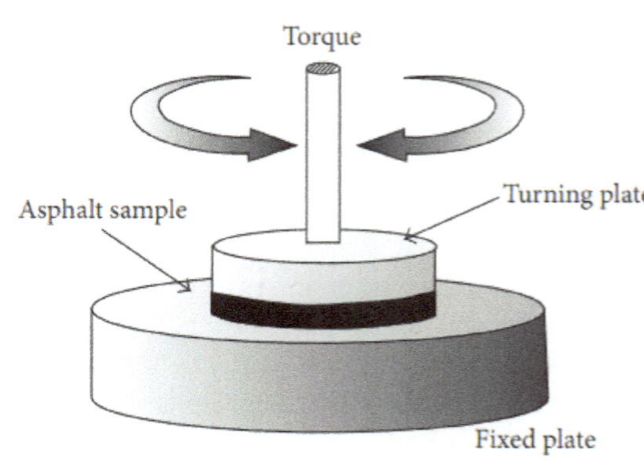

Figure 2: Asphalt sample loading way.

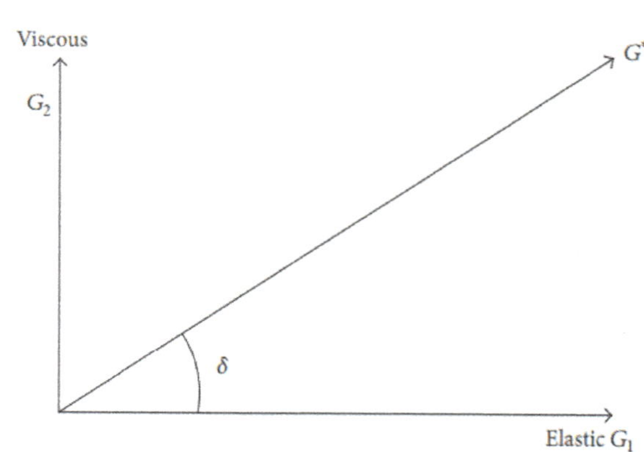

Figure 3: Viscoelastic complex modulus.

RESULTS AND DISCUSSION

Nanoscaled Results from AFM

Nanoscaled mechanical properties, that is, adhesion of asphalt binders before and after RTFO, were studied by using AFM. Nanoadhesion and nanoforce were measured, respectively [9]. AFM test probe of the elastic coefficient of cantilever beam is 0.4 N/m and correction of the cantilever beam of K value was 0.4840 N/m, with sweeping range of 20 microns × 20 microns. The results of the nanoadhesion and nanoforce of the asphalt before and after RTFO aging are shown in Figures 4–7, respectively.

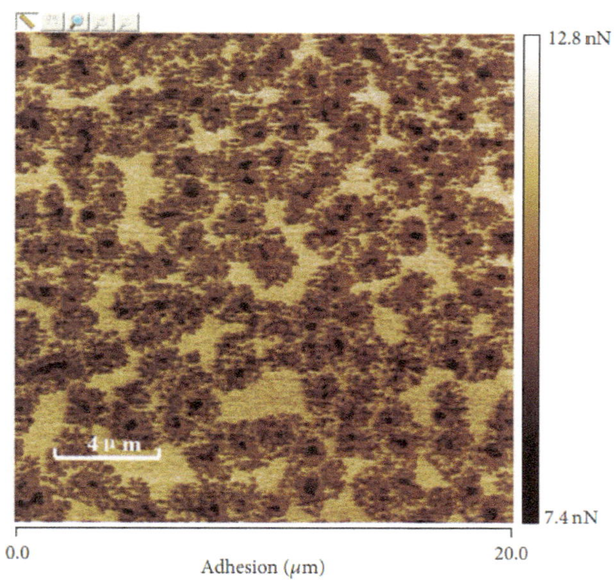

Figure 4: Nanoadhesion of asphalt.

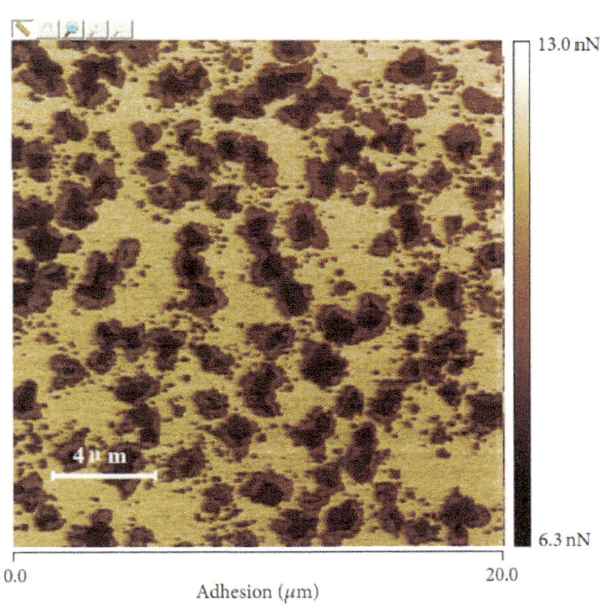

Figure 5: Nanoadhesion of RTFO residual.

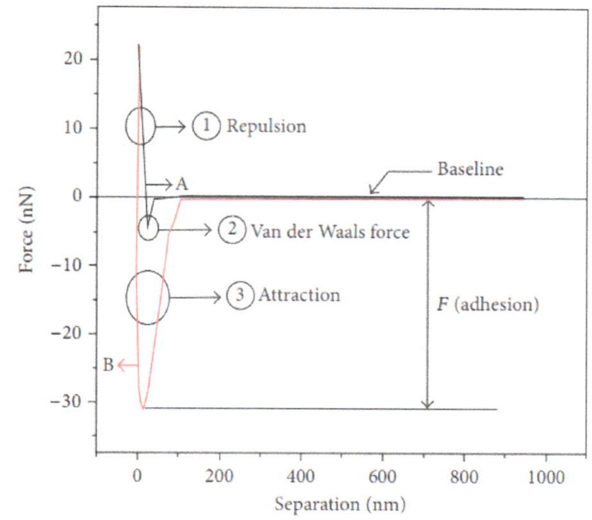

Figure 6: Nanoforce of asphalt.

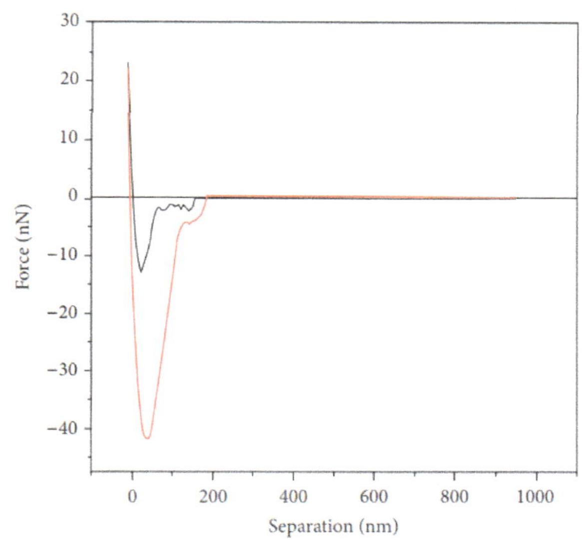

Figure 7: Nanoforce of asphalt RTFO residual.

Figures 4 and 5 are the results of nanoadhesion taken for asphalt before and after aging. These figures showed the nanoadhesion distribution on the area of asphalt of 20 microns × 20 microns. The darker the color, the smaller the adhesion. It is clear that the dark area was smaller and light color area was increased after RTFO aging, indicating that the whole area of adhesion was increased.

Figures 6 and 7 are the nanoforces curves of the binder before and after aging, a typical result taken from asphalt; both had the same trend. Here are the exploration and explanation for Figure 6. A nanoforce curve result can be divided into two parts—engagement part (curve A, abscissa from right to left), retraction part (curve B, abscissa from left to right). Curve A is characterization of the process that needle got closing to the sample, concave valley of curve A was caused by the van der waals force (②part) between probe and sample molecules, it mainly showed the attraction force at this time. As the probe kept getting close to the sample, the nanorepulsive force between tip and sample increased (①part represented repulsion part above the baseline). When the nanorepulsive force reached the maximum, the distance between the sample and probe became the closest, while the probe began to retract, that is, leaving the sample. Curve B in Figure 6 showed that the probe began to leave the sample (③part represented attraction part below the baseline). The adhesion of the sample can be reflected by the difference between the lowest point of retract part of the force curve (Curve B) and the baseline. The lowest point meant the largest attraction when the probe retracted out in the process of the sample of the binder sample. The point was often used to reflect the information of binding force of the asphalt binder sample. Work of adhesion is associated with the energy necessary to promote adherence. The behavior of the work of adhesion is similar to that observed for the adhesive force [10]. In principle, the value of separation cannot be negative, which was negative in some of Figures 4–7. The reason may be caused by the overloaded ramp force that can be solved by reducing the value of trigger threshold during the process of AFM experiment.

Figures 6 and 7 indicated that the adhesion can be reflected by retraction of nanoforce curve (i.e., the difference between the lowest part of the valley in the force curve B and the baseline). The point of valley means the largest attraction at the process of probe exiting samples. After that point, the probe can get rid of attraction and the force curve reaches the baseline. It can be seen from Table 2 that the adhesion values were 30.7 nN and 41.5 nN for asphalt and RTFO residuals, respectively. Adhesion force increased after RTFO aging.

Table 2: Adhesion force of different asphalt aging stage.

Degree of aging	Adhesion force of asphalt (nN)
Original	30.7
RTFO residual	41.5

Rheological Results from DSR

DSR measured rheological properties of asphalt before and after aging under temperatures from 52°C to 70°C, respectively, in accordance with the American AASHTOTP-93 requirements. DSR test results of asphalt are shown in Table 3. The relationship of different temperatures with G^*, G_1, G_2, and $G^*/\sin(\delta)$ was shown in Figures 8, 9, 10, and 11, and respectively.

Table 3: Rheological properties of asphalt before and after RTFO tests.

Test project	Test temperature/°C	Asphalt	RTFO aged asphalt residual
$G^*/\sin(\delta)$ (KPa)	52	5.10	13.70
	58	2.25	5.77
	64	1.05	2.46
	70	0.50	1.44
Phase angle (°)	52	85.8	82.5
	58	86.7	84.5
	64	87.2	86
	70	86.3	84.4
G^* (KPa)	52	5.10	13.58
	58	2.25	5.74
	64	1.05	2.45
	70	0.50	1.43
Elastic G_1 (KPa)	52	0.37	1.77
	58	0.13	0.55
	64	0.05	0.17
	70	0.03	0.14
Viscous G_2 (KPa)	52	5.09	13.47
	58	2.25	5.72
	64	1.05	2.45
	70	0.50	1.43

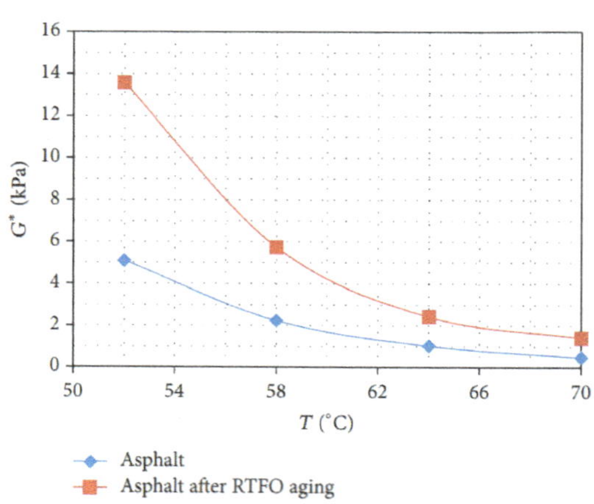

Figure 8: G^*, temperature of asphalt before and after RTFO aging.

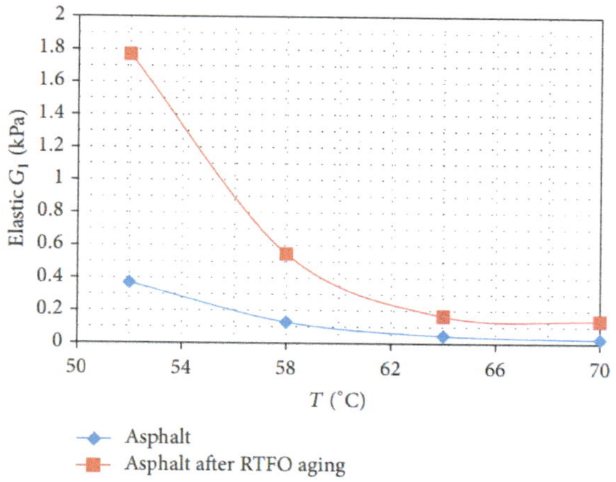

Figure 9: G_1, temperature of asphalt before and after RTFO aging.

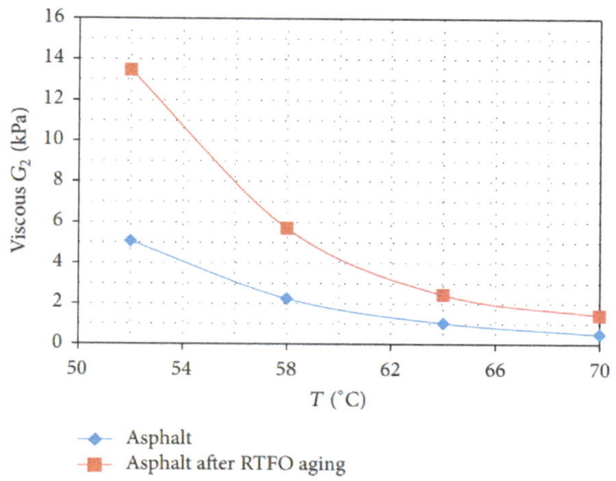

Figure 10: G_2, temperature of asphalt before and after RTFO aging.

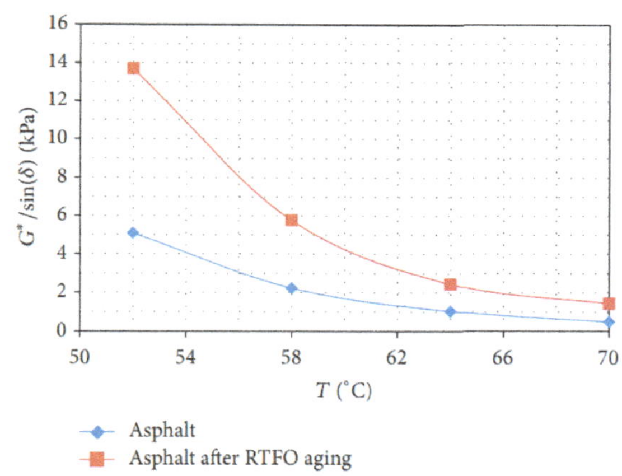

Figure 11: $G^*/\sin(\delta)$, temperature of asphalt before and after RTFO aging.

Figures 8, 9, 10, and 11 presented the complex shear modulus G^*, elastic modulus G_1, viscous modulus G_2, and $G^*/\sin(\delta)$ of asphalt binders before and after RTFO aging. The greater the $G^*/\sin(\delta)$, a rutting factor, the better the asphalt material resistant to high temperature deformation. A higher G^* value indicates that the asphalt binder has higher stiffness to resist rutting, and low phase angle (δ) value ensures that the asphalt becomes more elastic to recover part of deformation [11]. Overall, the G^* decreased as the testing temperature increased for both the original and RTFO residual. Moreover, G^* of the asphalt after RTFO had a higher value than that of the original. However, the difference between the G^* of RTFO residual and that of the original one decreased as the temperature increased; see Figure 8. It can be concluded that RTFO aging had much more influence on the G^* when the temperature is low. The G_1, G_2, and $G^*/\sin(\delta)$ changed with the temperature in a similar way to that of G^*; that is, the trend of the change of G^* with the temperature is true for the other properties presented (see Figures 8–11 again).

Apparent changes in viscoelastic properties (e.g., G^* , $G^*\sin(\delta)$) with number of cycles of loading are used to define fatigue performance of asphalt binder. However, the mechanisms responsible for apparent changes in these viscoelastic properties were previously not well understood [12]; it is necessary to make further study of micromechanical mechanism of asphalt binders. Nanoadhesion of the asphalt binder from both the image figure and curve was higher for RTFO residual than the original one. This was consistent with the increase of G^* of the binder before and after RTFO aging.

SUMMARY AND CONCLUSIONS

In the study, the changes in nanomechanical properties were measured on base binder of both original and RTFO residuals using atomic force microscope (AFM), and the rheological properties were tested by dynamic shear rheometer (DSR). Conclusions can be drawn as follows:

(1) The nanoadhesion was successfully measured from AFM of the picture and curve and it was found that it increased after RTFO aging for base binders. Nanoforce curve from AFM showed that adhesion force is 30.7 nN and 41.5 nN for asphalt and RTFO residuals, respectively.

(2) $G^*$of the binders was higher for RTFO residual than for the original one. The influence between the G^* of the RTFO residual and that of the original was temperature-dependent and decreased as the temperature increased.

(3) The rheological properties of the asphalt changed consistently with the nanomechanical properties with the aging.

(4) From the tests of AFM and RTFO, the results had a good correlation of base asphalt with the aging; SBS modified asphalt and other asphalt with different modifying agents should be tested in the future study to understand the aging mechanism of different asphalt better.

Conflict of Interests

The authors declare that there is no conflict of interests regarding the publication of this paper.

ACKNOWLEDGMENT

This research was supported financially by Chinese National Science Foundation with Grant no. 51378328.

REFERENCES

1. C. Joshi, A. Patted, M. R. Archana, and M. S. Amarnath, "Determining the rheological properties of asphalt binder using dynamic shear rheometer (DSR) for selected pavement stretches," International Journal of Research

in Engineering and Technology, no. 11, pp. 192–196, 2013.

2. D.-W. Cho and H. U. Bahia, "New parameter to evaluate moisture damage of asphalt-aggregate bond in using dynamic shear rheometer," Journal of Materials in Civil Engineering, vol. 22, no. 3, pp. 267–276, 2010.
3. C. M. Johnson, H. U. Bahia, and H. Wen, "Evaluation of strain-controlled asphalt binder fatigue testing in the dynamic shear rheometer," in Proceedings of the 4th International SIIV Congress, pp. 1–11, Palermo, Italy, 2007.
4. T. R. Hoare and S. A. M. Hesp, "Low-temperature fracture testing of asphalt binders," Transporation Research Record, vol. 1782, pp. 36–42, 2000.
5. D. A. Anderson, L. Lapalu, M. O. Marasteanu, Y. M. Le Hir, J.-P. Planche, and D. Martin, "Low-temperature thermal cracking of asphalt binders as ranked by strength and fracture properties,"Transportation Research Record, vol. 1766, no. 1, pp. 1–6, 2001.
6. F. Olard and H. Di Benedetto, "Fracture toughness and fracture energy of bituminous binders at low temperature," in Proceedings of the 5th International RILEM Conference on Reflective Cracking in Pavements, vol. 37, pp. 359–366, Limoges, France, May 2004.
7. B. McCarron, X. Yu, and M. Tao, The Investigation of "Bee-Structures" in Asphalt Binders, Department of Physics, Worcester Polytechnic Institute, 2011.
8. T. Khedaywi, G. Al-Khateeb, and M. Irfaeya, "Effect of medical ash on shear properties of asphalt binder using superpave dynamic shear rheometer (DSR)," Journal of Solid Waste Technology and Management, vol. 38, no. 1, pp. 19–27, 2012.
9. R. G. Allen, D. N. Little, and A. Bhasin, "Structural characterization of micromechanical properties in asphalt using atomic force microscopy," Journal of Materials in Civil Engineering, vol. 24, no. 10, pp. 1317–1327, 2012.
10. E. R. Dourado, R. A. Simao, and L. F. M. Leite, "Mechanical properties of asphalt binders evaluated by atomic force microscopy," Journal of Microscopy, vol. 245, no. 2, pp. 119–128, 2012.
11. G. G. Al-Khateeb and K. Z. Ramadan, "Investigation of the effect of rubber on rheological properties of asphalt binders using superpave DSR," KSCE Journal of Civil Engineering, vol. 19, no. 1, pp. 127–135, 2015.
12. C. Hintz and H. Bahia, "Understanding mechanisms leading to asphalt binder fatigue in the dynamic shear rheometer," Road Materials and Pavement Design, vol. 14, supplement 2, pp. 231–251, 2013.

CHAPTER 11

STRENGTH MECHANISM AND INFLUENCE FACTORS FOR COLD RECYCLED ASPHALT MIXTURE

Tao Ma, Hao Wang, Yongli Zhao, Xiaoming Huang, and Yuhui Pi

INTRODUCTION

Recycling technology has become one of the most promising approaches for asphalt pavement rehabilitation and maintenance with considerable savings in natural resource, cost, and energy [1–3]. It has been proved that hot recycled asphalt mixture with controlled RAP content between 10% and 30% performs similarly to virgin hot-mix asphalt mixture [4]. However, with the continuously increasing energy costs and environmental awareness, central-plant hot recycling technologies have difficulties meeting the need of using high-content RAP in asphalt pavements. Compared to hot recycling technology, cold recycling technology can use much higher RAP contents and bring more benefits [5, 6]. Therefore, cold recycling technology has attracted significant attentions in recent years.

The overall process of cold recycling includes milling, processing, mix design, binder addition, placement, compaction, and quality assurance [7]. To successfully use cold recycling for asphalt pavements, a lot of researches have been conducted on the mix design method and evaluation of mechanical properties of cold recycled asphalt mixtures. Based on previous research findings [8–13], the "black rock" theory is the basis for current design of cold recycled asphalt mixtures. The asphalt binder in RAP is usually ignored and RAP plays a single role of aggregate. Emulsified asphalt or foamed asphalt is used as new binder in cold recycled asphalt mixtures. However, it was found that the cold recycled asphalt mixture that is composed of RAP and emulsified or foamed asphalt may not perform well to carry heavy traffic loading. Thus, additives such as Portland cement, lime, and fly ash are suggested to be used to improve the performance of cold recycled asphalt mixture. Among these additives, adding Portland cement is an efficient way to consume water in the emulsified or foamed asphalt and form a stronger and denser skeleton in the cold recycled asphalt mixture [14]. In this case, the cold recycled asphalt mixture is regarded as semiflexible material different from the normal asphalt mixture and cement concrete [15, 16].

Studies have shown that many agencies used cold recycled asphalt mixtures in the base layer. However, some problems with low early strength, extended curing time, cracking, and raveling were commonly observed from field experience [17–20]. One of the most important reasons for causing these problems is the strength forming mechanism in the cold recycled asphalt mixture. Although RAP may look like "black" aggregate, the contribution of RAP to the cold recycled asphalt mixture may not be the same as real aggregate due to the existence of aged asphalt binder. If RAP does not play the same role as real aggregate, the mix design that is based on "black rock" for RAP may lead to improper mixture design and poor pavement performance. As Portland cement is

Citation: Tao Ma, Hao Wang, Yongli Zhao, Xiaoming Huang, and Yuhui Pi, "Strength Mechanism and Influence Factors for Cold Recycled Asphalt Mixture," Advances in Materials Science and Engineering, vol. 2015, Article ID 181853, 10 pages, 2015. doi:10.1155/2015/181853.

added to improve strength of the cold recycled asphalt mixture, the cement hydration could be affected by the existence of asphalt binder in RAP and the emulsified or foamed asphalt. Meanwhile, it is not clear if the aggregate gradation and the type or content of asphalt emulsion have significant influences on the performance of cold recycled asphalt mixtures.

The main objective of this study is to understand the strength forming mechanism and its key influence factors for the cold recycled asphalt mixture with cement and emulsified asphalt. To analyze the strength mechanism of cold recycled asphalt mixture, the inner structure and strength between RAP and real aggregate were compared and the interaction between the emulsified asphalt and cement was investigated. The influences of RAP, emulsified asphalt, and cement on the tensile strength of cold recycled asphalt mixture were evaluated.

MATERIALS AND METHODS

Materials

RAP was obtained from milling materials of asphalt layers of an expressway which had been on service for 8 years. To control the variability of gradation design for the cold recycled asphalt mixture, RAP was crushed and then sieved into three grades which are coarse RAP bigger than 16 mm, median RAP between 16 mm and 2.36 mm, and fine RAP smaller than 2.36 mm. The RAP particles for which the size is bigger than 30 mm were excluded for use. The gradations of coarse RAP, median RAP, and fine RAP are shown in Table 1.

Table 1: Gradations for coarse, median, and fine RAP

RAP	Passing ratios/% for different sieving sizes/mm												
	31.5	26.5	19	16	13.2	9.5	4.75	2.36	1.18	0.6	0.3	0.15	0.075
Coarse	100	77.9	38.0	0									
Median		100	100	100	90.4	69.2	29.1	0.0					
Fine		100	100	100	100	100	100	100	53	29	8.4	3.6	0.6

New fine aggregates and mineral fillers with qualities meeting the requirement by Chinese Technical Specification for Construction of Highway Asphalt Pavement [21] were used to improve the gradation design. The gradations of fine aggregates and mineral fillers are shown in Table 2.

Table 2: Gradations for new aggregates and mineral fillers

Material	Passing ratios/% for different sieving sizes/mm							
	9.5	4.75	2.36	1.18	0.6	0.3	0.15	0.075
New aggregates	100	99.5	75.7	55.5	35.7	26.5	13.8	4.5
Mineral fillers						100	90	80

Emulsified asphalt, cement, and potable water were used as binder in the cold recycled asphalt mixture. Three different types of emulsified asphalt, normal slow-setting emulsified asphalt, normal medium-setting emulsified asphalt, and high-viscosity slow-setting emulsified asphalt, were used in this study. Table 3 shows the basic rheological properties of the emulsified asphalt. Ordinary Portland cement with grade 32.5 was used and its physical properties are shown in Table 4. All three types of emulsified asphalt and the Portland cement meet the requirement by Chinese Technical Specification for Construction of Highway Asphalt Pavement [21].

Table 3: Basic properties of emulsified asphalt

Indexes	Emulsified asphalt			Specification
	High-viscosity	Slow-setting	Medium-setting	
Demulsification speed	Slow	Slow	Medium	Slow or medium
Particle charge	(+)	(+)	(+)	(+)
Residual stay on 1.18 mm/%	0.05	0.04	0.08	≤0.1
Englar viscosity	22	9	10	2–30
Residual after distillation				
Content/%	66	60	62	≥55
Penetration (25°C)/0.1 mm	55	69	72	50–300
Ductility (15°C)/cm	60	110	95	≥40
Solubility/%	98.5	98.1	99.8	≥97.5

Table 4: Basic properties of cement

Indexes	Testing value	Specification
Initial setting time/min	147	≥45
Final setting time/min	326	≤600
Compressive strength/MPa		
3 d	13.2	≥11.0
28 d	33.8	≥32.5
Flexural strength/MPa		
3 d	3.9	≥2.5
28 d	6.2	≥5.5

One controlling cold recycled mixture was prepared following Chinese Technical Specification for Highway Asphalt Pavement Recycling [22]. The aggregate mix of cold recycled asphalt mixture includes 75% RAP, 20% new aggregates, and 5% mineral fillers for gradation. The new binder added to the mixture was composed of 4% emulsified asphalt, 1.5% cement, and 5.4% water by the weight of the aggregate mix. Based on the mix design of the controlling cold recycled asphalt mixture, different gradations, emulsified asphalt types and contents, and cement contents were used to prepare cold recycled asphalt mixtures for analysis.

Laboratory Testing

Specific surface area is defined as the total surface area per unit mass of material. The specific surface area of aggregates has significant influence on the binder content and the adhesion between binder and aggregate. Specific surface area can be calculated using the gradation and the specific area coefficient for different sieving sizes, as shown in (1). The specific surface area coefficients (FA_i) for different sieving sizes of aggregates are shown in Table 5. The specific surface area of both RAP and the aggregates extracted from RAP was analyzed in this study:

$$\text{SA} = \sum P_i \times FA_i, \tag{1}$$

where SA is the total specific surface area of different sieving sizes, m^2 /kg; P_i is the passing ratio of different sieving sizes, %; and FA_i is the specific surface area coefficient for different sieving sizes.

Table 5: Specific surface area coefficients for different sieving sizes of aggregates

Sieving size (i)/mm	≥4.75	2.36	1.18	0.6	0.3	0.15	0.075
FA_i (10^{-2})/m^2/kg	0.4	0.8	1.64	2.87	6.14	12.3	32.77

Scanning electron microscope (SEM) can be used to observe the microstructure of material by scanning the material with electron and analyzing the secondary electron promoted on the surface of the material. Cement and emulsified asphalt were mixed based on quality ratio of 1 : 1 and then cured for 7 days at ambient temperature to form the cement-asphalt mix. Then asphalt was solved and eliminated from the cement-asphalt mix to obtain the cement hydrates. In this study, SEM was used to observe the microstructures of both the cement-asphalt mix and the cement hydrates extracted from the cement-asphalt mix.

The tensile strength is the most important mechanical property for design of cold recycled asphalt mixtures. Splitting test can be used to determine the tensile strength of mixtures at biaxial stress states. Samples with diameter of 101.6 mm and height of 63.5 mm were prepared. Splitting test was conducted with a loading rate of 50 mm/min at 15°C. The tensile strength of samples was calculated based on (2). There are two different testing conditions that can be used for splitting test. One is called dry splitting test in which the testing specimen is conditioned at 15°C in the air bath for 6 h or in the water bath for 2 h before splitting test. The other one is called immersion splitting test in which the testing specimen is conditioned at 25°C in the water bath for 22 h and then at 15°C in the water bath for another 2 h before splitting test. The tensile strength by the splitting test can be determined by (2). And the tensile strength ratio defined by (3) can be used to evaluate the moisture stability of test specimen:

$$R_T = \frac{0.006287 P_T}{h}, \tag{2}$$

where R_T is the splitting tensile strength, MPa; P_T is the maximum loading when sample is split, MPa; and h is the sample height, mm. Consider

$$\text{TSR} = \frac{R_2}{R_1} \times 100, \tag{3}$$

where TSR is the tensile strength ratio, %; R_2 is the tensile strength from immersion splitting test, MPa; and R_1 is the tensile strength from dry splitting test, MPa.

RESULTS AND DISCUSSION

Effect of RAP

Structure and Strength Analysis of RAP

Compared to real aggregates, RAP material may have similar shapes and sizes. However, many coarse particles in RAP are actually agglomerates of fine RAP particles bonded by the aged asphalt. This can be proved by the significant gradation difference between RAP and the extracted aggregates from RAP, as shown in Figure 1. It clearly shows that the gradation of extracted aggregates from RAP is much finer than the gradation of RAP. The biggest particle size of the extracted aggregates is 19 mm while it is 26.5 mm for RAP. In addition, there are much fewer RAP particles passing through the sieve of 0.3 mm compared to the extracted aggregates from RAP. To quantify the influence of agglomeration in RAP, the specific surface areas of RAP and the extracted aggregates from RAP were calculated using (1). The results are shown in Figure 2.

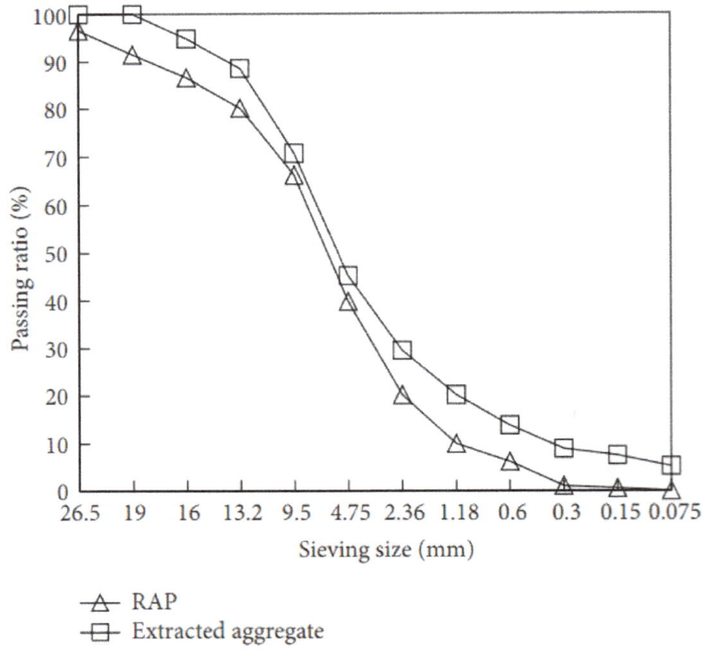

Figure 1: Gradations of RAP and aggregates.

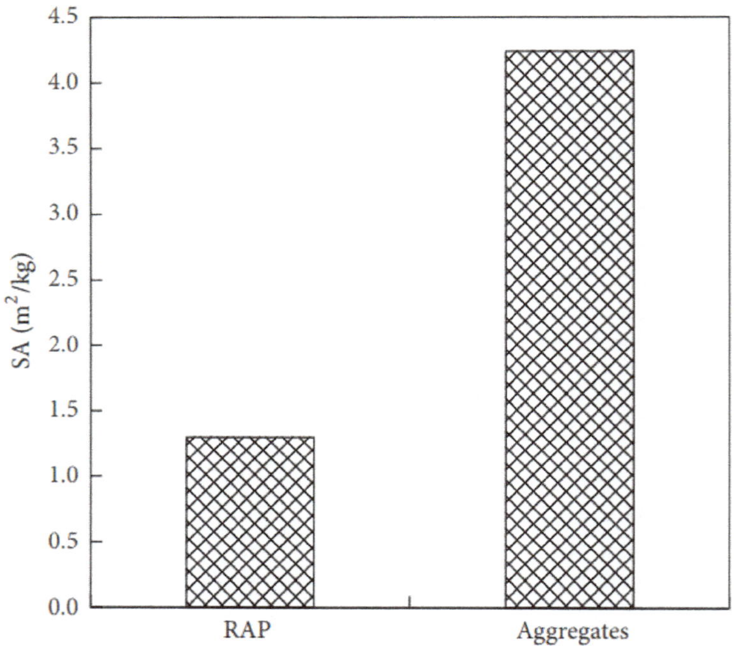

Figure 2: SA of RAP and aggregates.

From Figure 2, it can be seen that the specific surface area of the extracted aggregates is about three times the specific surface area of RAP. It indicates that when the new binder is added into RAP, only about one-third of the total surface of the aggregates in RAP can be covered and bonded by the new binder. The remaining surface of the aggregates in RAP is still covered by the aged asphalt in RAP. This explains why the asphalt content of cold recycled asphalt mixture is usually lower than hot-mix asphalt mixture. More importantly, the adhesion strength of aged asphalt binder inside RAP is much weaker than new binder due to long-term aging. Thus, when subjected to mechanical and environmental loading, the internal surface inside the RAP particles can be weak spots in the cold recycled asphalt mixture.

Splitting test was conducted to estimate the tensile strength of RAP. Since the splitting test cannot be used directly to determine the tensile strength of RAP, testing samples were prepared by mixing RAP with cement mortar. Cement mortar was prepared with water, cement, and sand according to a mass ratio of 1 : 2 : 3. Testing samples were prepared by mixing cement mortar and RAP at different RAP contents (0, 10, 20, 30, 40, and 50% by the weight of the total mix). The relationship between the tensile strength of cement-mortar RAP samples and the RAP content is summarized in Figure 3.

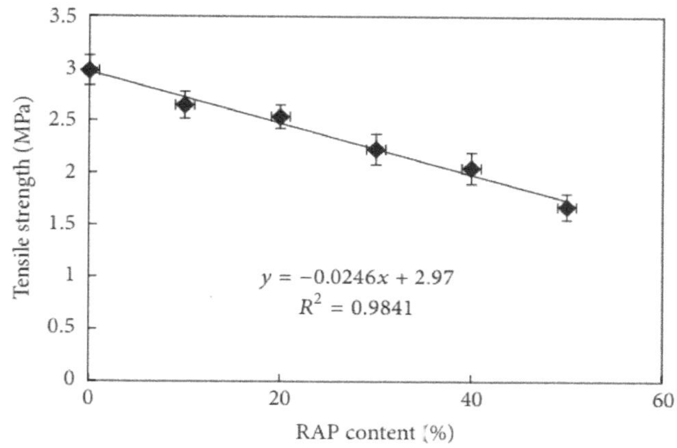

Figure 3: Tensile strength versus RAP content.

It can be seen that the tensile strength decreases as the percentage of RAP increases. A linear relationship was obtained with a high R^2 value. According to the linear equation, the tensile strength of pure RAP was estimated to be 0.51 MPa when the percentage of RAP is 100%. The value is within the range of 0.4 and 0.6 MPa for tensile strength of cold recycled asphalt mixtures observed from real projects. Combined with the previous specific surface area analysis, it indicates that RAP particles are much different with real aggregates, as the main composition material in cold recycled asphalt mixture; the strength of RAP has important influences on the strength of cold recycled asphalt mixture.

Influence of RAP Gradation

To analyze the influences of gradation on the tensile strength of cold recycled asphalt mixtures, three different gradations named as fine gradation, median gradation, and coarse gradation were used to prepare cold recycled asphalt mixtures, as shown in Figure 4. The median gradation is the base gradation used for the controlling cold recycled asphalt mixture as described in Section 2.1. During the preparation of cold recycled asphalt mixtures, the contents of emulsified asphalt, cement, and water were 4%, 1.5%, and 5.4%, respectively. Figure 5summarizes the tensile strength and tensile strength ratio from splitting tests for cold recycled asphalt mixtures with different gradations.

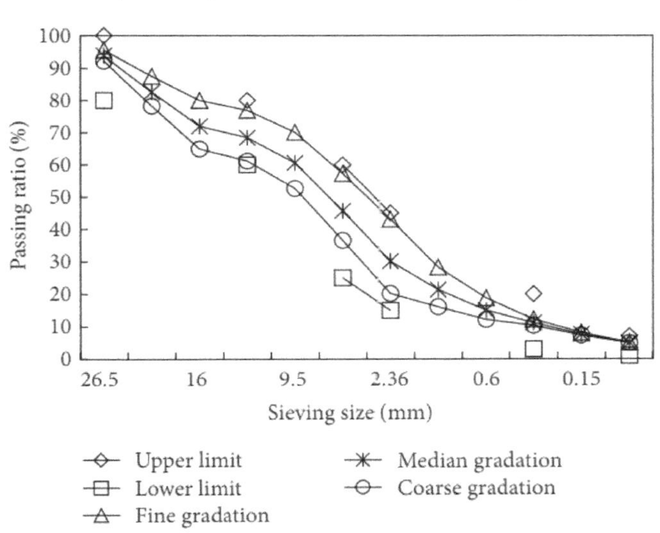

Figure 4: Different gradations used for cold recycled asphalt mixtures.

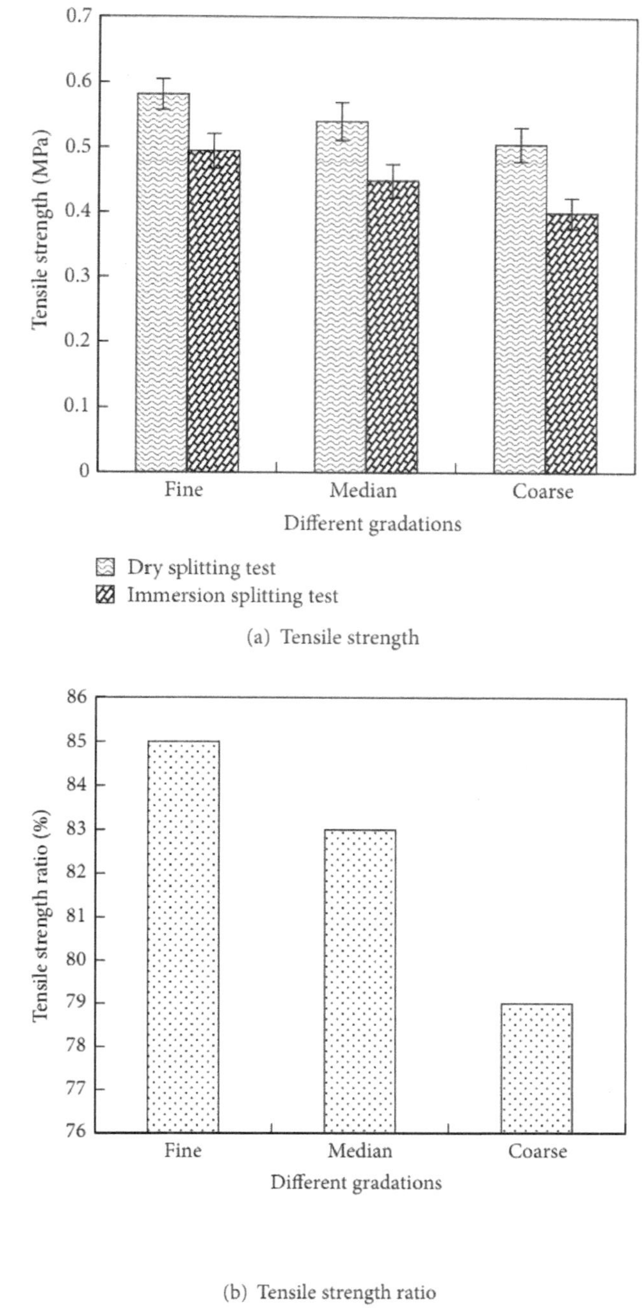

Figure 5: Splitting test results for cold recycled asphalt mixtures with different gradations.

The results show that gradation can affect the tensile strength and moisture stability of cold recycled asphalt mixtures. Both tensile strength and tensile strength ratio of cold recycled asphalt mixtures increase as gradation becomes finer. When the fine gradation is used, there are fewer coarse RAP particles in the mixture and the specific surface area of aggregate mix increases. This could improve the bonding between new binder and RAP and thus increase the tensile strength of cold recycled asphalt mixtures. This conforms with the previous analysis of RAP.

Effect of Emulsified Asphalt

Influence of Asphalt Content

Three different emulsified asphalt contents, 3%, 4%, and 5%, were used to analyze the influence of asphalt content on the tensile strength of cold recycled asphalt mixtures. All of the three gradations described in Figure4 were used to prepare the cold recycled asphalt mixtures. For the cold recycled asphalt mixtures with different emulsified asphalt contents of 3%, 4%, and 5%, the cement content was 1.5% and the water contents were 5.8%, 5.4%, and 5%, respectively. Figure 6 summarizes the tensile strength and tensile strength ratio from splitting tests for cold recycled asphalt mixtures with different gradations and asphalt contents. It is shown that, for each gradation, as the asphalt content increases from 3% to 5%, both the tensile strength and tensile strength ratio increase first and then decrease. Therefore, there is optimal asphalt content to guarantee the best tensile strength and moisture stability for cold recycled asphalt mixture. For instance, the optimal asphalt content is determined to be 4% for the specimens tested in this study.

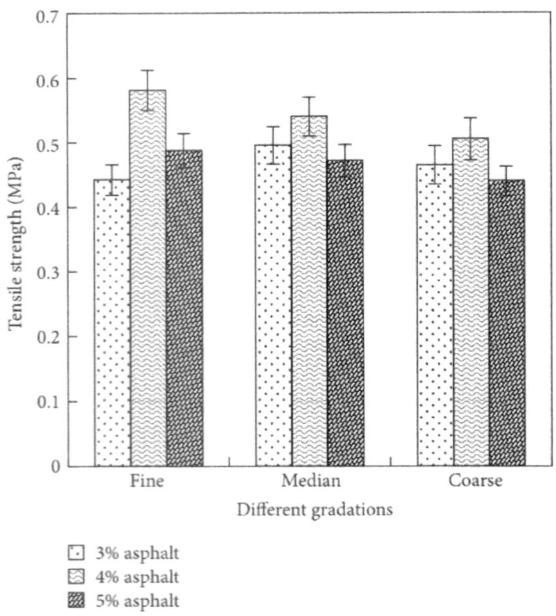

(a) Dry splitting test

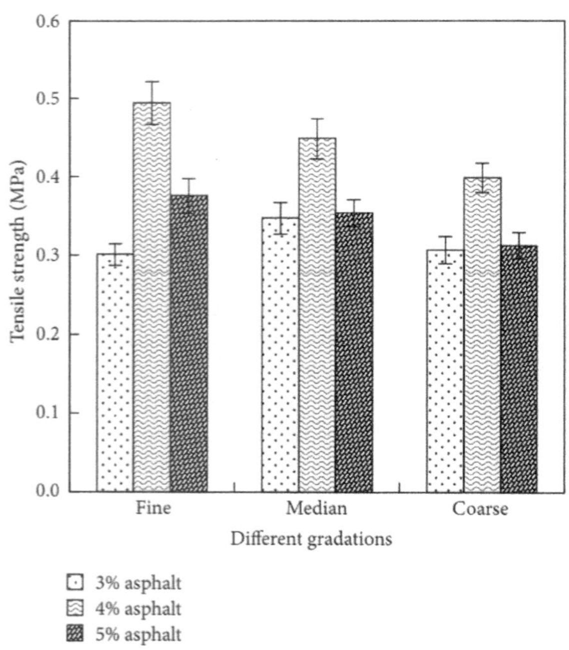

(b) Immersion splitting test

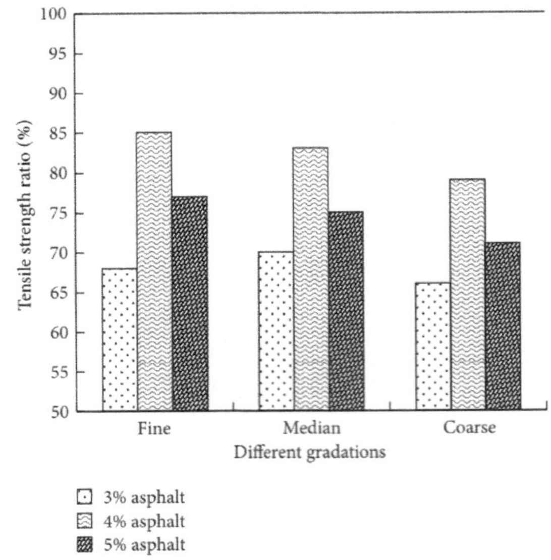

(c) Tensile strength ratio

Figure 6: Tensile strength of cold recycled asphalt mixtures with emulsified asphalt contents.

Influence of Emulsified Asphalt Type

Three types of emulsified asphalt, including high-viscosity slow-setting emulsified asphalt, normal slow-setting emulsified asphalt, and medium-setting emulsified asphalt, were compared in this study. The median gradation described in Figure 4 was used to prepare the cold recycled asphalt mixture. During the preparation of cold recycled asphalt mixtures, the contents of emulsified asphalt, cement, and water were 4%, 1.5%, and 5.4%, respectively. To simulate the increasing of tensile strength of cold recycled asphalt mixtures over time, samples were prepared and cured at room temperature for 2, 3, 4, and 7 days, respectively. The tensile strength by dry splitting test after different curing days is shown in Figure 7. To simulate the final tensile strength of cold recycled asphalt mixture, samples were prepared and cured at room temperature for 7 days and then at 60°C for another 48 h. The final tensile strength by dry and immersion splitting tests for cold recycled asphalt mixtures with different types of emulsified asphalt is shown in Figure 8.

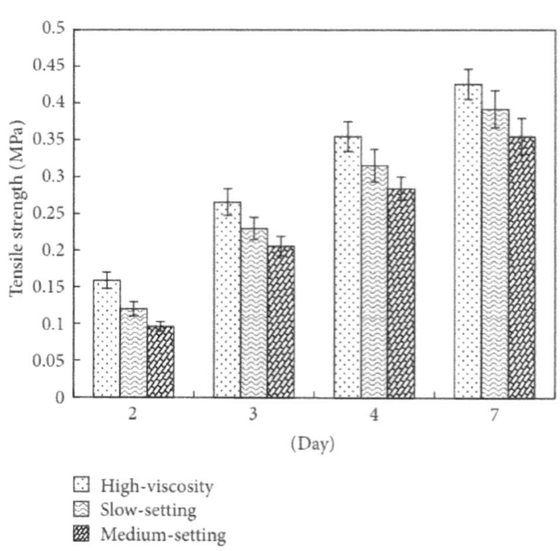

Figure 7: Tensile strength versus curing days for cold recycled asphalt mixtures with different types of emulsified asphalt.

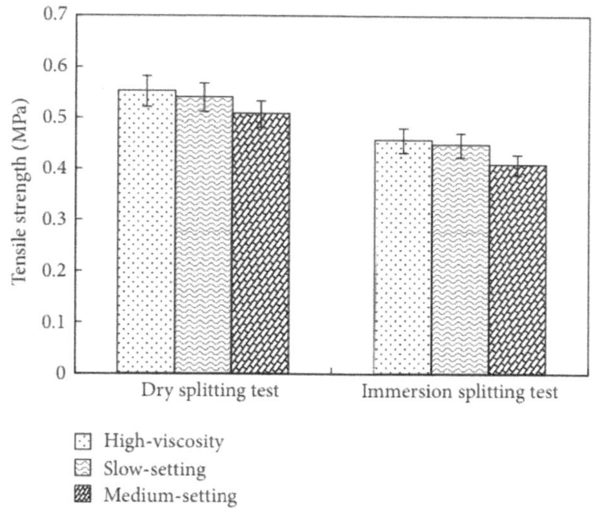

Figure 8: Final tensile strength for cold recycled asphalt mixtures with different types of emulsified asphalt.

The results in Figure 8 show that the final tensile strengths of cold recycled asphalt mixtures with high-viscosity slow-setting emulsified asphalt and normal slow-setting emulsified asphalt are similar to each other. But Figure7 shows that the tensile strengths of cold recycled asphalt mixture with the high-viscosity emulsified asphalt at different curing times are always greater than the tensile strength of cold recycled asphalt mixtures with low-viscosity emulsified asphalt. It means the viscosity of emulsified asphalt has no significant influence on the final tensile strength but does have influences on the early strength of cold recycled asphalt mixture which are important for traffic opening in the field. Meanwhile, the tensile strength of cold recycled asphalt mixtures with slow-setting emulsified asphalt is greater than that of cold recycled asphalt mixture with medium-setting emulsified asphalt. The amounts of water loss at different curing time for cold recycled asphalt mixtures with different types of emulsified asphalt are shown in Figure 9. It shows that the water loss of medium-setting emulsified asphalt is much greater than the other two types of slow-setting emulsified asphalt during the first three days. The greater water loss means the faster setting rate of emulsified asphalt, which can cause insufficient hydration of cement and thus the smaller strength of cold recycled asphalt mixture.

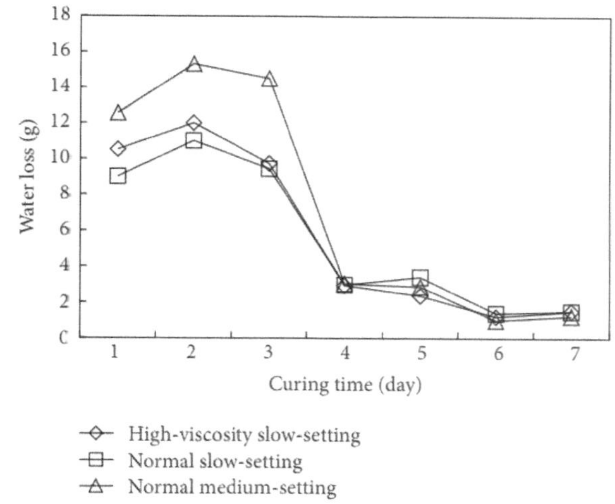

Figure 9: Water loss at different curing time.

Effect of Cement

SEM Analysis of Cement-Asphalt Mix

The microstructure of cement-asphalt mix and cement hydrates after elimination of asphalt was observed using scanning electron microscope (SEM), respectively, as shown in Figures 10 and 11. Typical cement hydration products including acicular and reticular hydrated calcium silicate, columnar ettringite, and flaky calcium hydroxide can be observed in the SEM images of cement-asphalt mix. It means that cement gets hydrated with water in the emulsified asphalt, which can lead to demulsification of the emulsified asphalt. The hydrated cement products interweave with asphalt to form the new binder in the cold recycled asphalt mixture. This explains the semiflexibility of the cold recycled asphalt mixture with cement hydration providing the rigidity and asphalt binder providing the flexibility [23].

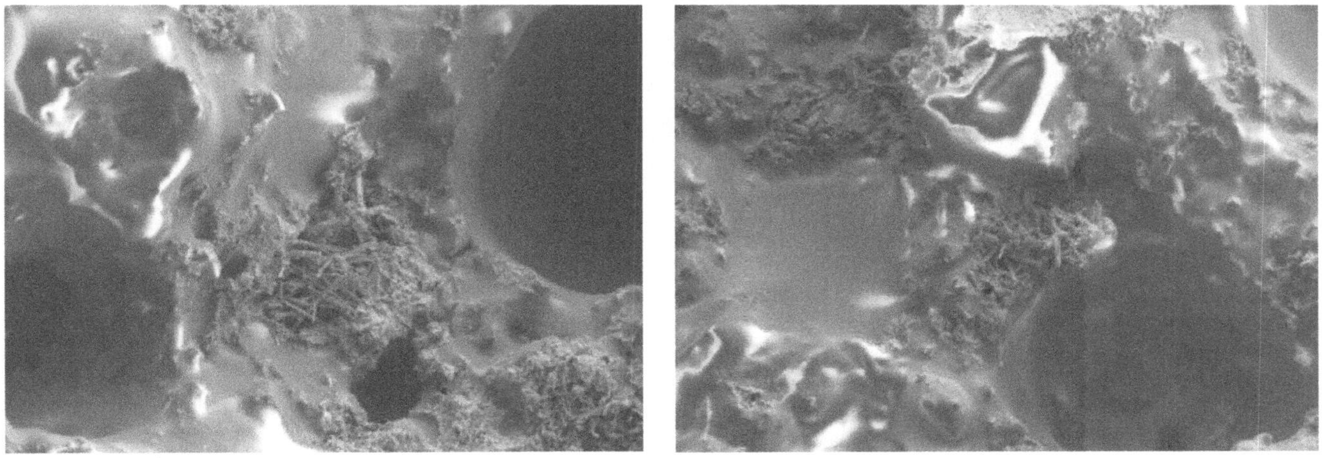

Figure 10: Microstructure of cement-asphalt mix.

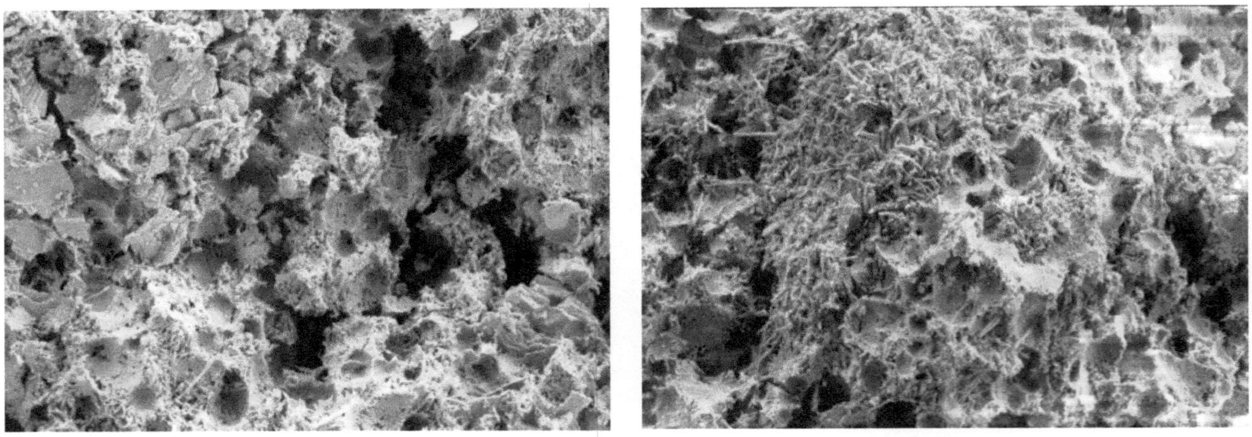

Figure 11: Microstructure of cement hydrates after elimination of asphalt.

However, it is observed that both asphalt and cement hydrates have disadvantageous influences on each other's microstructure integrity. As shown in Figure 10, micropores can be observed in the cement-asphalt mix that could result from water loss in the hydration reaction of cement and demulsification of emulsified asphalt. On the other hand, a number of micropores were found in the cement hydration after elimination of asphalt, as shown in Figure 11. The cement hydrates change from massive structure into powdery form by losing its integrity and strength. Therefore, the microflaw inside the cement-asphalt mix indicates weak bonding between asphalt and hydrated cement that can cause strength defects in the cold recycled asphalt mixture. This finding is consistent with the observation reported by previous researches on the strength of cement-asphalt mortar, which indicated that asphalt membrane could have negative effects on the hydration of cement and weaken the skeleton formed by cement hydrates, while cement hydrates could impale the asphalt membrane and weaken the integrity of asphalt membrane [24].

Influence of Cement Content

Different cement contents, 1%, 1.5%, 2%, 2.5%, and 3%, were used to prepare testing specimens for dry and immersion splitting test. The median gradation described in Figure 4 was used to prepare the cold recycled asphalt mixture. For the cold recycled asphalt mixtures with different cement contents of 1%, 1.5%, 2%, 2.5%, and 3%, the emulsified asphalt content was 4% and the water contents were 5.2%, 5.4%, 5.4%, 5.6%, and 5.7%, respectively. Both the dry and immersion tensile strength of cold recycled asphalt mixtures with different cement contents are shown in Figure 12(a) and the tensile strength ratios at different cement contents are shown in Figure 12(b). It is clearly shown that the tensile strength increases as cement content increases. However, the relationship between the tensile strength and cement content is nonlinear. It can be seen that the increasing rate decreases when the cement content exceeds 1% and the tensile strength barely changes when the cement content exceeds 2%. Compared to the tensile strength with 1.5% cement, 3% cement only improves the tensile strength by less than 10%. Similar relationship between tensile strength ratio and cement content was observed in Figure 12(b). It can be seen that after cement content exceeds 1.5%, the tensile strength ratio barely changes with the increasing of cement content.

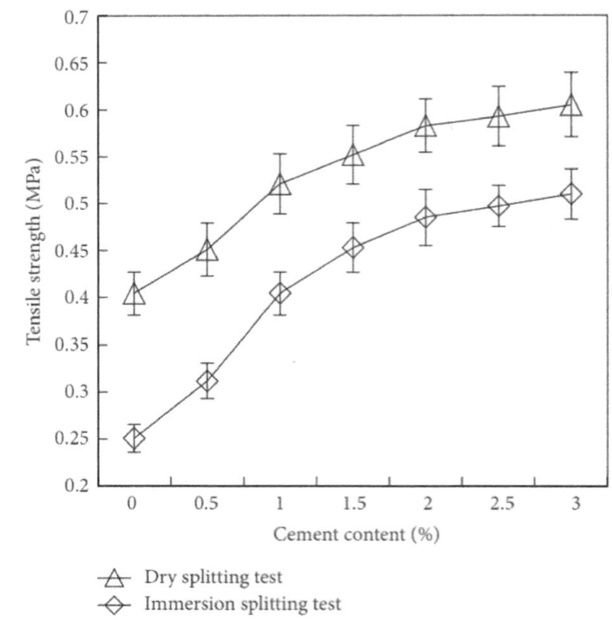

(a) Dry and immersion tensile strength

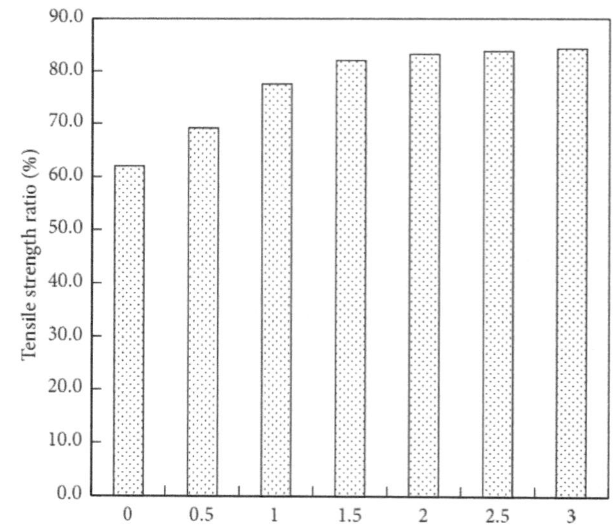

(b) Tensile strength ratio

Figure 12: Tensile strength versus cement content for cold recycled asphalt mixtures.

To further confirm the effect of cement without influence of asphalt, RAP and different contents of cement were used to prepare cement-RAP mixtures without emulsified asphalt. As comparison, new aggregate and different contents of cement were used to prepare cement-aggregate mixtures. The same gradation was used for both RAP and new aggregate. Splitting tests were conducted for both cement-RAP mixtures and cement-aggregate mixtures with cement content changing from 1% to 3% by the weight of RAP or aggregate. Figure 13summarizes the test results.

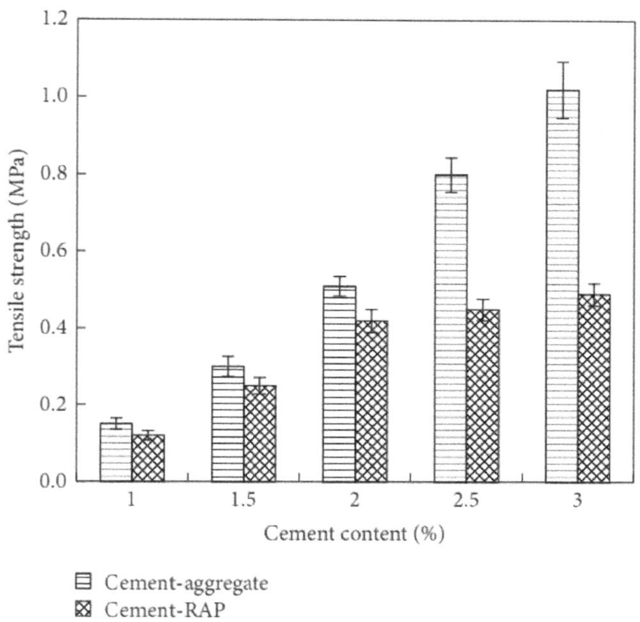

Figure 13: Tensile strength for cement-RAP mixture and cement-aggregate mixture.

As shown in Figure 13, although RAP and new aggregate have the same gradation, the relationships between the tensile strength and the cement content are much different for cement-RAP mixture and cement-aggregate mixture. It is clearly indicated that the tensile strength of the cement-aggregate mixture keeps growing linearly as the cement content increases. When the cement content is smaller than 2%, the tensile strength of the cement-RAP mixture grows linearly with the increasing of cement content. After the cement content exceeds 2%, the tensile strength of the cement-RAP mixture barely grows with the increase of cement content. The cement-RAP mixture has obviously smaller tensile strength compared to the cement-aggregate mixture with the same gradation and cement content. The tensile strength of the cement-RAP mixture with 3% cement is about 0.5 MPa which is only half of the tensile strength of the cement-aggregate mixture with 3% cement. Combined with previous analysis in Section 3.1, it proves that the strength of RAP has decisive influence on the tensile strength of mixtures and it can limit the effect of cement in cold recycled asphalt mixture.

CONCLUSIONS

Based on laboratory testing of cold recycled asphalt mixtures, the following conclusions can be made:

1. RAP cannot be simply treated as black aggregate in the cold recycled asphalt mixture. The specific surface area of RAP is much smaller than the aggregate in RAP that affects the bonding between RAP and new binder. Adequate crushing of RAP, addition of new aggregates, and fine gradation design can increase the tensile strength of cold recycled asphalt mixture by reducing agglomeration of RAP and improving the bonding between RAP and binder.
2. The optimum emulsified asphalt content can be determined to achieve the highest tensile strength of cold recycled asphalt mixtures. It is suggested that slow-setting emulsified asphalt can provide enough time for cement to hydrate which is beneficial for the final strength of the cold recycled asphalt mixture. The viscosity of emulsified asphalt has no significant influences on the final tensile strength but it does affect the early strength of the cold recycled asphalt mixture. The higher viscosity of emulsified asphalt can provide better early strength of the cold recycled asphalt mixture that is helpful to carry traffic loading during initial period of traffic opening.
3. Cement can improve the tensile strength of cold recycled asphalt mixtures by providing cement hydrates and promote demulsification of emulsified asphalt. However, microstructure flaws were observed in the mix of asphalt and cement hydration products, which could reduce the bonding between asphalt and cement and the strength of cement hydrates. Meanwhile, the influences of cement and its content on the strength of cold recycled asphalt mixture are not as significant as that on the strength of cement concrete. The effect of cement is limited by the structure and strength of RAP.
4. It is found that factors related to emulsified asphalt, cement, and gradation design can affect the tensile strength of cold recycled asphalt mixture. But the inner structure and tensile strength of RAP have decisive influence on the tensile strength of cold recycled asphalt mixture.

ACKNOWLEDGMENTS

The study is financially supported by the National Natural Science Foundation of China (no. 51378006), National Science and Technology Support Program (2014BAG05B04), Huoyingdong Foundation of the Ministry of Education of China (no. 141076), Excellent Young Teacher Program of Southeast University (2242015R30027), and State Key Laboratory of High Performance Civil Engineering Materials.

REFERENCES

1. F. P. Xiao, S. N. Amirkhanian, B. J. Putman, and H. Juang, "Feasibility of Superpave gyratory compaction of rubberized asphalt concrete mixtures containing reclaimed asphalt pavement," Construction and Building Materials, vol. 27, no. 1, pp. 432–438, 2012.
2. F. Hong, D.-H. Chen, and M. M. Mikhail, "Long-term performance evaluation of recycled asphalt pavement results from Texas: pavement studies category 5 sections from the long-term pavement performance program," Transportation Research Record, vol. 2180, pp. 58–66, 2010.

3. H. Wen, J. Lu, and T. M. VanReken, "Modeling the effects on energy and carbon dioxide from the use of recycled asphalt pavement in hot mix asphalt," International Journal of Sustainable Transportation, vol. 8, no. 4, pp. 249–261, 2014.
4. X. Yu and Y. Li, "Optimal percentage of reclaimed asphalt pavement in central plant hot recycling mixture," Journal of Wuhan University of Technology-Mater. Sci. Ed., vol. 25, no. 4, pp. 659–662, 2010.
5. C. Plati, A. Loizos, V. Papavasiliou, and A. Kaltsounis, "Investigating in situ properties of recycled asphalt pavement with foamed asphalt as base stabilizer," Advances in Civil Engineering, vol. 2010, Article ID 565924, 10 pages, 2010.
6. G. Thenoux, Á. González, and R. Dowling, "Energy consumption comparison for different asphalt pavements rehabilitation techniques used in Chile," Resources, Conservation and Recycling, vol. 49, no. 4, pp. 325–339, 2007.
7. Wirtgen Cold Recycling Manual, Wirtgen GmbH, Windhagen, Germany, 2010.
8. J. H. Yan, H. R. Zhu, Z. X. Zhang, L. Gao, and S. Charmot, "The theoretical analysis of the RAP aged asphalt influence on the performance of asphalt emulsion cold recycled mixes," Construction and Building Materials, vol. 71, pp. 444–450, 2014.
9. Y. Kim and H. D. Lee, "Performance evaluation of Cold In-Place Recycling mixtures using emulsified asphalt based on dynamic modulus, flow number, flow time, and raveling loss," KSCE Journal of Civil Engineering, vol. 16, no. 4, pp. 586–593, 2012.
10. S. Charmot and P. Romero, "Assessment of fracture parameters to predict field cracking performance of cold in-place recycling mixtures," Transportation Research Record, vol. 2155, pp. 34–42, 2010.
11. F. Long and H. Theyse, "Mechanistic-empirical structural design models for foamed and emulsified bitumen treated materials," in Proceedings of the 8th Conference on Asphalt Pavements for Southern Africa (CAPSA ‹04), Sun City, South Africa, September 2004.
12. T. S. Ozsahin and S. Oruc, "Neural network model for resilient modulus of emulsified asphalt mixtures,"Construction and Building Materials, vol. 22, no. 7, pp. 1436–1445, 2008.
13. J. H. Yan and Z. X. Zhang, "The correlativity analysis between the aggregates characteristics of reclaimed asphalt pavement and the performance of emulsion cold recycling mixture," in Proceedings of the International Symposium on Heavy Duty Asphalt Pavements and Bridge Deck Pavements, Nanjing, China, May 2012.
14. R. Issa, M. M. Zaman, G. A. Miller, and L. J. Senkowski, "Characteristics of cold processed asphalt millings and cement-emulsion mix," Transportation Research Record, vol. 1767, pp. 1–6, 2001.
15. B. Gómez-Meijide and I. Pérez, "A proposed methodology for the global study of the mechanical properties of cold asphalt mixtures," Materials and Design, vol. 57, pp. 520–527, 2014. ·
16. M. A. T. Alsheyab and T. S. Khedaywi, "Effect of electric arc furnace dust (EAFD) on properties of asphalt cement mixture," Resources, Conservation and Recycling, vol. 70, pp. 38–43, 2013.
17. A. Modarres, M. Rahimzadeh, and M. Zarrabi, "Field investigation of pavement rehabilitation utilizing cold in-place recycling," Resources, Conservation and Recycling, vol. 83, pp. 112–120, 2014.
18. A. Forsberg, E. Lukanen, and T. Thomas, "Engineered cold in-place recycling project: Blue Earth County State Aid Highway 20, Minnesota," Transportation Research Record, vol. 1813, pp. 111–123, 2002.
19. Y. Kim, S. Im, and H. D. Lee, "Impacts of curing time and moisture content on engineering properties of cold in-place recycling mixtures using foamed or emulsified asphalt," Journal of Materials in Civil Engineering, vol. 23, no. 5, pp. 542–553, 2011.
20. E. S. Peter, B. Gabriel, H. Edgard, D. Weitzel, and S. Bemanian, "Performance of cold in-place recycling in Nevada," Transport Research Record, vol. 1896, pp. 162–169, 2007.
21. Ministry of Transport of the People›s Republic of China, Technical Specification for Construction of Highway Asphalt Pavement, 2006.
22. Ministry of Transport of the People›s Republic of China, Technical Specification for Highway Asphalt Pavement Recycling, 2008.
23. L. He, Research on constructing semi-flexible pavement by cold in-place recycling of asphalt pavement [Ph.D. thesis], Chongqing Jiaotong University, Nan›an, China, 2008.
24. W. Qiang, Y. Peiyu, A. Ruhan, Y. Jinbo, and K. Xiangming, "Strength mechanism of cement-asphalt mortar," Journal of Materials in Civil Engineering, vol. 23, no. 9, pp. 1353–1359, 2011.

CHAPTER 12

INVESTIGATION ON USING SBS AND ACTIVE CARBON FILLER TO REDUCE THE VOC EMISSION FROM BITUMINOUS MATERIALS

Peiqiang Cui, Shaopeng Wu, Fuzhou Li, Yue Xiao and Honghua Zhang

Keywords: Volatile organic compounds (VOC), bituminous materials, ultraviolet-visible spectroscopy test, thermogravimetric analysis-mass spectrometry

INTRODUCTION

During the construction process and service period, bituminous materials will release volatile organic compounds (VOC), which are harmful for the environment and contractors [1]. Such VOC emission would cause environmental pollutions and toxicity to humans and aquatic organisms [2,3]. An example cross-sectional study conducted by Gamble [4] and Karakaya [5] proved that workers exposed weekly to bitumen fume had more symptoms, such as eye irritation and abnormal fatigue.

VOC emission is not only harmful to pavement workers and the environment, but it also degrades the material behavior of bituminous materials. Therefore, innovative technologies of detection and analysis technology for VOC and the development of VOC inhibitors to reduce the emission speed and amount are becoming more and more important to achieve green asphalt pavement. Nowadays, a number of construction guidelines [6,7] are requiring bituminous materials with a lower VOC content. This is a common requirement, especially in regards to construction for environmentally-sensitive areas (e.g., living area, schools and wetlands). Gasthauer [8,9] developed a process to identify the VOC from bituminous materials with gas chromatography coupled with mass spectrometry. It was reported that the VOC emission depended on various parameters, such as the asphalt temperature, the asphalt oxidation, the humidity of air, and so on.

Styrene butadiene styrene (SBS) has very good compatibility with bitumen and was added into bituminous materials to improve the material behaviors for pavement construction and roofing purposes [10,11]. The SBS molecules dissolved in bituminous materials will swell and absorb the soft components of bitumen. This swell and absorption process not only makes it an excellent property modifier [12], but also it has a potential function to make it difficult for the smaller molecules inside of the bituminous materials to escape from the surface of the bitumen. This means that SBS might be used as an inhibitor to prevent the release of VOC. Activated carbon has extremely high surface energy and has a porous structure that offers a large specific surface. It is commonly used for removing organic constituents and residual disinfectants in water supplies. It is a favored water treatment technique because of its multifunctional nature. Research has also proven that active carbon filler can be added into bituminous materials to improve the material behaviors [13,14]. Therefore, active carbon filler also might have a strong capacity of adsorption to achieve the purpose of preventing VOC from releasing.

The aim of this work was to understand the possibility of using SBS and active carbon filler to reduce the VOC emission from bituminous materials, as well as the material behaviors of such VOC-inhibited bituminous materials. The VOC emission speed and amount from bituminous materials were studied by means of thermal gravimetric analysis-mass

Citation: Cui, P.; Wu, S.; Li, F.; Xiao, Y.; Zhang, H. Investigation on Using SBS and Active Carbon Filler to Reduce the VOC Emission from Bituminous Materials. Materials 2014, 7, 6130-6143. doi:10.3390/ma7096130.

spectrometer testing and ultraviolet-visible spectroscopy testing, respectively. The rheological properties of bituminous materials modified by VOC-inhibitors, such as SBS and active carbon filler, were characterized with temperature sweep and frequency sweep analyses.

MATERIALS AND TEST METHODS

Raw Materials

Two kinds of bituminous materials, PJ-90 bituminous materials and modified bituminous materials, were used. A combination of 4 wt% of SBS and 4 wt% of active carbon filler were added into the bituminous materials to investigate the influence on the materials' properties and VOC emission characteristics by adding such inhibitors. The optimum added contents used in this research were concluded from the investigations on the rheology property, ageing resistance, crack resistance, *etc.* Table 1 presents the basic characteristics of the active carbon filler, and Table 2 concludes on the fundamental properties of the studied materials.

Table 1. Basic characteristics of active carbon

Items	Values
PH	5.0–7.0
Surface area	1100 m^2/g
Residual after dried	90%
Dissolved by hydrochloric acid	Less than 0.8%
Dissolved by ethanol	Less than 0.2%

Table 2. The general performance of the evaluated materials

Property	PJ90	PJ90 (4% SBS + 4% C)
Softening point (°C)	45	70
Ductility (15 °C) (cm)	>120	>120
Penetration (25 °C) (0.1 mm)	75.6	56.7

Principle of TG-MS Test

Thermogravimetry analysis studies the mass changes of a sample while it is subjected to a controlled temperature program [14]. However, it is only a quantitative thermal technique. It gives no direct chemical information. Mass spectrometry is an analytical technique that produces the spectrum of the masses of the molecules [15,16]. They might be combined together to characterize the additional chemical information. The spectrum can be used to illustrate the masses of molecules and to distinguish the structure and chemical properties of different molecules.

Thermal gravimetric analysis-mass spectrometry (TG-MS) is an analytical method that combines the features of thermogravimetry and mass spectrometer to identify different volatile products during a weight lost process with one test sample. Applications of TG-MS include weight detection, temperature investigation and identification of volatile emissions. TG-MS is therefore employed to analyze the bitumen VOC emissions, both qualitatively and quantitatively, in this research.Figure 1 introduces the principal of the TG-MS test and analysis program, and Figure 2 presents the TG-MS instrument.

When the VOC emissions go through a mass spectrometer, it then separates the ions according to their specific mass-to-charge ratio and records the relative amount of each ion type. Thus, a mass spectrum of the molecule can be produced. AsFigure 1 shows, the collected VOC source ions will be resolving into their characteristics mass components according to their mass-to-charge ratio at the analyzer step. Then, the ion detector will detect the ions and record the relative amount of each of the resolved ionic species.

The STA 449 F3 Jupiter (TG) and QMS403C quadrupole mass spectrometer (MS), produced by NETZSCH (Altenstadt, Germany), were employed to analyze the chemical components during VOC emission from bituminous materials. The specimen was heated up to 300 °C in the TG instrument with the temperature increasing at a rate of 10 °C/min. In the meantime, MS was conducted to characterize the molecular weight and volatile speed of VOC emissions.

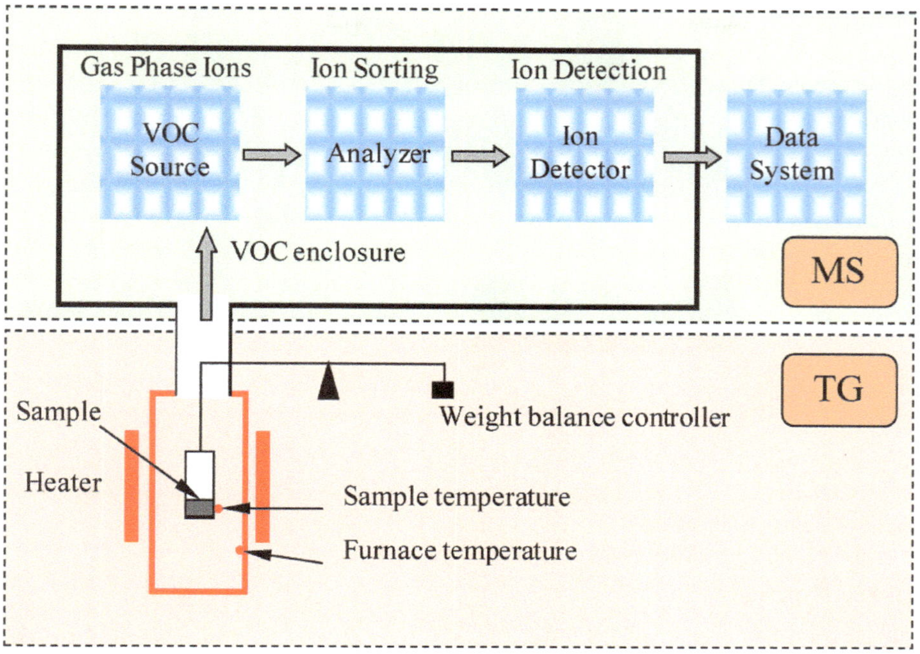

Figure 1. Principal of thermal gravimetric analysis-mass spectrometry (TG-MS) test and analysis program.

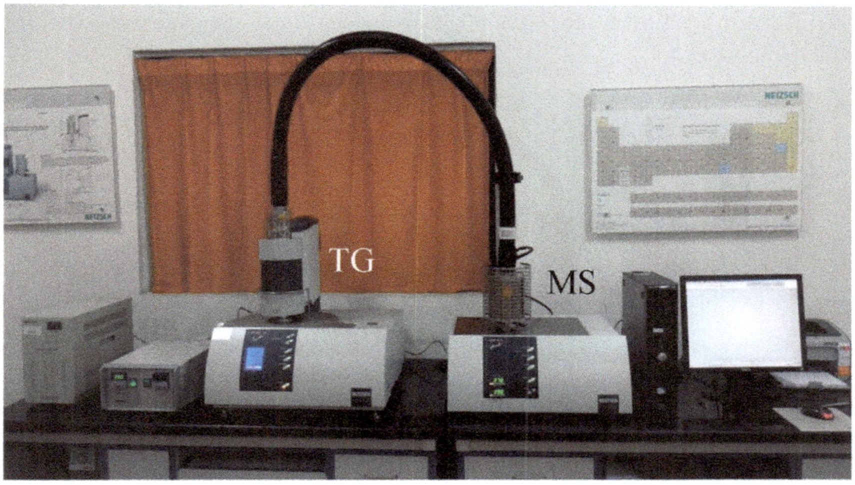

Figure 2. The TG-MS instrument used for characterizing volatile organic compound (VOC) emissions.

Ultraviolet-Visible Spectroscopy Test

Ultraviolet-visible spectrophotometry is a method to formulate the absorption spectroscopy in the ultraviolet-visible spectral region. It is widely employed to conduct the identification and measurement of organic and inorganic compounds in a wide range of products and processes. The Beer–Lambert law states that the absorbance of a solution is directly proportional to the concentration of the absorbing species in the solution and the path length. Thus, for a fixed path length, UV-Vis spectroscopy can be used to determine the concentration of the absorber in a solution, when the calibration curve has been determined to express the relation between absorbance and concentration.

The calibration curve can be determined by detecting the absorbance of incident monochromatic light at different wavelength when passing through solutions with a fixed concentration of the absorbing specimens. Based on our previous research [17], ultraviolet light with a wavelength of 288 nm was used to characterize the VOC emissions. The calibration curve was first defined and then used to determine the emission amount from bituminous materials.

A diagrammatic sketch of the VOC generating and collecting device for UV-Vis analysis is shown in Figure 3. The air pump (Airchek 2000 sampler, SKC Inc., Eighty Four, PA, USA) was connected at the end of the experimental system to create negative air pressure, leading to air flow from the specimen container to the solution container. The glass fiber filter membrane is first to filter the heavy components from the bituminous materials before the air goes through the cyclohexane, which was used as the solution to collect VOC emissions.

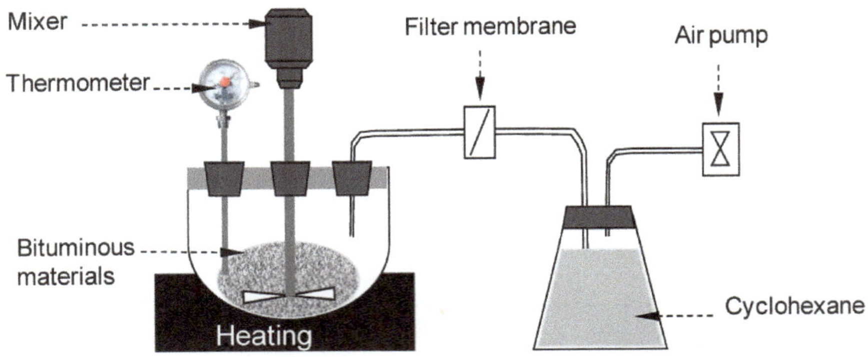

Figure 3. VOC generating and collecting device for the ultraviolet-visible spectroscopy test (UV-Vis) analysis.

The bituminous material was heated to 180 °C to accelerate the VOC emission speed. The sampling flow was 500 mL/min, and the sampling time was 15 min. A UV-1601 visible spectrometer, produced by Japanese Shimadzu, was used in this research to investigate the amount of VOC components.

Rheological Property Analysis

A dynamic shear rheometer (DSR), produced by Anton Paar Austria (Graz, Austria), was used to evaluate the rheological properties at varied temperatures and frequencies. Temperature sweep tests within a wide range of frequencies and frequency sweep tests at different temperatures were conducted separately. The sinusoidal signal as Figure 4 shows was applied on the bituminous specimen. A spindle with an 8-mm diameter was used for lower temperature ranges from 5 °C to 30 °C, while a 25-mm spindle was used for higher temperature ranges from 30 °C to 80 °C. The phase angle and complex shear modulus response of tested specimen were then analyzed.

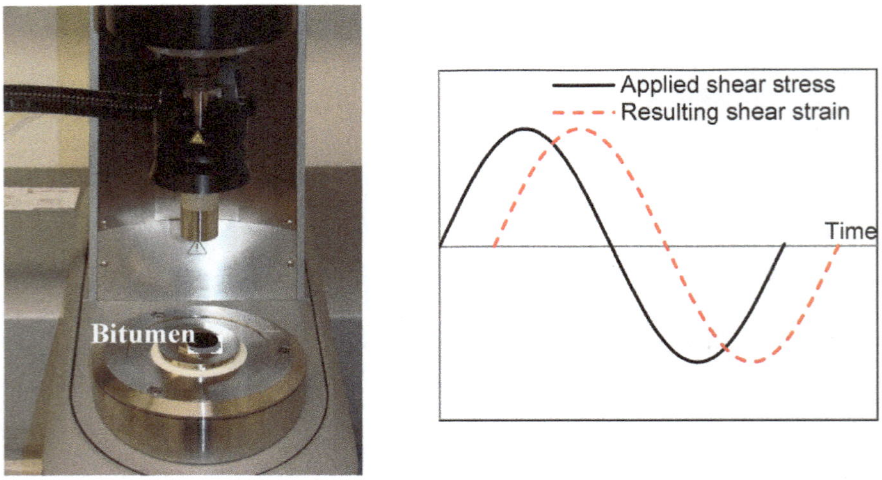

Figure 4. The dynamic shear rheometer (DSR) instrument and the applied sinusoidal signal.

RESULTS AND DISCUSSIONS

Residual Volatile Speed

Residual volatile speed was first introduced to characterize the changing of VOC emission before and after adding inhibitors into bitumen. The residual volatile speed was defined as the ratio between the VOC volatile speed of bituminous material modified with inhibitors and the VOC volatile speed of non-modified bituminous material:

$$G^* = \frac{E_1}{E_2} \times 100(\%) \tag{1}$$

where G^* is the residual volatile speed, E_1 is the VOC volatile speed of bituminous material modified with inhibitors, whileE_2 is the VOC volatile speed of bituminous material without modified with inhibitors.

When $G^* > 1$, the added VOC inhibitor is unsteady and cannot inhibit the VOC emission at the evaluated temperature. Instead of decreasing the volatile speed, the speed would be increased by using such an inhibitor. When $G^* < 1$, the contribution of the added inhibitor was positive, and the VOC volatile speed can be decreased to a certain level.

Normal octane and dichloroethane, which can cause damage to the nervous system, liver, kidneys and lungs and may cause cancer, are reported as the two main volatile components from bituminous materials during their heating and service period. Naphthalene is concerned as the most volatile PAH (Polycyclic aromatic hydrocarbon) from bituminous materials [18]. It is also well known that the naphthalene compounds and phenanthrene compounds are the main components of VOC, which are hazardous to human health and the environment. Therefore, these four types of gas in the VOC emissions from bituminous materials, including dichloroethane (molecular weight 98, $C_2H_4Cl_2$), normal octane (molecular weight 114, C_8H_{18}), naphthalene (molecular weight 128, $C_{10}H_8$) and methylphenanthrene (molecular weight 191, $C_{15}H_{12}$), were collected and analyzed. Table 3 summarizes the fundamental information for these four analyzed VOC gas.

Table 3. Information for the four analyzed VOCs

Volatile Type	Molecular Type	Main Characteristics
Naphthalene	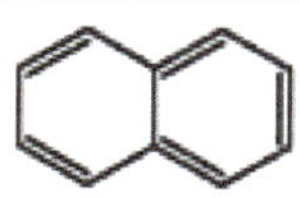	The most volatile components from bituminous materials, may cause cancer
Dichloroethane	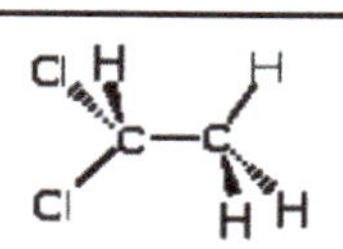	The two main volatile components from bituminous materials, toxicity to humans and aquatic organisms
Normal octane		
Methylphenanthrene	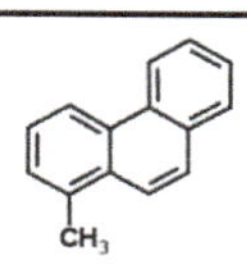	Main components of VOC, toxicity to humans and aquatic organisms

Figure 5 illustrates the VOC volatile speed of non-modified bituminous material. The emission of naphthalene and dichloroethane mainly happened at the beginning stage of the test, although at the beginning, the conditional temperature was lower.

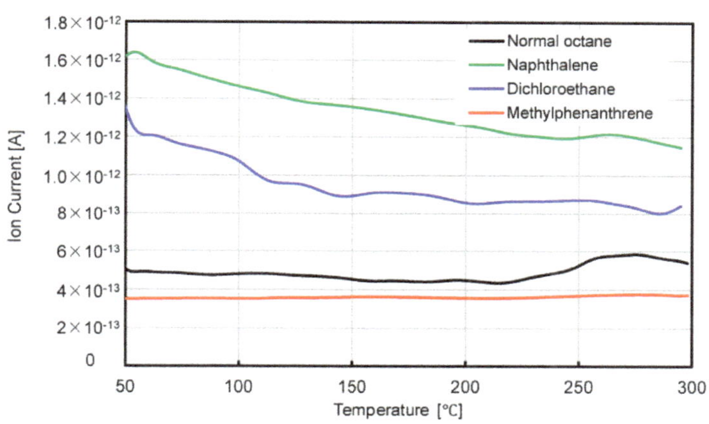

Figure 5. VOC volatile speed of non-modified bituminous material.

Figure 6 compares the residual volatile speed of the four analyzed VOC gases. It is clear that the combined introduction of SBS and active carbon filler into bituminous materials can significantly decrease the VOC emission speed, while the decreasing influence is different on different types of gas. At the very beginning stage of the test, the volatile speed of naphthalene can be decreased by more than 70% when the SBS and active carbon filler were added. The emission speed of dichloroethane decreased by more than 60% and about 20% for normal octane. With the increasing test temperatures, the inhibitory effect changed. With naphthalene and dichloroethane, the inhibitory effect decreased with the increasing of temperature; while with normal octane and methylphenanthrene, the influence of temperature on the inhibitory effect is not as significant as that compared to the other two types of VOC gas.

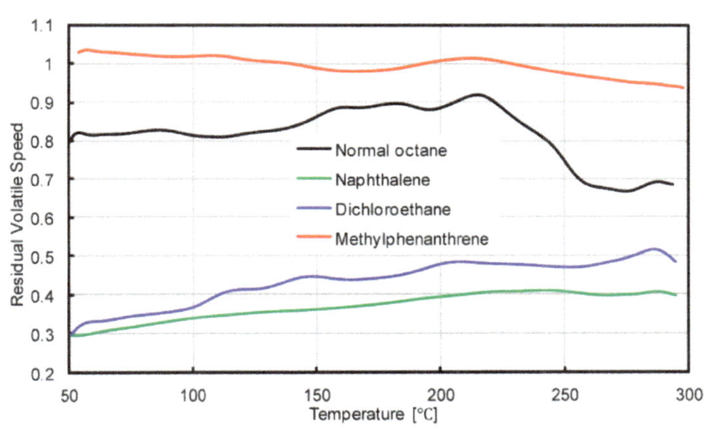

Figure 6. Residual volatile speeds of four types of gas.

Ultraviolet-Visible Spectroscopy Test (UV-Vis) Test Analysis

The evaluation of VOC amount was based on the linear relationship between the VOC quality of the absorption solution and its absorbance value. Therefore, the relationship between the absorbance of the standard bituminous specimen and the amount of VOC emission was first finalized.

Calibration Curve

A certain amount of cyclohexane solution was used to absorb different known amounts of VOC, resulting in solutions with a variety of VOC concentrations. These treated solutions were than tested with UV-Vis. At the end, the relationship between VOC concentration and absorbance can be concluded, as shown in Table 4. Figure 7 presents the calibration curve.

Table 4. Relationship between the absorbance and concentration of VOC emission

No.	Absorbance (Y)	Concentration of VOC (μg/mL, X)
1	0	0
2	0.0686	9.4949
3	0.1573	28.4848
4	0.3897	47.4747
5	0.4813	66.4646
6	0.6254	75.9596
7	0.7192	94.9495
8	0.8242	123.4343
9	0.9594	142.4242
Mathematical Solver		$Y = 0.005X + 0.0078$ ($Y = aX + b$, $R^2 = 0.998$)

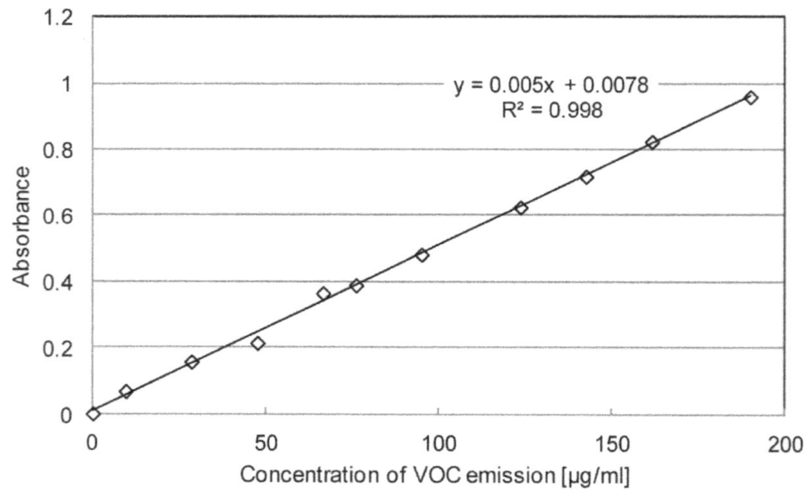

Figure 7. Mathematical solvers for the calibration curve.

Figure 7 illustrates that the relation between absorbance and the amount of VOC emission shows a good linear correlation. Such a linear correlation can be expressed with the function presented in Table 4. Two constant parameters, "a = 0.005" and "b = 0.0078", were used to define the calibration curve.

UV-Vis Data

Based on the calibration curve, the UV-Vis test was conducted with PJ90 and modified bituminous material. The quantity of VOC emissions from every gram of bituminous material can be calculated according to this equation (μg/g):

$$w = \frac{\frac{Y-b}{a} \times V}{M} \tag{2}$$

where M is the weight of tested bituminous material (g), a and b are the constant parameters of the calibration curve, Y is the absorbance value of the test sample and V is the volume of cyclohexane that was used to dissolve VOC (mL).

In this research, this volume was fixed at 18 mL, while only 3.5 mL of cyclohexane with dissolved VOC were used for UV-Vis evaluation. The test results were presented in Table 5. The VOC was collected within 15 min when the bituminous material was heated at 180 °C.

It is obvious from Table 5 that under the same temperature condition, the amount of VOC emission from PJ90 bituminous material was much higher than that from bituminous material modified with inhibitors. The analysis shows that 7.886 μg/g of VOC emissions volatilized from PJ90, while only 3.63 μg/g of VOC emissions were found from the modified bituminous material. The combination of using SBS and active carbon filler as inhibitors can prevent more than 50% of VOC molecules from volatilization.

It is believed that the SBS modifier with a physical cross-linking structure can prevent the small molecules in the bituminous materials from escaping to the surface. In the meantime, an ultra-fine granular additive, such as activated carbon, has a very remarkable higher surface energy than traditional fillers. Such characteristics could absorb small molecules, resulting in lower VOC emission.

Table 5. Quantity values of VOC emission from bituminous materials

Parameter	PJ-90	PJ90 (4% SBS + 4% C)
M (g)	935.48	1190.94
Y	2.0569	1.2087
V (mL)	18	18
Concentration of VOC in cyclohexane (μg/mL)	409.82	240.18
Total amount of VOC (μg)	7376.76	4323.24
W (μg/g)	7.886	3.63
VOC decrement percentage (%)	53.96	

Rheological Behaviors

The TG-MS test results strongly proved that adding functional modifiers into bituminous materials can prevent the VOC from emission. Besides this remarkable achievement of inhibiting VOC, if these modifiers can maintain the material behavior of bituminous materials for pavement purposes in the meantime, it would be very helpful to design environmentally-friendly bituminous materials.

Temperature Sweep

Figure 8 and Figure 9 compare the differences of phase angle and the shear complex modulus between bituminous materials with and without additives of the VOC inhibitor, at a loading frequency of 10 rad/s.

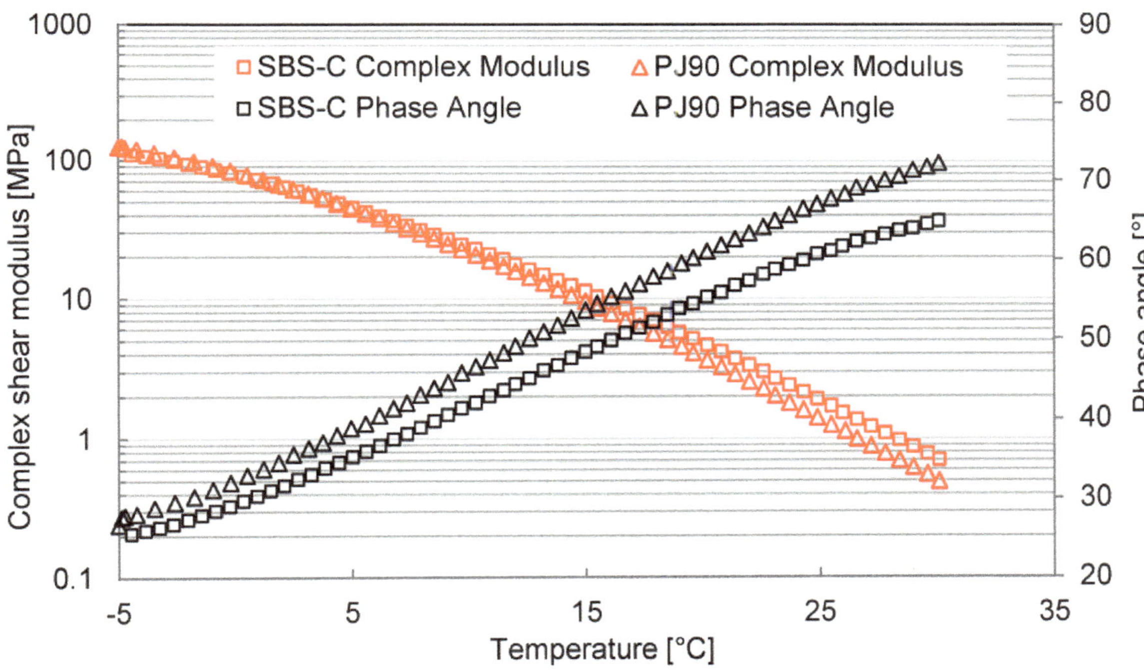

Figure 8. Phase angle and modulus curves at the lower temperature range.

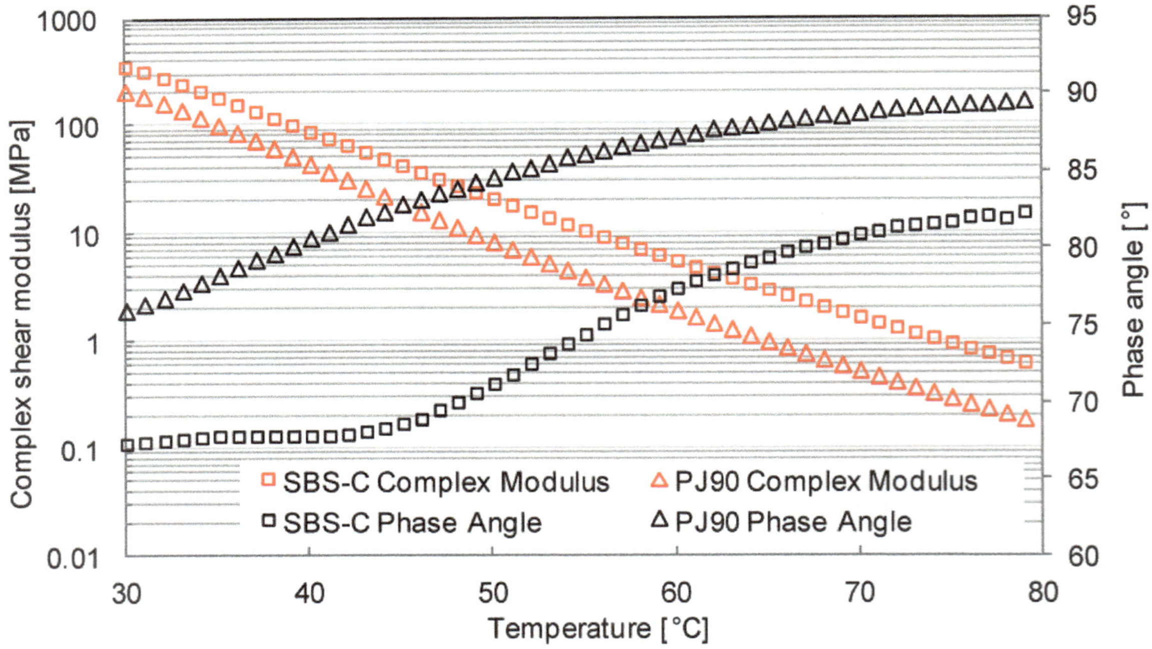

Figure 9. Phase angle and modulus curves at the higher temperature range.

With the increasing of the test temperature, the phase angle of both bituminous materials increased, while the complex shear modulus decreased. This is because the bituminous material is a viscoelastic material. It has a viscous domain behavior at higher temperature, while at a lower temperature, it has an elastic domain behavior. The amount of viscous components in the bituminous material was increasing, resulted from the increase of temperature. The state of bituminous materials will transform from an elastic state to a viscous state when the temperature is increased. This process will then accelerate the movement of small molecules.

The additives of SBS and active carbon filler will decrease the values of the phase angle, while increasing the values of the complex shear modulus. This indicates that the added inhibitors can enhance the material behavior contributed by the elastic component, as well as improve the high-temperature resistance. Furthermore, both the slopes of the modulus curve and the phase angle curve for modified bituminous materials are smaller than those values for the PJ90 bituminous material. Such a slope represents the temperature susceptibility of the tested materials. Therefore, it can be concluded from Figure 8and Figure 9 that the modified bituminous material is not as sensitive to temperature as the PJ90 material is. The combination of adding 4% SBS and 4% active carbon filler into bituminous materials could decrease the temperature susceptibility.

Under a lower temperature condition, the differences of the phase angle or the complex shear modulus of two kinds of bituminous materials were smaller than the differences at higher temperature conditions. This proved that the influence of SBS and active carbon filler on the bituminous behavior is more significant at higher temperature ranges than at lower temperature ranges. The VOC emission speed increases with the increase of the conditioning temperature. Such a phenomenon will make the significant influence of SBS and active carbon filler at a higher temperature range greater.

Frequency Sweep

Frequency sweep tests were performed by means of an applied sinusoidal strain signal on bituminous materials at fixed testing temperatures under a wide frequency range from 0.1 rad/s to 400 rad/s. The fixed test temperatures were −10, 0, 10, 20, 30, 40, 50 and 60 °C. Master curves of the complex shear modulus and the phase angle of the two evaluated bituminous materials were constructed at the reference temperature of 20 °C, with the data from the frequency sweep test.

The master curves can be used to analysis the material behavior at a wide range of loading frequencies. Based on the time-temperature superposition principle [19], the material behavior at a lower frequency range can be theoretically transferred to represent the material behavior at a higher temperature condition, and *vice versa*. Figure 10 proved that the introduction of SBS and active carbon filler can slightly increase the modulus values, but significantly decrease the values of the phase angle. Furthermore, such influences resulting from SBS and active carbon filler are more obvious at a lower frequency range than at a higher frequency range. At a frequency higher than 200 rad/s, the influence on the modulus is negligible, while the differences on the phase angle are also very small.

Research results from both the temperature sweep test and frequency sweep test indicated that the combination of adding 4% SBS and 4% active carbon filler into bituminous materials has a significant contribution to improving the deformation resistance behavior at higher temperatures.

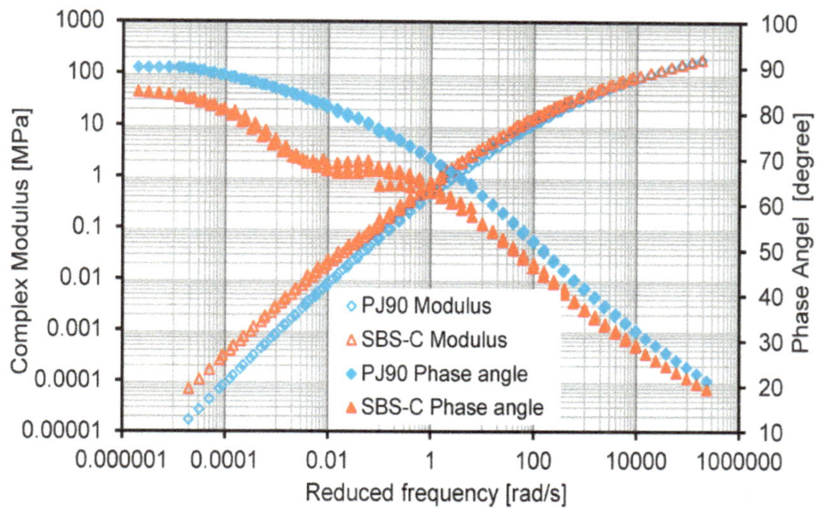

Figure 10. Master curves of two kinds of bituminous materials.

CONCLUSIONS

The VOC emission and material behaviors were analyzed and compared in this research to investigate the possibility of adding SBS and active carbon filler into bituminous materials to develop environmentally-friendly materials. Thermal gravimetric analysis-mass spectrometry and ultraviolet-visible spectroscopy testing were employed to characterize the VOC emission process. The temperature sweep test and frequency sweep test were conducted to evaluate the rheological properties of bituminous materials with and without SBS and active carbon filler as VOC inhibitors. Based on the research results discussed, the following conclusions can be finalized:

1. TG-MS, a combination of thermogravimetry and mass spectrometry can be used to identify different VOC components from bituminous materials during a weight lost process.
2. The combined introduction of SBS and active carbon filler into bituminous materials can significantly decrease the VOC emission speed, while the decreasing influence is different on different types of gas. Such an inhibitory effect resulting from SBS and active carbon filler is temperature related.
3. Under the same temperature condition, the amount of VOC emission from PJ90 bituminous material was much higher than that from bituminous material modified with inhibitors. The combination of using SBS and active carbon filler as inhibitors can prevent more than 50% of VOC molecules from volatilization.
4. The combined introduction of SBS and active carbon filler into bituminous materials can not only lower the VOC emission speed and amount, but also improve the deformation resistance behavior at a higher temperature, which was proven by means of a temperature sweep test and a frequency sweep test.

ACKNOWLEDGMENTS

This work is supported by the Natural Science Foundation of China (No. 51278392) and the National Key Scientific Apparatus Development Program from the Ministry of Science and Technology of China (No. 2013YQ160501). The authors gratefully acknowledge their financial support.

AUTHOR CONTRIBUTIONS

Peiqiang Cui contributed with the TG-MS and UV-Vis tests and wrote the corresponding part of the manuscript. Shaopeng Wu, Yue Xiao and Honghua Zhang took charge of the rheological behavior analysis and wrote the corresponding part of the manuscript. Fuzhou Li controlled the whole project, participated in the discussion of the results and wrote and revised the manuscript.

REFERENCES

1. Hicks, J.B. Asphalt industry cross-sectional exposure assessment study. *Appl. Occup. Environ. Hyg.* **1995**, *10*, 840–848.
2. Boffetta, P.; Burstyn, I. Studies of carcinogenicity of bitumen fume in humans. *Am. J. Ind. Med.* **2003**, *43*, 1–2.
3. Kitto, A.M.; Pirbazari, M.; Badriyha, B.N.; Ravindran, V.; Tyner, R.; Synolakis, C.E. Emissions of volatile and semi-volatile organic compounds and particulate matter from hot asphalts. *Environ. Technol.* **1997**, *18*, 121–138.
4. Gamble, J.F.; Nicolich, M.J.; Barone, N.J.; Vincent, W.J. Exposure-response of asphalt fumes with changes in pulmonary function and symptoms. *Scand. J. Work Environ. Health* **1999**, *25*, 186–206.
5. Karakaya, A.; Yücesoy, B.; Turhan, A.; Erdem, O.; Burgaz, S.; Karakaya, A.E. Investigation of some immunological functions in a group of asphalt workers exposed to polycyclic aromatic hydrocarbons. *Toxicology* **1999**, *135*, 43–47.
6. Xiao, Y.; van de Ven, M.F.C.; Molenaar, A.A.A.; Wu, S.P. Possibility of using epoxy modified bitumen to replace tar-containing binder for pavement antiskid surfaces. *Constr. Build. Mater.* **2013**, *48*, 59–66.
7. Xiao, Y.; van de Ven, M.F.C.; Molenaar, A.A.A.; Wu, S.P. Environmental concern of using coal tar in road

engineering and its possible alternatives. *J. Wuhan Univ. Technol.* **2010**, *32*, 1–7.

8. Gasthauer, E.; Mazé, M.; Marchand, J.P.; Amouroux, J. Characterization of asphalt fume composition by GC/MS and effect of temperature. *Fuel* **2008**, *87*, 1428–1434.
9. Lange, C.; Stroup-Gardiner, M. Quantification of potentially odorous volatile organic compounds from asphalt binders using head-space gas chromatography. *J. Test. Eval.* **2005**, *33*.
10. Yu, J.; Wang, X.; Kuang, D.; Zhang, H. Effect of organophilic montmorillonite on thermal-oxidative aging behavior of SBS modified bitumen crack filling material. *J. Wuhan Univ. Technol. Mater. Sci. Ed.* **2009**, *24*, 673–676.
11. Zhang, F.; Yu, J.; Wu, S. Effect of ageing on rheological properties of storage-stable SBS/sulfur-modified asphalts. *J. Hazard. Mater.* **2010**, *182*, 507–517.
12. Wu, S.P.; Pang, L.; Mo, L.T.; Chen, Y.C.; Zhu, G.J. Influence of aging on the evolution of structure, morphology and rheology of base and SBS modified bitumen. *Constr. Build. Mater.* **2009**, *23*, 1005–1010.
13. Chaala, A.; Roy, C.; Ait-Kadi, A. Rheological properties of bitumen modified with pyrolytic carbon black. *Fuel* **1996**,*75*, 1575–1583.
14. Gentzis, T.; Rahimi, P.; Malhotra, R.; Hirschon, A.S. The effect of carbon additives on the mesophase induction period of Athabasca bitumen. *Fuel Process. Technol.* **2001**, *69*, 191–203.
15. Boczkaj, G.; Przyjazny, A.; Kamiński, M. Characteristics of volatile organic compounds emission profiles from hot road bitumens. *Chemosphere* **2014**, *107*, 23–30.
16. Flego, C.; Carati, C.; Gaudio, L.D.; Zannoni, C. Direct Mass Spectrometry of tar sands: A new approach to bitumen identification. *Fuel* **2013**, *111*, 357–366.
17. Yu, M.; Wu, S.P.; Chen, M.Z.; Zhang, H.H. Evaluation of volatile organic compounds from asphalt using UV-visible spectrometer. *Adv. Mater. Res.* **2012**, *472–475*, 432–436.
18. Jia, C.; Batterman, S. A critical review of naphthalene sources and exposures relevant to indoor and outdoor air. *Int. J. Environ. Res. Public Health* **2010**, *7*, 2903–2939.
19. Xiao, Y.; van de Ven, M.F.C.; Molenaar, A.A.A.; Su, Z.; Zandvoort, F. Characteristics of two-component epoxy modified bitumen. *Mater. Struct.* **2011**, *44*, 611–622.

CHAPTER 13

ANALYSIS ON ADHESIVELY-BONDED JOINTS OF FRP-STEEL COMPOSITE BRIDGE UNDER COMBINED LOADING: ARCAN TEST STUDY AND NUMERICAL MODELING

Xu Jiang, Xuhong Qiang, Henk Kolstein and Frans Bijlaard

Keywords: FRP deck, adhesively-bonded joint, combined loading, finite element analysis, failure criterion

INTRODUCTION

Fiber Reinforced Polymer (FRP) bridge decks are increasingly used in practice for rehabilitation of existing bridges and also for new bridge constructions, which are due to the remarkable advantages of FRP decks: lightweight of bridge superstructures, ease of installation, minimum traffic disturbance, large tolerance for environmental corrosion, long service life, as well as low maintenance cost. To be cost-effective, FRP decks are commonly supported by longitudinal main girders made of steel, prestressed concrete or FRP composites [1,2,3,4]. Between FRP decks and steel girders, adhesive bonding is usually employed as a preferable connection method. Compared with welding and bolted connections, it can reduce constructional time, save weight by eliminating fasteners, introduce more uniform load transfer and provide better long-term performance. Adhesively-bonded connections have been intensively investigated over the past 70 years. Most researches were related to the aerospace and aircraft engineering industry [5]. During the last two decades, the adhesive bonding technique was increasingly used in civil Engineering field, especially for repairing the deteriorated concrete components (beam and column) and structures using Carbon Fiber Reinforced Polymer (CFRP) sheets [6]. As FRP profiles are increasingly used in civil infrastructures, researches focused on the mechanical performance of FRP adhesively-bonded single-lap joints and double-lap joints were conducted [7,8,9,10,11]. These adhesive joints composed of pultruded Glass Fiber Reinforced Polymer (GFRP) composite profiles glued by epoxy adhesives. Parametric studies were conducted experimentally and numerically on the overlap length, adhesive layer thickness, adherend thickness and degree of chamfering of adherends. The results indicated that the combination of local through-thickness tensile (peeling) stress and shear stress was the most severe stress-state and usually initiated the failures in the adhesive fillet and in the outer fiber-mat layers of the adherends below the joint edges. A probabilistic strength prediction method on the adhesive joints was also investigated, considering the scale sensitivity by using Weibull statistical distribution [11].

Recently, for traditional concrete-steel composite bridges, adhesively-bonded connections started to get more attention to be utilized between concrete deck and steel girder [12,13,14,15,16,17]. The opportunity to substitute the traditional metallic connection (e.g., studs) with the adhesive bonding connection can be due to the advantages, including the reduction of construction time, continuity of force transmission and consequently a reduction in the risk of transversal cracks in the concrete, elimination of welding in connectors, more effective protection against steel girder corrosion and more possibility to use precast concrete for reducing the concrete shrinkage cracks [14]. Investigations on the adhesively-

Citation: Jiang, X.; Qiang, X.; Kolstein, H.; Bijlaard, F. Analysis on Adhesively-Bonded Joints of FRP-steel Composite Bridge under Combined Loading: Arcan Test Study and Numerical Modeling. Polymers 2016, 8, 18. doi:10.3390/polym8010018.

bonded connection between steel and concrete indicated that the adhesive bonding definitely enhanced the stiffness and strength of adhesive connections to levels almost as high as or higher than those of the traditional steel stud connection [12,13,16,17]. In Berthet *et al.*'s research [12], by employing push-out tests, the average measured shear stress of adhesive connections was found to be between 3.4 and 5.4 MPa, which was larger than the shear strength of concrete (3 MPa). The peeling stress was evident due to the additional bending moment, which influenced the ultimate load of the composite structures. Further researches were performed on the laboratory tests of 3–5 m span beams [13].The study showed that bonding concrete to steel resulted in less slip between the two components comparing with steel stud connections. The brittle and unforeseen failure mode of specimens may be explained by combined peeling and shear stress in the concrete. To better understand the influence of combined peeling and shear stress on the strength of concrete-steel adhesively-bonded joints, a further research was conducted [14]. Finally, the quadratic tensile-shear interaction failure criterion was established and validated by experimental steel-concrete composite beams, which was recommended for the design of steel-concrete bonded composite bridges.

For adhesively-bonded joints between the FRP bridge deck and the longitudinal steel girder, the technical background and researches have not been documented adequately in the literature. To fill in this gap, in this paper, the mechanical behavior of FRP-to-steel adhesively-bonded joints is experimentally and numerically investigated. A tensile-shear loading device is designed with the adaptability to provide the combination of tensile and shear loads in six different ratios. The quadratic failure criterion of the adhesively-bonded joint is obtained according to the test results. To characterize the stress distribution throughout the adhesively-bonded joint, Finite Element modeling method is employed.

TENSILE-SHEAR LOADING DEVICE (ARCAN TEST DEVICE)

Generally, there are three typical stress states for the adhesive joint between FRP deck and steel girder:

1. Shear stress τ: due to the composite action between FRP deck and steel girder in the longitudinal direction of bridge, the deck and steel girder trend to bend together to carry the traffic load. Thus, the adhesive joints are under the shear stress condition to transfer the load from the FRP deck to the steel girder, as shown in Figure 1a;
2. Tensile stress σ: in the transverse direction of bridge, loading on other traffic lanes causes up-lift forces on adhesive joints, which results in the tensile stress (peeling stress), as shown in Figure 1b;
3. Combination of shear stress and tensile stress with different contribution ratios.

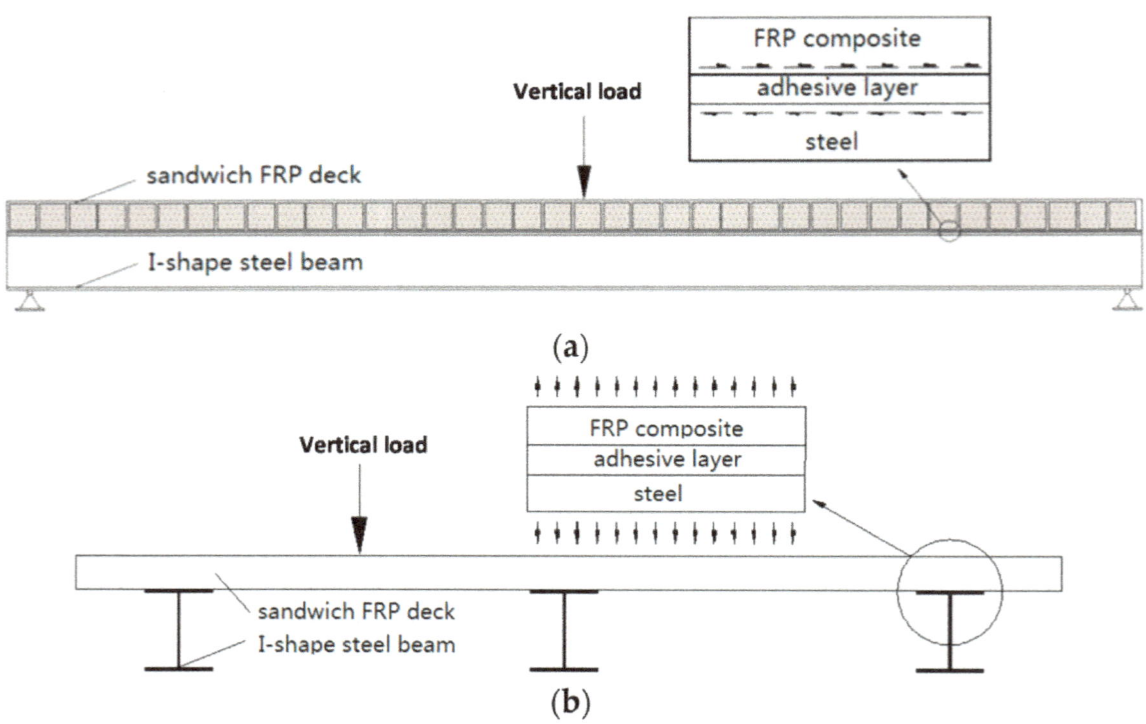

Figure 1. Typical stress states of the adhesive joint: (**a**) shear stress in the longitudinal direction; and (**b**) tensile stress in the transverse direction.

According to the above three stress states, a specific loading device is needed for providing tensile load, shear load and the combination of both simultaneously.

The adhesively-bonded joint between FRP deck and steel girder is extracted for experimental investigation as shown in Figure 2 190 mm × 90 mm sandwich bridge deck element is adhesively bonded to a convex shape steel support. The steel used herein is the commercial S355 grade steel. In the middle of sandwich deck element is the 38.1 mm Balsa SB150 (3A Composites [18]), a core material produced from certified kiln-dried balsa wood in the "end-grain" configuration. The surface layers are three layers of 0.94 mm EQX1200 (OCV Technical

Fabrics [19]), which are the glass-fiber reinforced laminated polymer composites (54% glass content by weight). Properties of EQX1200 are shown in Table 1. The sandwich profiles are manufactured by resin vacuum infusion (Infra Composite B.V. [20]). The thickness of the 90 mm × 90 mm adhesive layer is 6 mm. Dimensions of adhesive joints are determined depending on the actual application of composite bridges as well as limitations of the loading equipment. In order to fix the adhesive joint to the loading system, other accessorial components are designed as shown in Figure 3. The steel support is drilled with 4 holes to connect to steel blocks via bolts. For the sandwich deck element, no hole is made, since the discontinued part in the FRP material causes more stress distribution distortion, which is not actual in the application of composite bridges. All the accessorial components are made of steel. Compared with the FRP composites and adhesive materials, the deformation of steel components can be neglected during tests, due to the relatively high stiffness of the steel material. To fix the sandwich deck part, it is designed to be fastened by two purple L-shape steel plates through four bolts to the top steel block, as shown in Figure 3a. The steel support is fastened directly through four bolts to the bottom steel block, as shown in Figure 3b. And then, the two steel blocks are fastened to circular steel plates, as shown in Figure 4. The circular steel plates are composed of two single pieces. Three bolts are employed to transfer the load uniformly. By loading the different angles of the circular steel plates, the proposed stress-state can be achieved in the adhesive joint, such as pure tension, pure shear and combinations of both. Correspondingly, six loading conditions can be realized through this loading system. Similar Arcan loading devices have been employed by other researches [21,22,23].

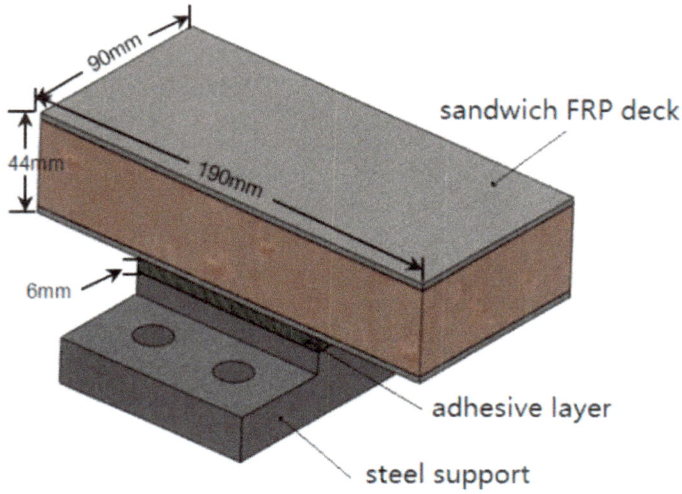

Figure 2. Adhesively-bonded joint.

Table 1. Properties of EQX1200 (supplied by manufacturer [19])

Product name	Total weight (g/m²)	Weight uniformity (g/m²)				
		Yarn roving				Knit yarn
		0°	+45°	90°	−45°	
EQX 1200	1,193	283	300	300	300	10

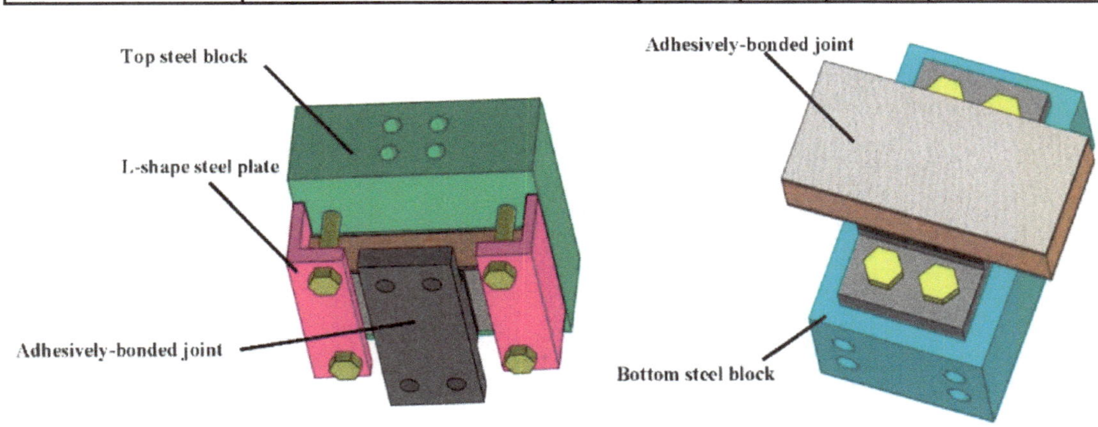

(a) **(b)**

Figure 3. Joint fix configuration. (**a**) Deck fix configuration and (**b**) steel support fix configuration.

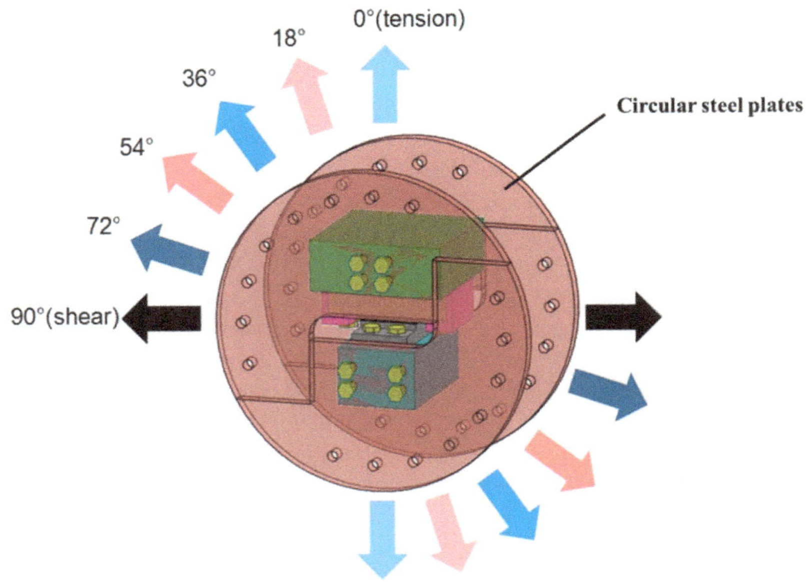

Figure 4. Tensile/shear loading device.

The sand blasting method is employed for the surface pretreatment on FRP laminates and steel supports. It starts from wiping specimen surfaces with acetone to eliminate any presence of oil used during the machining process. Then, the surfaces are abraded using a sand blasting machine to remove the impurities and oxide layer which can potentially exist. After that, re-degreasing and re-cleaning of the surface are done with acetone to remove the particles that can remain after sanding. Three replicated specimens are employed for each loading angle. Thus, in total, 18 pieces of specimens are prepared.

EXPERIMENT

The test set-up is shown in Figure 5. A SCHENCK Hydropuls test machine with loading capacity of 600 kN in tension is employed and controlled by the INSTRON 8400 controller, which can provide both load control and displacement control. The whole tensile/shear loading device is loaded by jacks through two hinged joints, which avoid the additional bending moment due to the eccentric loading from the specimen misalignment. Two LVDTs (linear variable differential transformer) are fixed on each side of the loading system, as shown in Figure 5, to measure the displacement between the top and bottom semi-circle loading device, for checking equal distribution of the load. The measure range of the LVDT is 0–10 mm. The quasi-static experiments of adhesively-bonded joints are displacement controlled by LVDTs at a rate of 0.001 mm/s. When the adhesive joint specimen is installed in the loading device, the four bolts connecting the L-shape steel profiles (see Figure 3a) are not fully fastened. A preload of 1kN is applied to ensure every loading component contact each other. In this way, the load can transfer smoothly from the loading device to the adhesive joint. After that, these four bolts are fully fastened and then the preload is unloaded. The tests start from 0 kN.

Figure 5. Test set-up.

RESULTS AND DISCUSSION

Table 2 lists failure loads of the adhesive joints under six load angles (0°, 18°, 36°, 54°, 72° and 90°). The lowest load-bearing capacity of the adhesive joints is obtained under the 18° angle loading condition, with an average failure load of 11.9 kN. The load-bearing capacity gradually increases from 11.9 to 23.27 kN as the loading angle increases from 18° to 72°. The largest deviation obtained is 34.0% under the shear loading condition, while the smallest deviation obtained is 5.55% under the 54° loading angle.

To easily recognize the load combination effects, the total failure load is vectorially separated into shear and tensile loads, with regard to the loading angle of each loading condition, as shown in Figure 6. The horizontal axis represents the shear load applied to the adhesive joint, while the vertical axis is the tensile load. It is apparent that, under the combination of tensile and shear loads, the load-bearing capacity of adhesive joints decreases, compared with that of pure tensile and pure shear loading conditions. From Table 2, the failure load of adhesive joints under the pure shear loading (69.3 kN) is considerably higher than that of the other loading conditions, due to the different failure modes, which are discussed in detail in the following sections. It is manifest that the failure load of an adhesive joint under different combined loading conditions is more sensitive to the percentage of tensile load vectorially separated from the resultant force. For the loading conditions of 18°, 36° and 54° angles, the vectorially separated tensile loads are close to each other, even though the deviations of failure loads are big. It means that, for a certain ratio limitation of the tensile and shear combination, the tensile load dominates the failure of adhesive joints more than the shear load. However, for the pure tensile loading condition (without any combination of shear load), the tensile failure load (17.43 kN) is much larger than the vectorially separated tensile loads (about 10kN) of the 18°, 36° and 54° angle loading conditions. It reconfirms that the combination of tensile and shear loads significantly degrades the load-bearing capacity of the adhesive joints.

Table 2. Ultimate failure loads of adhesive joints under six loading conditions

Loading angle	Failure load (MPa)			Average	Deviation
	Specimen01	Specimen02	Specimen03		
Pure tensile	17.53	16.05	18.72	17.43	7.94%
18° angle	12.7	11.2	11.8	11.9	6.72%
36° angle	11.2	14.6	11.8	12.53	16.5%
54° angle	16.5	17.7	16.1	16.77	5.55%
72° angle	26.6	20.0	23.2	23.27	14.3%
Pure shear	51.0	92.9	64.1	69.3	34.0%

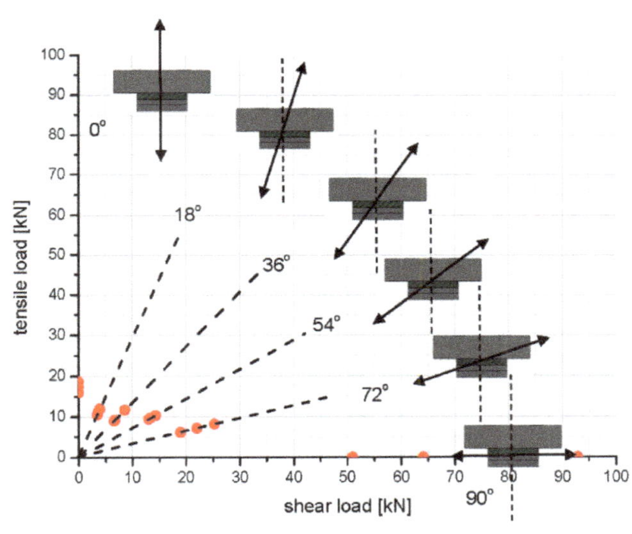

Figure 6. Failure loads of adhesive joints under six loading angles.

To investigate the failure criterion of adhesive joints, the value of average stress is considered, which is obtained using the tensile/shear load divided by the adhesive bonding area. An ellipsoid function is employed for curve fitting by the least square method, which was successfully used to establish the failure criteria of the concrete-steel adhesive joint under combined shear and peeling stresses [14]. Firstly, considering all the test results of six loading conditions, the predictive curve Equation (1) is obtained and indicated as the solid line in Figure 7. It can be found that the agreement between the test results and the predictive equation is not very well. All the test data of four combined loading conditions are below the predictive curve, which implies that the predictive equation is not conservative for the design of these adhesive joints. To solve this issue, only the test results of four combinations of shear and tensile loads are selected to perform as the basic curve fitting data. The modified predictive equation is expressed by Equation (2), and indicated as the dashed line in Figure 7. It can be found that the modified predictive equation is more conservative and practical, since in the real application of an adhesive joint between FRP decks and steel girders, the joint mainly carries the combined loading of tension and shear, but not only the pure tension or pure shear load.

Predictive curve:

$$\left(\frac{\sigma}{1.54}\right)^2 + \left(\frac{\tau}{9.59}\right)^2 = 1 \quad (1)$$

Modified predictive curve:

$$\left(\frac{\sigma}{1.41}\right)^2 + \left(\frac{\tau}{3.18}\right)^2 = 1 \quad (2)$$

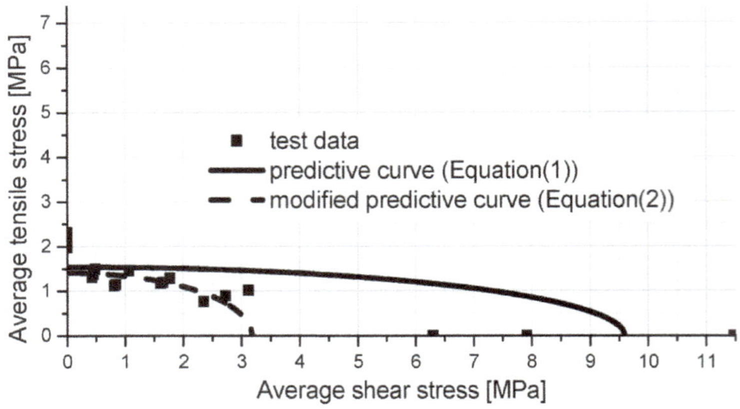

Figure 7. Shear–tensile failure criterion for the adhesively-bonded joint.

The failure modes of all the test specimens are shown in Figure 8. For adhesive joints under the shear loading, the ultimate failure of the adhesive joints always occurs in a brittle and sudden manner. The failure mode is the cohesive fracture (near the interface between adhesive layer and steel support) in the adhesive layer, as shown in Figure 9. It is visible that some residual adhesive material is left on the surface of the steel support. It can be predicted from the fracture plane (see Figure 9), cracks initiate at the edge of the adhesive bonding area and steadily propagate towards the middle of the adhesive layer. The stress re-distribution occurs throughout the bonding area and continuously carries the load. When the rest adhesive bonding area is too limited to carry the load, the cracks rapidly develop through the whole section and the adhesive joint immediately loses its load-bearing capacity.

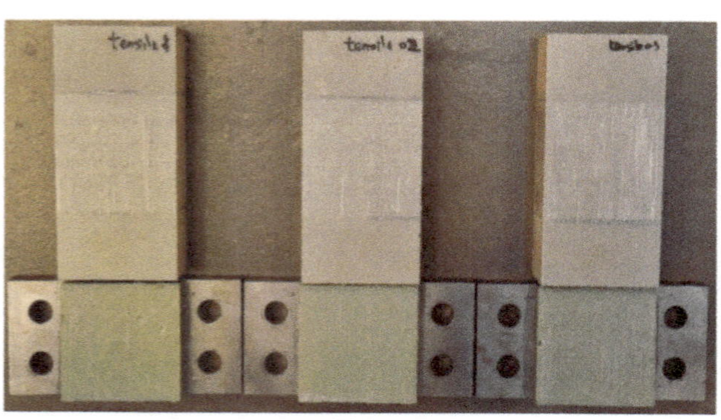

(a)

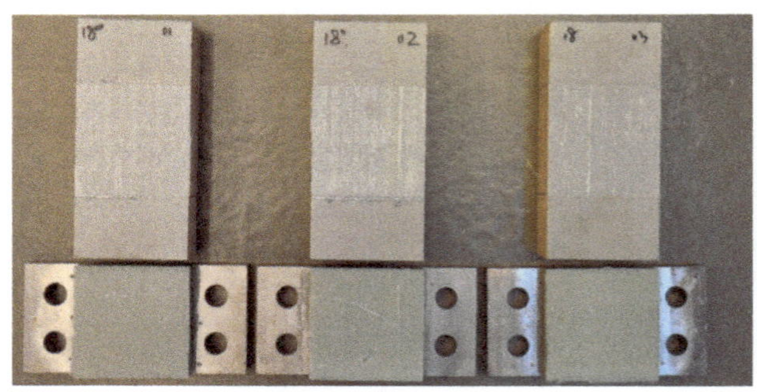

(b)

(c)

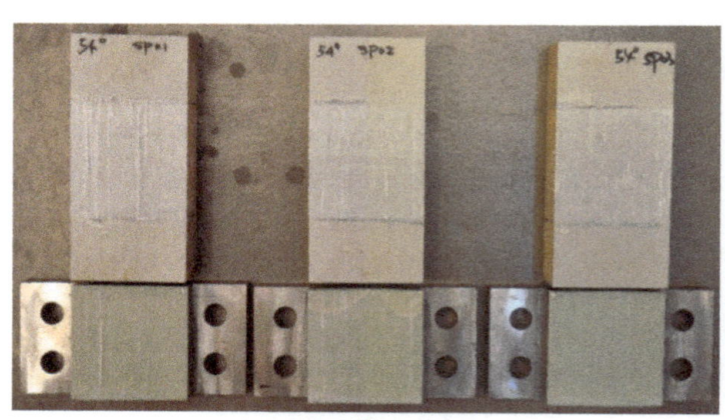

(d)

(e)

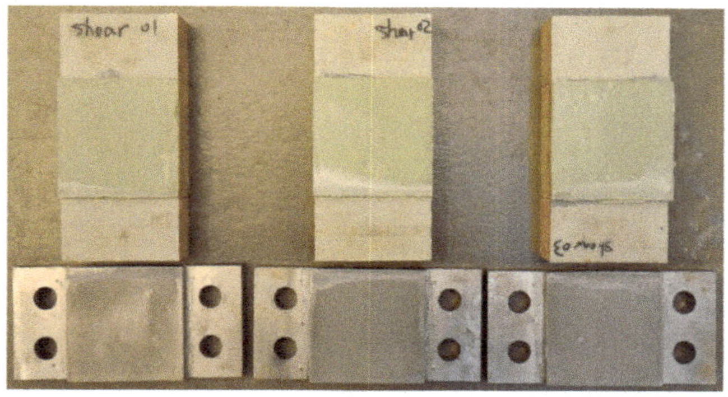

(f)

Figure 8. Failure modes of adhesive joints under six loading angles:(**a**)pure tension; (**b**) 18° angle; (**c**) 36° angle; (**d**) 54° angle; (**e**) 72° angle; and (**f**) pure shear.

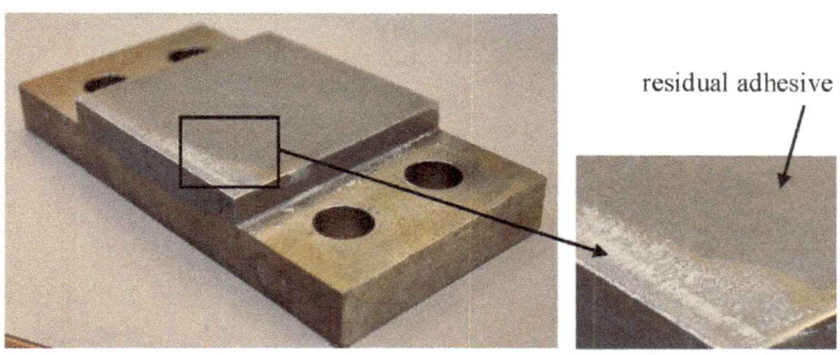

Figure 9. Cohesive failure of the adhesive joint under the pure shear loading condition.

For the other five angle loading conditions, all the fracture planes are through the interface between the FRP sandwich deck and the adhesive layer. Some areas of fiber breaking or FRP delamination are evident from the view of the FRP laminate failure surface, as shown in Figure 10. But these areas do not fully cover the adhesively-bonded area. This failure mode can be defined as the combination of fiber breaking, FRP delamination and interfacial adhesion failure between the FRP sandwich deck and the adhesive layer. Furthermore, multiple failure modes under the pure shear loading and the other five loading conditions indicate that the shear load-bearing capacity of the adhesive layer is much lower than that of adhesive-steel and adhesive-FRP interfaces and the failure of the interface between the FRP sandwich deck and adhesive layer is dominated by the tensile load.

It is worthwhile mentioning that, for the 72°-angle loading condition, there are some cracks observed in the adhesive layer of the 72°-01 specimen and propagate through the interface between the adhesive layer and the steel support, as illustrated in Figure 11. It indicates that the failure plane almost switches to the interface between the adhesive layer and the steel support, which occurred for the adhesive joints under the pure shear loading condition. This phenomenon suggests that the upper and lower interfaces between the adhesive layer and the FRP laminate or the steel support almost achieve the failure homogenously. However, for the other two specimens under the 72° loading angle, cracks in the adhesive layer are not visible. Instead, a large portion of FRP delamination or fiber breaking area is observed. Although all the specimens were prepared and cured in the same condition, multiple failure modes under the 72° angle loading condition are obtained. This phenomenon is attributed to the different bonding qualities of specimens, which could be induced by the non-uniform distribution of voids in the adhesive layer or the different interfacial adhesion strengths between the FRP laminate and the adhesive layer. It indicates that the controllable adhesive bonding technique is essential to guarantee the reliable mechanical performance of the adhesive joints.

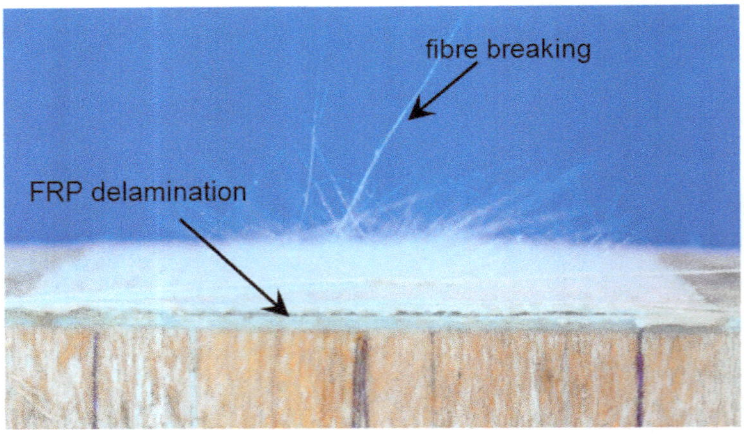

Figure 10. FRP delamination and fiber breaking in the FRP sandwich deck.

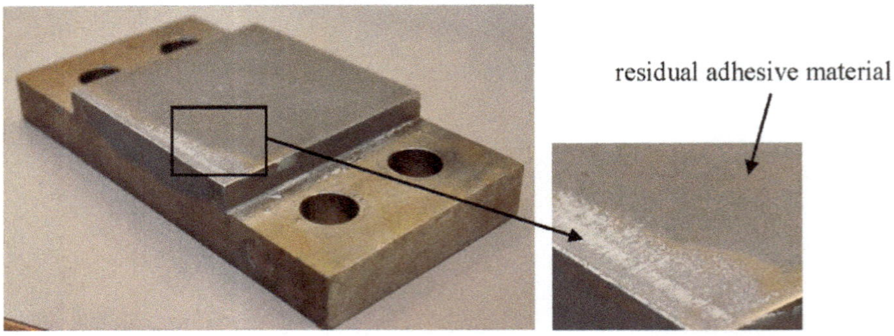

Figure 11. Cracks in the adhesive layer of the 72°-01 specimen under the 72° angle loading condition.

FE ANALYSIS

FE Model

The FE model of adhesive joints under the combination of tensile and shear loads is developed using ABAQUS 6.8 software (Dassault Systèmes Americas Corp., Waltham, MA, USA). In order to save computational time, the FE model is simplified, as shown in Figure 12. The shear and tensile loads are applied via the surface tractions on the loading area respectively, while all the degrees of freedom are restricted for the reaction area. The center of shear and tensile loading area is exactly through the centroid of the adhesive layer, in such a way to confirm that the resultant force is also through the centroid of the adhesive layer and no additional bending moment is involved. Subsequently, four loading angles can be realized by varying the ratio between shear load and tensile load with specific tangent values. Depending on the investigations of mesh dependence of an FE model under the shear loading condition [24], the FE model with the 1.50 mm mesh scale and six-layer discretization through the thickness of the adhesive layer is preferable to achieve reasonable accuracy as well as to save computational time. Thus, this mesh configuration is employed for FE analysis in this research, as shown in Figure 13 and Figure 14. All the elements used are C3D8R, which are eight-node brick elements with reduced integration. No geometric non-linearity or elastic-plastic material properties are involved in the FE analysis. The input of material properties is listed in Table 3, which are supplied by the manufactures (3A Composites [18], OCV Technical Fabrics [19] and Infra Composites B.V. [20]) and the three-point bending test results of FRP laminates [25]. The lacking properties of balsa wood are referred to the Wood Handbook [26], and the material properties of steel are referred to the Chinese Design Code for Steel Structures [27]. The resultant force of 15 kN (vectorial combination of tensile and shear loads) is applied on the FE model. The linear FE results can be amplified by any ratio in order to be comparable with test results.

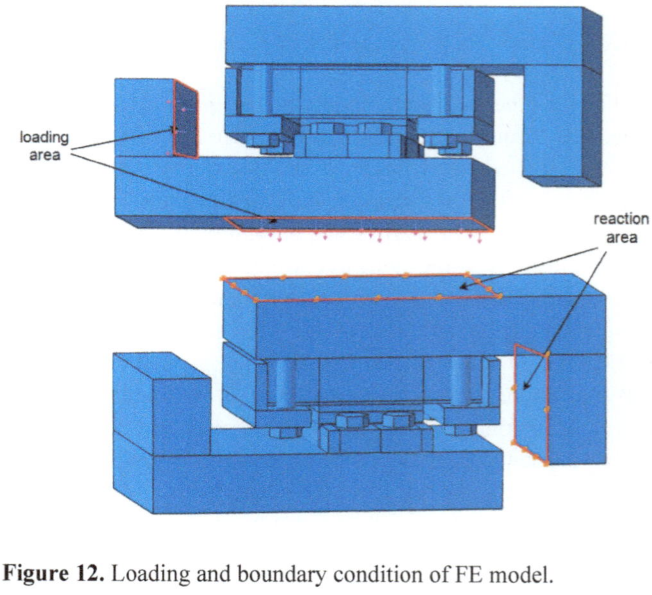

Figure 12. Loading and boundary condition of FE model.

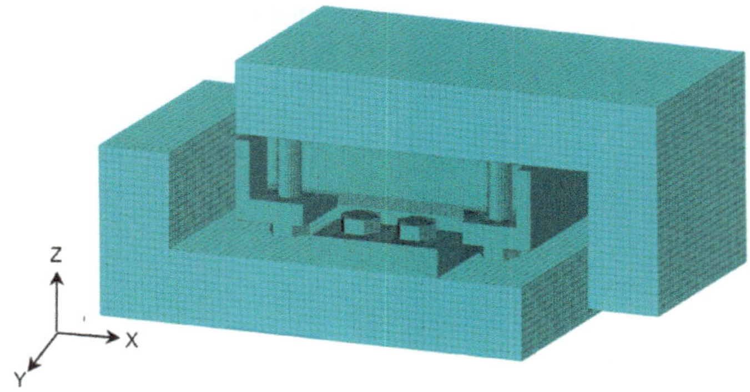

Figure 13. FE model.

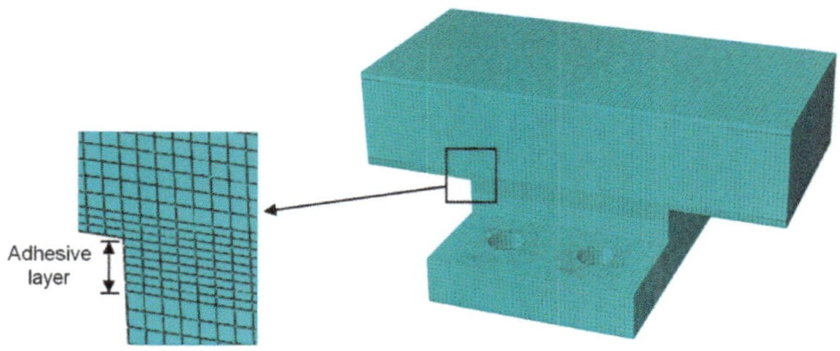

Figure 14. Mesh configuration of adhesive joint.

Table 3. Mechanical properties of materials for FE model

Property	FRP Laminates
Elastic modulus 11 (MPa)	16,609
Elastic modulus 22 (MPa)	16,609
Elastic modulus 33 (MPa)	11,000
Poisson ratio 12	0.33
Poisson ratio 23	0.3
Poisson ratio 13	0.18
Shear modulus 12 (MPa)	6,986
Shear modulus 23 (MPa)	1,200
Shear modulus 31 (MPa)	1,200
Adhesive	
Elastic modulus (MPa)	3,400
Poisson ratio	0.37
Core material	
Elastic modulus 11 (MPa)	86
Elastic modulus 22 (MPa)	265
Elastic modulus 33 (MPa)	5,759
Poisson ratio 12	0.23
Poisson ratio 23	0.23
Poisson ratio 13	0.49
Shear modulus 12 (MPa)	29
Shear modulus 13 (MPa)	213
Shear modulus 23 (MPa)	309
Steel	
Elastic modulus (MPa)	206,000
Poisson ratio	0.3

FE Results and Discussion

Figure 15 and Figure 16 illustrate the contour map of tensile and shear stress on the interface between the FRP laminates and the adhesive layer under six loading angles, where the failure plane locates. For all the six angle loading conditions, in the transverse direction (the *Y* axis as indicated in Figure 13), tensile and shear stresses distribute more uniformly. The maximum absolute stress values are not located exactly at the ends of the interface, but at 9 mm distance from the end. To further investigate the stress distribution in the longitudinal direction (the *X* axis as indicated in Figure 13), the tensile and shear stresses are extracted through the longitudinal path at the location of 9mm distance from the end, as shown in Figure 17 and Figure 18.

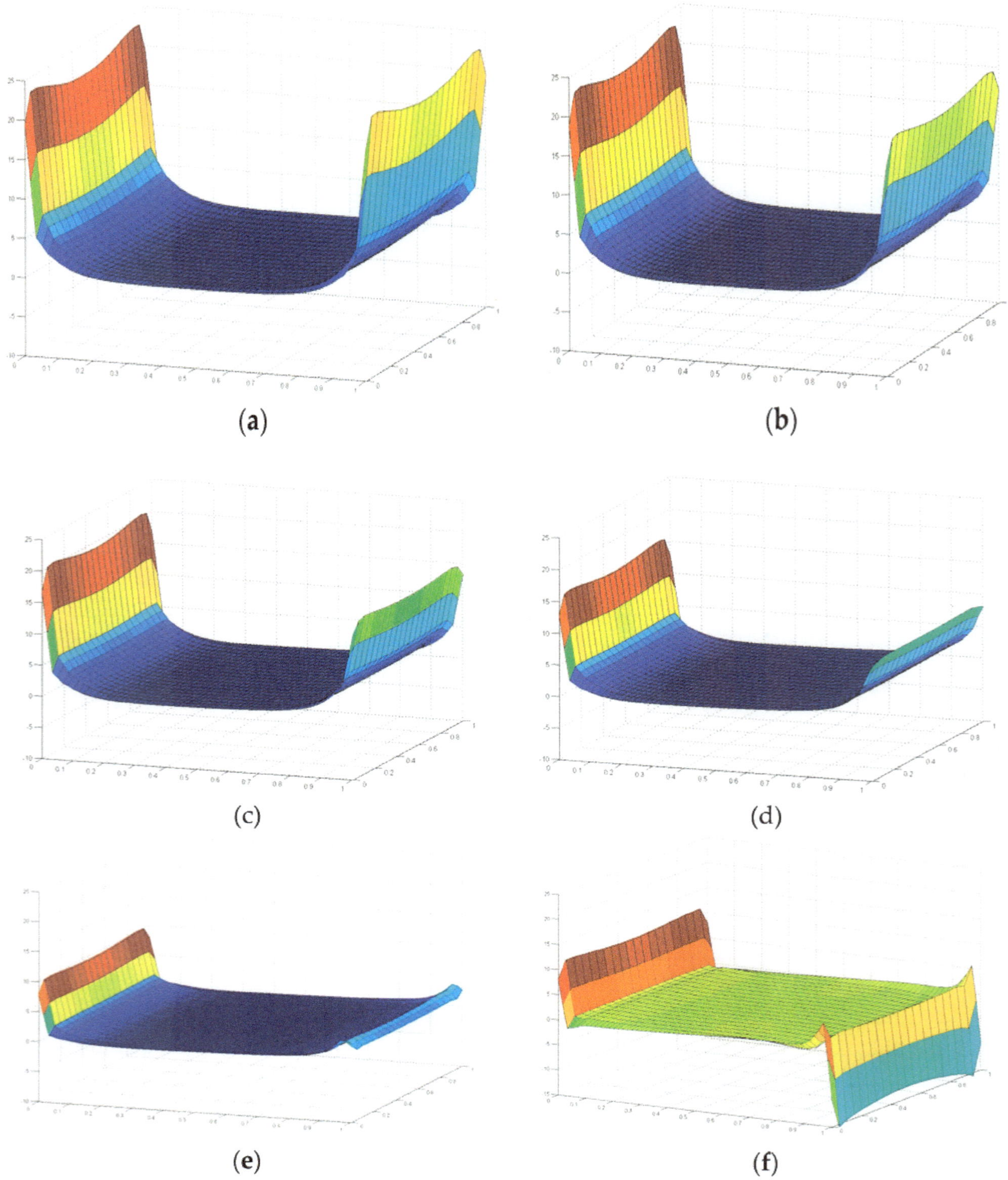

Figure 15. Contour maps of tensile stress on the interface between FRP laminates and adhesive layer under six loading conditions: (**a**) 0° angle loading (tensile); (**b**) 18° angle loading; (**c**) 36° angle loading; (**d**) 54° angle loading; (**e**) 72° angle loading; and (**f**) 90° angle loading (shear).

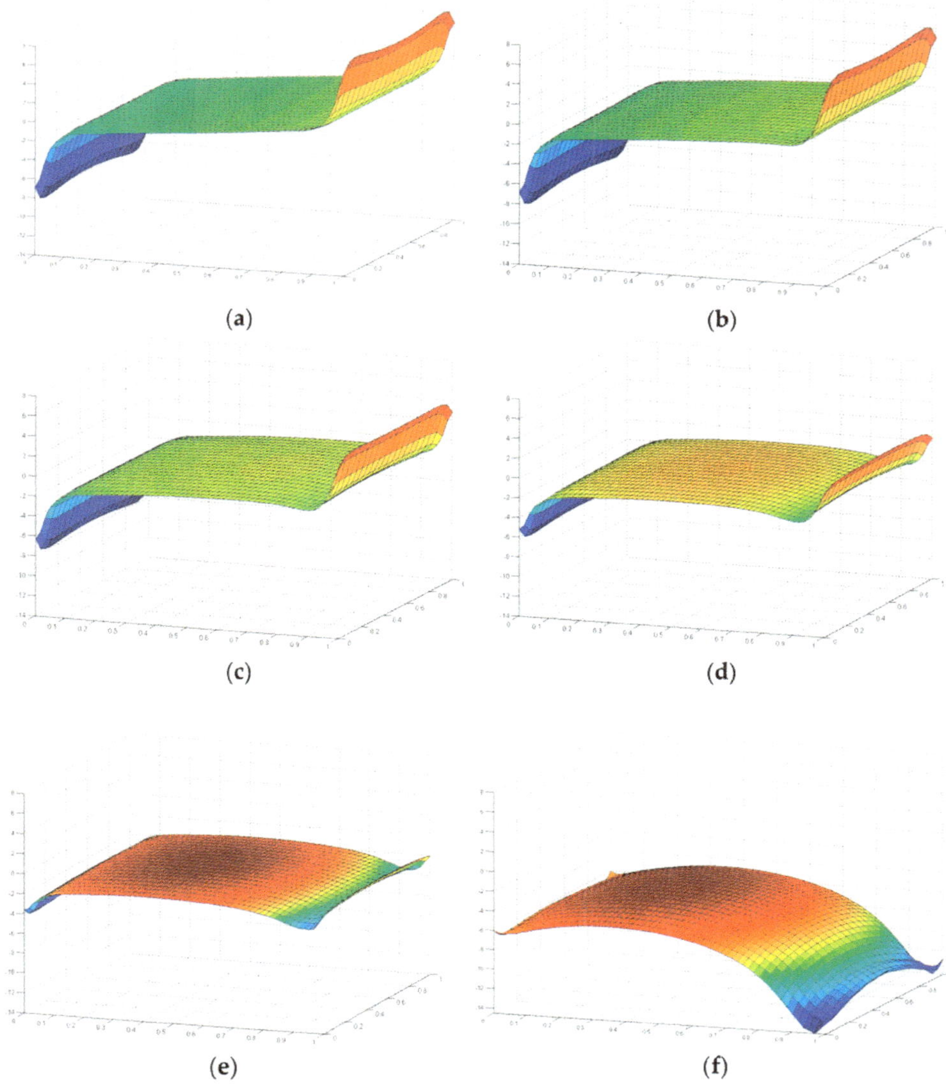

Figure 16. Contour maps of shear stress on the interface between FRP laminates and adhesive layer under six loading conditions:(**a**) 0° angle loading (tensile); (**b**) 18° angle loading; (**c**) 36° angle loading; (**d**) 54° angle loading; (**e**) 72° angle loading; and (**f**) 90° angle loading (shear).

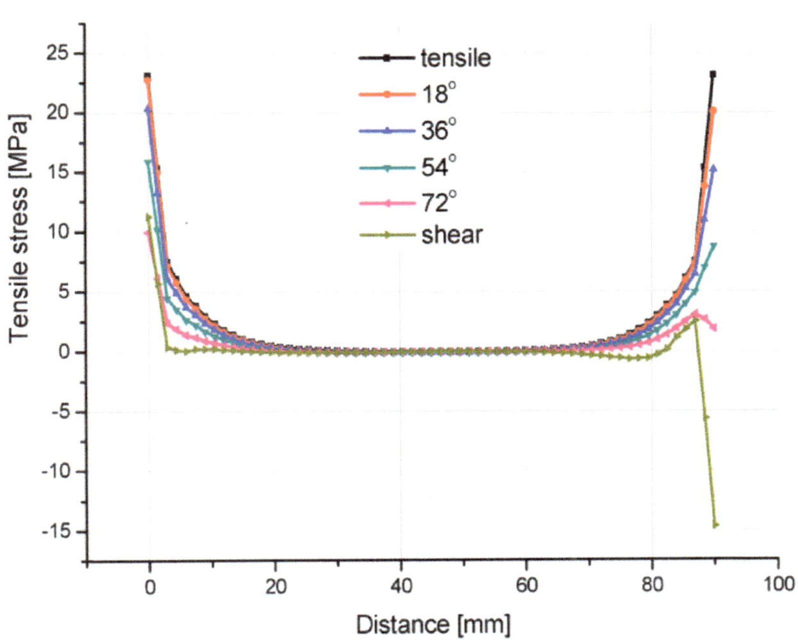

Figure 17. Tensile stress distribution in the longitudinal path at the location 9mm away from the end of the interface.

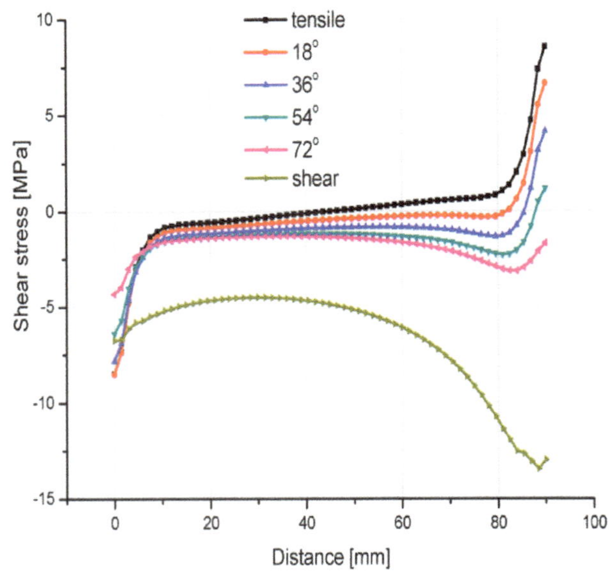

Figure 18. Shear stress distribution in the longitudinal path at the location 9mm away from the end of the interface.

From Figure 17, it can be found that for all six loading conditions the tensile stress is approximately zero in the central part of the interface in the longitudinal direction (the X axis as illustrated in Figure 13). However, in the vicinity of adhesive layer ends, the tensile stress concentration is evident, where the cracks can initiate and develop. As the loading angle rotates from 0° (tensile) to 90° (shear), the tensile stress singularity decreases at both ends of the interface, since the vectorially separated tensile load applied on the FE model of adhesive joint decreases. For the right end (shear load applied side), the extent of tensile stress decreasing is more significant than that on the left side, which introduces an asymmetrical tensile stress state throughout the longitudinal path. Subsequently, the tensile stress singularity at the right end drops to the negative zone for the shear loading condition.

Figure 18 illustrates the shear stress distribution along the longitudinal path. The tensile and shear loading conditions achieve the two extreme shear stress states. The shear stress distributions of the other four angle loading conditions regularly transit from the tensile loading condition to the shear loading condition. In the central part of the interface, the shear stress values of four combined loading angles are not zero anymore. The absolute values are increasing with the vectorially separated shear load which increases from the 18° angle loading to the 72° angle loading. Without any doubt, the absolute value of shear stress achieves the maximum (13.5 MPa) for the shear loading condition (90° angle), which is many times higher than that under other loading conditions. This could be the reason that the failure plane of the adhesive joint under the pure shear load switched to the cohesive failure in the adhesive layer but not the fiber breaking or FRP delamination in the FRP laminates. For the right edge (the side where shear load applied), the variation of the shear stress state is more evident than that on the left side, since the shear stress at the load directly forced side is more sensitively influenced. To the contrary, the shear stress distribution at the left side does not vary too much for the six loading conditions. The three curves of 0° (tensile), 18° and 36° loading conditions almost cover each other.

CONCLUSIONS

From the experimental and numerical investigation on mechanical behavior of adhesive joints under six angle loading conditions, some conclusions can be drawn as follows:

1. The load-bearing capacities of adhesive joints under combined loading are much lower than those of the pure tensile and pure shear loading conditions.
2. According to failure loads of adhesive joints under the four combined loading conditions, the modified failure criterion of the adhesive joint is addressed by using aquadratic ellipsoid function, which is conservative and practical for the real application of an adhesive joint between FRP decks and steel girders in bridges.
3. For adhesive joints under the shear loading condition, the failure mode is the cohesive failure (near the interface between the adhesive layer and the steel support) in the adhesive layer. However, for adhesive joints under the tensile loading and four combined loading conditions, the fracture plane is through the interface between the FRP sandwich deck element and the adhesive layer, combining with fiber breaking or FRP delamination in the FRP laminates.
4. The controllable adhesive bonding technique is essential to guarantee the reliable mechanical performance of adhesive joints.
5. The stress states on the interface between the FRP laminate and the adhesive layer under six loading angles are characterized using Finite Element modeling method, which gives a brief indication on crack initiation location and failure mode of the tested adhesive joints.

ACKNOWLEDGMENTS

The authors would like to thank the National Natural Science Foundation of China (Program No. 51408424), the Fundamental Research Funds for the Central Universities (2014KJ001) and Infra Composites B.V. for financial support and supplying test materials of this research.

AUTHOR CONTRIBUTIONS

Xu Jiang conceived and designed the test setup and experiments. Xu Jiang performed the experiments. Xu Jiang and Xuhong Qiang analyzed the test data and did the finite element analysis. Xu Jiang and Xuhong Qiang co-wrote the paper. Henk Kolstein and Frans Bijlaard supervised the whole research process, and organized the works related to the experiments.

REFERENCES

1. Knippers, J.; Pelke, E.; Gabler, M.; Berger, D. Bridges with glass fibre-reinforced polymer decks: The road bridge in friedberg. *Struct. Eng. Int.* **2010**, *20*, 400–404.
2. Sebastian, W.M.; Ross, J.; Keller, T.; Luke, S. Load response due to local and global indeterminacies of FRP-deck bridges. *Compos. Eng.* **2012**, *43*, 1727–1738.
3. Luke, S.; Canning, L.; Collins, S.; Knudsen, E.; Brown, P.; Taljsten, B.; Olofsson, I. Advanced composite bridge decking system—project ASSET. *Struct. Eng. Int.* **2002**, *12*, 76–79.

4. Keller, T.; Rothe, J.; de Castro, J.; Osei-Antwi, M. GFRP-balsa sandwich bridge deck: Concept, design, and experimental validation. *J. Compos. Construct.* **2014**, *18*, 04013043-1-6.
5. Adams, R.D. *Adhesive Bonding: Science, Technology And Applications*; Woodhead Publishing Limited and CRC Press LLC: New York, NY/Washington, DC, USA, 2005.
6. Hollaway, L.C. A review of the present and future utilization of FRP composites in the civil infrastructure with reference to their important in-service properties. *Construct. Build. Mater.* **2010**, *24*, 2419–2445.
7. Keller, T.; Vallee, T. Adhesively bonded lap joints from pultruded GFRP profiles. Part I: stress-strain analysis and failure modes. *Compos. Eng.* **2005**, *36*, 331–340.
8. Keller, T.; Vallee, T. Adhesively bonded lap joints from pultruded GFRP profiles. Part II: Joint strength prediction.*Compos. Eng.* **2005**, *36*, 341–350.
9. Vallee, T.; Keller, T. Adhesively bonded lap joints from pultruded GFRP profiles. Part III: Effects of chamfers. *Compos. Eng.* **2006**, *37*, 328–336.
10. Vallee, T.; Correia, J.R.; Keller, T. Probabilistic strength prediction for double lap joints composed of pultruded GFRP profiles part I: Experimental and numerical investigations. *Compos. Sci. Technol.* **2006**, *66*, 1903–1914.
11. Vallee, T.; Correia, J.R.; Keller, T. Probabilistic strength prediction for double lap joints composed of pultruded GFRP profiles—Part II: Strength prediction. *Compos. Sci. Technol.* **2006**, *66*, 1915–1930.
12. Berthet, J.F.; Yurtdas, I.; Delmas, Y.; Li, A. Evaluation of the adhesion resistance between steel and concrete by push out test. *Int. J. Adhes. Adhes.* **2011**, *31*, 75–83.
13. Bouazaoui, L.; Perrenot, G.; Delmas, Y.; Li, A. Experimental study of bonded steel concrete composite structures. *J. Constr. Steel Res.* **2007**, *63*, 1268–1278.
14. Larbi, A.S.; Ferrier, E.; Hamelin, P. Concrete to steel lap joint failure criteria under combined shear and peeling stress.*J. Construct. Steel Res.* **2009**, *65*, 386–394.
15. Jurkiewiez, B.; Meaud, C.; Michel, L. Non linear behavior of steel-concrete epoxy bonded composite beams. *J. Construct. Steel Res.* **2011**, *67*, 389–397.
16. Bouazaoui, L.; Jurkiewiez, B.; Delmas, Y.; Li, A. Static behavior of a full-scale steel-concrete beam with epoxy-bonding connection. *Eng. Struct.* **2008**, *30*, 1981–1990.
17. Meaud, C.; Jurkiewiez, B.; Ferrier, E. Steel-concrete bonding connection: An experimental study and non-linear finite element analysis. *Int. J. Adhesion Adhes.* **2014**, *54*, 131–142.
18. 3A Composites product data sheet. Available online: http://www.3acomposites.com/ (accessed on 3 February 2013).
19. Nauticexpo Product Data Sheet. Available online: http://pdf.nauticexpo.com/pdf/vetrotex-ocv-reinforcements/quadraxial-fabrics/27891–13674-_2.html (accessed on 3 February 2013).
20. Composite bridge decks. Available online: http://www.infracomposites.com (accessed on 3 February 2013).
21. Créac'hcadec, R.; Sohier, L.; Cellard, C.; Gineste, B. A stress concentration-free bonded arcan tensile compression shear test specimen for the evaluation of adhesive mechanical response. *Int. J. Adhes. Adhes.* **2015**, *61*, 81–92.
22. El-Hajjar, R.; Haj-Ali, R. In-plane shear testing of thick-section pultruded FRP composites using a modified Arcan fixture. *Compos. Eng.* **2004**, *35*, 421–428.
23. Dos Santos, D.J.; Batalha, G.F.; Polimeros-Ciencia, E. Failure criterion for adhesively bonded joints using Arcan's experimental method. *Tecnologia* **2014**, *24*, 441–445.
24. Jiang, X.; Kolstein, M.H.; Bijlaard, F.S.K. Experimental and numerical study on mechanical behavior of an adhesively-bonded joint of FRP-steel composite bridge under shear loading. *Compos. Struct.* **2014**, *108*, 387–399.
25. Qiang, X.; Jiang, X.; Kolstein, H.; Bijlaard, F.S.K. Mechanical degradation on flexural property of glass-fiber-reinforced polymer laminates under hot/wet environment. *Compos. Eng.* **2015**. Under review.
26. Wood Handbook, Wood as an Engineering Material. Available online: http://www.woodweb.com/Resources/wood_eng_handbook/wood_handbook_fpl_2010.pdf (accessed on 3 February 2013).
27. GB50017-2003 Code for Design of Steel Structures. Ministry of Construction of the People's Republic of China: Beijing, China, 2003. (In Chinese).

CHAPTER 14

INVESTIGATIONS OF PHYSICAL AND RHEOLOGICAL PROPERTIES OF AGED RUBBERIZED BITUMEN

Asim Hassan Ali, Nuha S. Mashaan, and Mohamed Rehan Karim

INTRODUCTION

Several road pavement distresses are related to bitumen properties. Rutting and fatigue cracking are the major distresses that lead to permanent failures in pavement construction. Bitumen is a viscoelastic material; its rheological properties are very sensitive to temperature and rate of loading. In general, road pavement performance properties are mainly affected by the bitumen binder properties; it is well known that the rheological properties and durability of conventional bitumen are not sufficient to resist pavement distresses. A conventional bitumen 80/100 penetration grade is commonly used in Malaysia, and moreover, it is subjected to the high traffic load and hot climate. The use of crumb rubber (waste tyre) in bitumen modification is considered as a sustainable technology which transforms unwanted residue into a new bituminous mixture. Aging is known as a factor influencing the performance and characteristics of bitumen binder. Many factors might contribute to this hardening of the bitumen such as oxidation, volatilization, polymerization, and thixotropy [1]. Field studies have shown that crumb rubber can improve the bitumen pavement performance [2–5].

Mohamed et al. [6] illustrated the influence of low CRM content on rheological properties of aged bitumen, reporting that the rheological changes in bitumen binder led to improving the mechanical properties of rubberized bitumen binder. In a joint experiment, Abdelaziz and Karim [7] conducted a study on rheological evaluation of aged rubberized bitumen. The results showed that CRM has an obvious significant effect on bitumen rheology by increasing complex shear modulus G^* and decreasing phase angle δ. Also, the RTFOT aging results of this study indicated that G^* was increased by 1 to 1.5 times for unmodified bitumen and 2.5 times for rubberized bitumen. The PAV test indicated an increase in complex shear modulus which was about 2 and 3 times for unmodified and rubberized bitumen, respectively.

The applications of rubberized bitumen binders have reported many advantages; for instance improved bitumen resistance to rutting due to high viscosity, high softening point and better resilience, reduction in road pavement maintenance cost, lower fatigue/reflection cracking, and less temperature susceptibility [8]. In general, the application of crumb rubber modified in bitumen binder can improve binder properties by lowering the inherent temperature susceptibility of binders [9, 10]. The use of crumb rubber modifier with bitumen binder led to enhancing the fatigue resistance [11, 12]. The improved performance of bitumen rubber pavements compared with conventional bitumen pavements is primarily a result of improved rheological properties of the rubberized bitumen binder.

A study by Aflaki and Memarzadeh [13] investigated the effect of rheological properties of crumb rubber on fatigue

Citation: Asim Hassan Ali, Nuha S. Mashaan, and Mohamed Rehan Karim, "Investigations of Physical and Rheological Properties of Aged Rubberized Bitumen," Advances in Materials Science and Engineering, vol. 2013, Article ID 239036, 7 pages, 2013. doi:10.1155/2013/239036.

cracking at low and intermediate temperatures using different shear methods. The results of this study showed that the high shear blending indicated more influence on improvement of low-temperature properties of rubberized binder than the low shear blending. The mechanism by which binder properties may change at low temperature was presented. This mechanism is referred to as physical hardening occurring at temperatures next to or lower than the glass transition temperature and causes significant hardening of the bitumen binder. Aging influenced the bitumen chemistry and rheology significantly; additionally, the use of the two aging conditions, that is, thin film oven test (TFOT) and rolling thin file oven test (RTFOT) showed a strong correlation and similar severity [14]. The ideal use of bitumen binders as paving materials is dependent on its resistance to the physical change across the range of temperatures encountered in a conventional pavement. Aging significantly changed both the chemical and physical properties of asphalts. Physical properties such as penetration, ductility, viscosity, and stiffness are all affected to some extent by aging [15].

Aging index measurement is one of the most popular procedures used to determine aging susceptibility and provides ranking of different bitumen binders. Aging index is defined as the ratio of a given binder property after aging to that property before aging as shown in (1):

$$\text{Aging index} = \frac{\text{Bitumen property after aging}}{\text{same property before aging}}. \quad (1)$$

Research Objective and Scope

The primary objective of this research is to evaluate and investigate the influence of crumb rubber content on the rubberized binder's rheological and physical changes subsequent to aging. The secondary objective is to identify the degree of aging in rheological and physical properties of rubberized bitumen and its effect on high-temperature properties (rutting) as well as intermediate temperature (fatigue cracking) through short-term aging test (rolling thin film oven test (RTFOT)) and long-term aging test (pressure aging vessel (PAV)).

EXPERIMENTAL PROGRAM AND MATERIALS

The base 80/100 penetration grade bitumen was used in this study. Tables 1 and 2 show the properties of base bitumen and the chemical components of original bitumen (before and after aging), respectively. Rubberized bitumen binder was prepared by mixing bitumen 80/100 penetration grade with CRM (8% and 16% by binder weight) passing the 30 mesh sieve as shown in Table 3, the propeller mixer used to mix the base bitumen, and the crumb rubber at constant blending conditions of 180°C temperature, time of 60 minutes, and velocity of 200 rpm.

Table 1: Physical and rheological properties of 80/100 PG bitumen

Test properties	Results
Viscosity at 135°C (Pa · s)	0.65
$G^*/\sin\delta$ at 64°C (kPa)	1.35
Ductility at 25°C (cm)	>100
Softening point (°C)	47
Penetration at 25°C (d-mm)	88
Ductility after RTFOT at 25°C (cm)	>100
RTFOT aged $G^*/\sin\delta$ at 64°C (kPa)	6.022
PAV aged $G^*\sin\delta$ at 25°C (kPa)	3122.5

Table 2: The chemical components of the original bitumen before and after ageing

	Unaged	After RTFOT	After PAV
Saturates (%)	15.1	13.88	12.47
Aromatics (%)	48.55	44.19	38.55
Resins (%)	26.22	28.94	31.21
Asphaltenes (%)	10.13	12.99	17.77

Table 3: (CRM gradation)

Sieve n (size, mm)	Particle size distribution
No. 10 (2.00)	100
No. 16 (1.180)	100
No. 20 (0.850)	100
No. 30 (0.600)	97.7
No. 40 (0.425)	61.1
No. 50 (0.300)	33.9
No. 80 (0.180)	12.5
No. 100 (0.150)	7.5
No. 200 (0.075)	0.0

The samples were aged using the standard rolling thin file oven test (RTFOT) (ASTM D-2872) at a temperature of 163°C for 85 min as short-term aging and pressure aging vessel (PAV) (AASHTO PP1) at 100°C and an air pressure of 2.1 Mpa for 20 hours to simulate the long-term aging. To investigate the influence of oxidative aging on rheological changes of rubberized bitumen, the age hardening is evaluated by measuring Brookfield viscosity (135°C) ASTM D4402, penetration test (25°C) ASTM D5, softening point test (25°C) ASTM D36, and dynamic shear (DSR) ASTM D-4 Proposal P246. In this study, the condition test of DSR test of parallel plate (diameter and gaps), frequency sweep, and the testing temperature for samples during RTFOT and PAV aging are listed in Table 4.

Table 4: DSR condition test used in the current study

Rheological parameter	Plate Dia. (mm)	Gap (mm)	Sample	T (°C)
$G^*/\sin\delta$	25	1	RTFOT	76
$G^*\sin\delta$	8	2	PAV	31

RESULTS AND DISCUSSION

Performance Properties at High Temperature

Brookfield Viscosity at 135°C

The viscosity of bitumen binders at high temperatures is an important property as it is a good indicator of the binder's ability of pumping through bitumen plant [16]. Figures 1 and 2 show the viscosity results of the rubberized bitumen tested in this study for various rubber contents and aging phases/conditions. As expected, the increase in crumb rubber content led to high viscosity of the rubberized binder. Additionally, the increase in viscosity of aged rubberized bitumen was about 36% and 44% for rubber contents of 8% and 16%, respectively, under RTFOT aging conditions, while the increase in the viscosity under PAV aging values was about 78% and 82% for rubber content of 8% and 16%, respectively. Thus, the increase in crumb rubber content has an obvious effect on aged rubberized bitumen viscosity compared to aged unmodified bitumen with correlation coefficient $R^2 = 0.997$. The reason is the conversion to asphaltenes (the highest molecular component fraction) of the other lower molecular weight asphalt components through oxidation. This is clearly evidenced by molecular weight analysis of asphalts before and after aging as shown in Table 2.

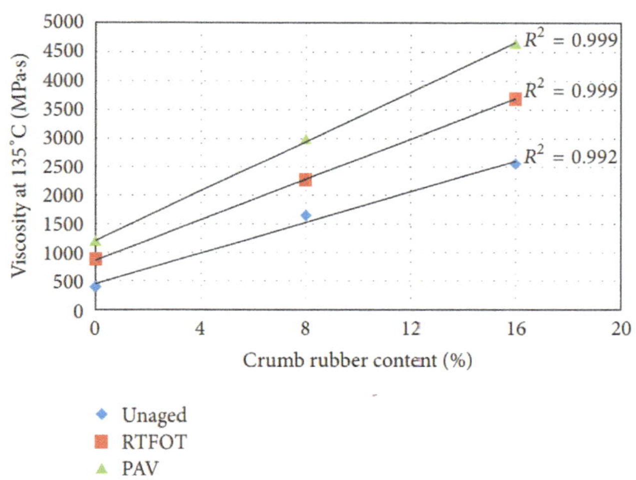

Figure 1: Brookfield viscosity results versus crumb rubber content.

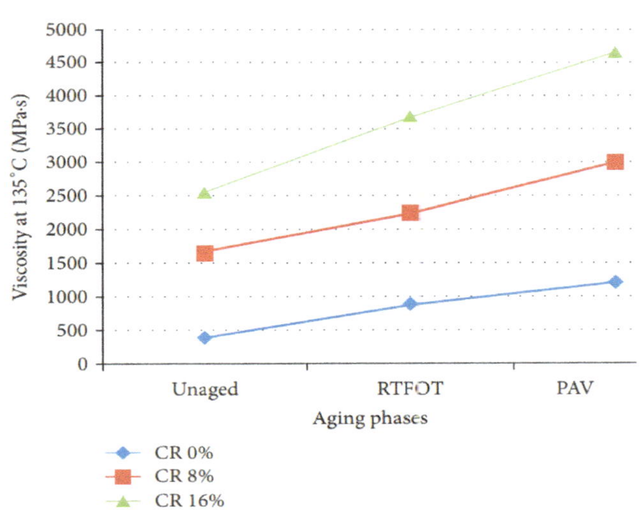

Figure 2: Brookfield viscosity results of CRM binders versus aging conditions.

Figure 3 shows VAI of the original bitumen and the rubberized binder after RTFOT and PAV aging. VAI is the ratio of aged viscosity of the binder to the unaged viscosity of same binder as illustrated in (2) [17]:

$$\text{VAI} = \frac{V_{\text{aged}}}{V_{\text{unaged}}}, \tag{2}$$

where VAI is the viscosity aging index, V_{unaged} and V_{aged} are the viscosities of unaged and aged values of the binder, respectively. This exhibits that viscosity aging index of rubber modified binder is significantly lower than that of unmodified bitumen, thus indicating that the rubberized bitumen would improve the thermooxidative aging properties of bitumen. This result can be explained by the chemical components change of the binder during the aging that led to the increase of the asphaltene content which was responsible for increasing the viscosity flow of the bitumen binder. The viscosity aging index reflected better resistance of crumb rubber modified bitumen to aging as it showed lower value after short-term aging, RTFOT, and long-term aging, PAV, respectively. Thus, the use of viscosity aging index determines the effect of aging of the original bitumen and modified bitumen which provides a much clearer effect of the hardening (aging) rate [17].

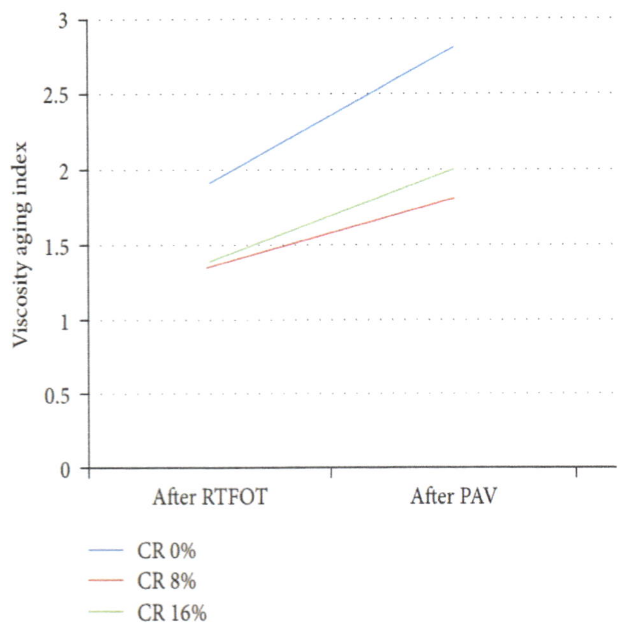

Figure 3: VAI of original and rubberized modified bitumen after RTFOT and PAV.

Penetration Aging Ratio Results

The penetration aging ratio is an alternative option to show the effects of aging on the physical properties of the bitumen binder. Penetration aging ratio (PAR) can be defined as the ratio of aged penetration value to unaged penetration as illustrated in (3):

$$\text{PAR} = \left(\frac{\text{penetration}_{\text{aged}}}{\text{penetration}_{\text{unaged}}}\right) \times 100. \tag{3}$$

Figure 4 shows the penetration aging ratio of original and rubberized bitumen after rolling thin film oven test and pressure aging vessel test. The results showed that the penetration of unmodified bitumen and rubberized bitumen binders decreased after the two different aging. In addition, the higher crumb rubber content, the lower penetration aging ratio which led to reducing the degree of aging of rubberized bitumen binder. Thus, the crumb rubber addition led to improving the binder resistance to oxidative aging.

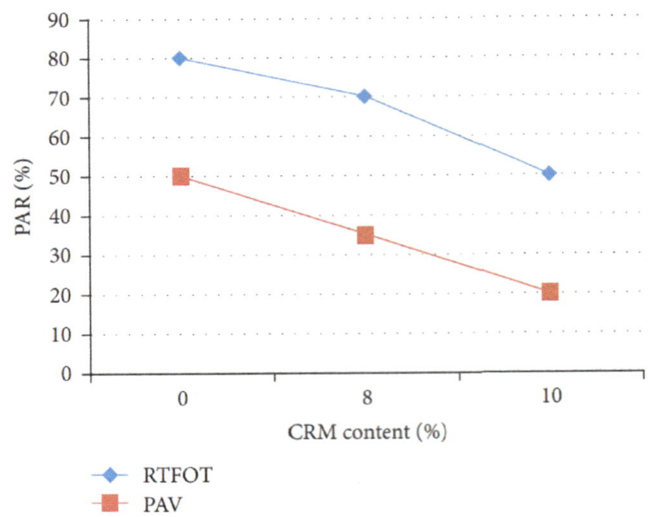

Figure 4: Effect of aging on penetration of bitumen binder.

Softening Point Increment (S)

The softening point increment can be used as a good indicator to measure the degree of aging, and it can be defined by ΔS and calculated as shown in (4):

$$\Delta S = \text{S.P}_{\text{aged}} - \text{S.P}_{\text{unaged}}, \tag{4}$$

where ΔS is the softening point increment, S.P_{aged} is aged softening point value, and $\text{S.P}_{\text{unaged}}$ is the unaged softening point value. Figure 5 shows the effect of crumb rubber modified binder bitumen on softening point changes after RTFOT aging and PAV aging. The softening point results of unmodified bitumen and rubberized bitumen showed an increase after both aging conditions of RTFOT and PAV, respectively, as the rubber content increased, indicating the hardening level of the materials during aging. In comparing the softening point increment (ΔS) of modified binder with bitumen without crumb rubber modifier, the crumb rubber modified bitumen showed lower (ΔS) as rubber crumb content was increased. This can be explained by the structural change of molecules in asphalt of fraction content during aging, which led to an increase in the softening point [17, 18]. Thus, the results of using rubberized bitumen binder have a significant effect on improving the aging resistance and show lower hardening aging which can reduce the oxidative aging resistance of the bitumen.

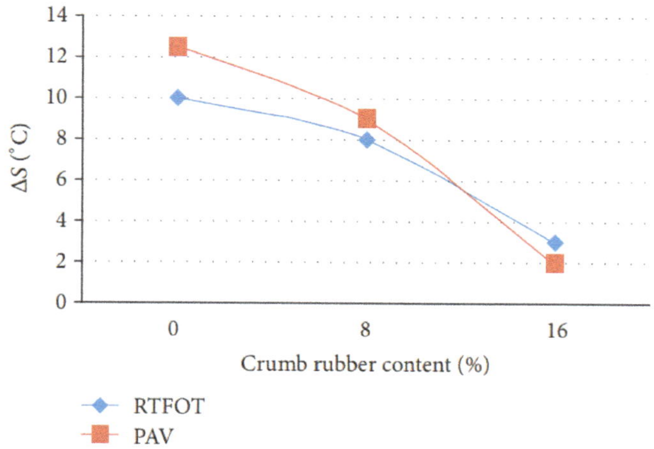

Figure 5: Effect of aging on softening point increment of bitumen binder.

Influence of Aging on Rheological Properties at 76°C

The dynamic shear rheometer test was conducted to select the rutting resistance parameter, $G*/ \sin \delta$, of the unmodified binder and rubberized bitumen binder were mastered at 76° C, and the results are shown in Figure 6. Generally, the higher rubber content led to high value of rutting factor, $G*/ \sin \delta$, for all combinations tested. Additionally, $G*/\sin(\delta)$ of aged rubberized bitumen was about 11% and 21% for rubber contents of 8% and 16% compared to unaged base binder using RTFOT aging conditions. The increase in the $G*/\sin(\delta)$ was about 38% and 60% for samples of 8% and 16% rubber content, respectively, compared to unaged base binder using PAV aging conditions. Thus, the increase in crumb rubber content has an obvious effect on rutting factor, $G*/\sin(\delta)$, with correlation coefficient $R^2 = 0.995$. In general, the results of rubberized binder are in tandem with the superpave binder specification of a minimum $G*/ \sin \delta$ requirement using RTFOT (2.2 Pa) which showed that the binder with higher $G*/ \sin \delta$ value is less susceptible to rutting deformation at high pavement temperatures [10, 19]. Figure 7 shows the aging index of rutting factor, $G*/ \sin \delta$, of original and crumb rubber modified bitumen binder at 76°C as calculated by (5) as follows:

$$\text{AIRF} = \frac{(G^* / \sin \delta)_{\text{aged}}}{(G^* / \sin \delta)_{\text{unaged}}}. \tag{5}$$

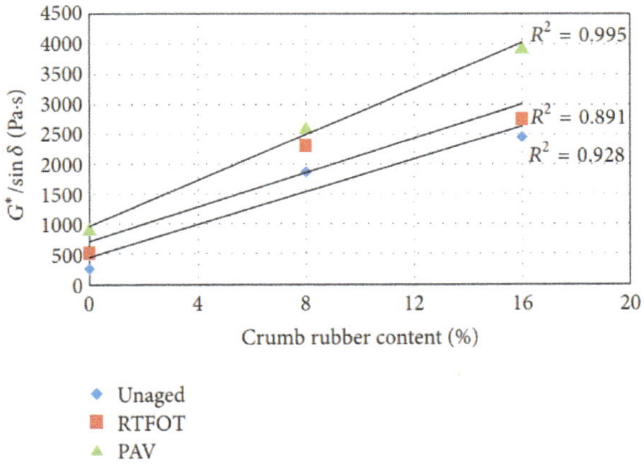

Figure 6: $G*/ \sin \delta$ results versus crumb rubber content.

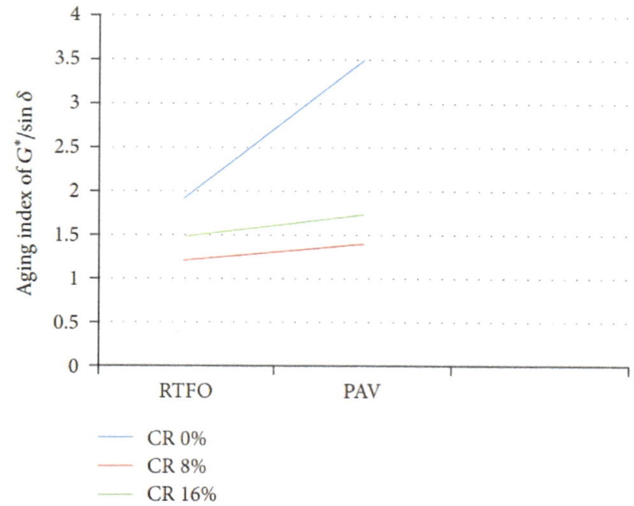

Figure 7: Aging index of $(G*/ \sin \delta)$ of bitumen binder versus CRM content.

In the case of AIRF as the aging index of rutting factor $(G*/ \sin \delta)_{aged}$ and $(G*/ \sin \delta)_{unaged}$ is the $G*/ \sin \delta$ value of bitumen binder after and before aging, respectively. The results showed that$G*/ \sin \delta$ of bitumen binder with different content of crumb rubber increased after different aging, obviously after PAV aging. Thus, crumb rubber modified bitumen binder showed better resistance to oxidation aging. As displayed in Figure 7, the unmodified binder was of higher AIRF which led the binder to being more susceptible to aging, while modified bitumen shows lower AIFR as crumb rubber increased in the specimen, leading to better aging resistance of modified binder.

Figure 8 shows the effect of aging on loss tangent of bitumen binder after RTFOT aging and PVA aging. The results displayed that the higher the crumb rubber content, the higher aging index of loss tangent, Tan δ, indicating better aging resistance. The aging ratio of loss tangent is defined as the ratio of aged Tan δ to unaged Tan δ of the bitumen binder as shown in (6):

$$\text{Tan}\,\delta\ \text{avg.} = \left[\frac{(\text{Tan}\,\delta)_{\text{aged}}}{(\text{Tan}\,\delta)_{\text{unaged}}}\right] \times 100, \tag{6}$$

where Tan δ avg. is the average of Tan δ. $(\text{Tan}\,\delta)_{aged}$ and/ $(\text{Tan}\,\delta)_{unaged}$ is the (Tan δ) value of bitumen binder after and before aging, respectively. The higher value of aging index Tan δ is related to higher elastic behavior (high in storage modulus G') since Tan $\delta = G''/G'$; thus the high-loss tangent of rubberized bitumen led to more elastic behavior and more solid-like material. This suggests that the rubberized bitumen binder can improve the resistance to oxidative aging.

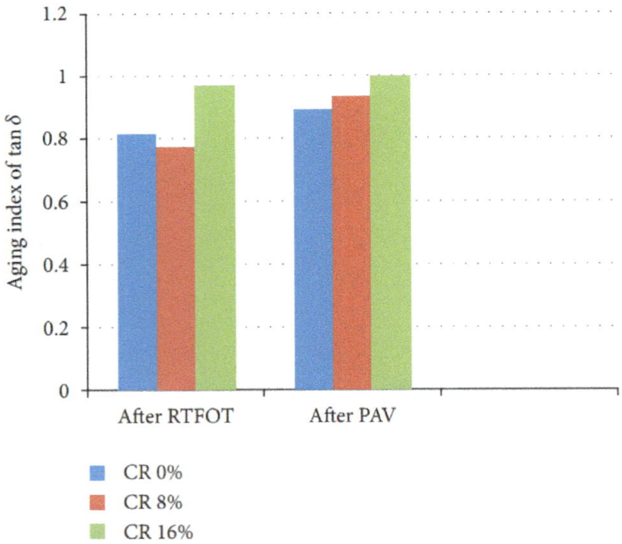

Figure 8: Effect of aging on Tan δ of bitumen binder.

Fatigue Cracking Properties at Intermediate Temperature

The Strategic Highway Research Program (SHRP) had a maximum value of 5000 kPa for $G*$ sin(δ), and low values of these parameters are considered good indicators of fatigue cracking resistance [16]. In this study, the fatigue resistance parameter, $G*$ sin(δ) values, of the unaged bitumen and rubberized binders after pressure aging vessel (PAV), were measured using the dynamic shear rheometer (DSR) at 31°C, and the results are illustrated in Figure 9. In general, the high crumb rubber content led to lower $G*$ sin(δ) values of the rubberized bitumen binders. The decrease in the $G*$ sin(δ) was about 13% and 25% for rubber content of 8% and 16%, respectively, after pressure aging vessel (PAV), aging conditions, respectively. Thus, the increase in crumb rubber content has an obvious effect on aged rubberized bitumen fatigue parameter, $G*$ sin δ, with correlation coefficient R^2 = 0.816 after PAV aging. Owing to the aging, the results showed that the lower the $G*$ sin δ value, the less shearing energy loss and the better ability of the fatigue resistance. Also,* sin δ of the modified bitumen is lower than the unmodified bitumen at temperature of 31°C, indicating that rubberized bitumen binder would improve the fatigue resistance of bitumen as a result of introduction of crumb rubber modifier (CRM).

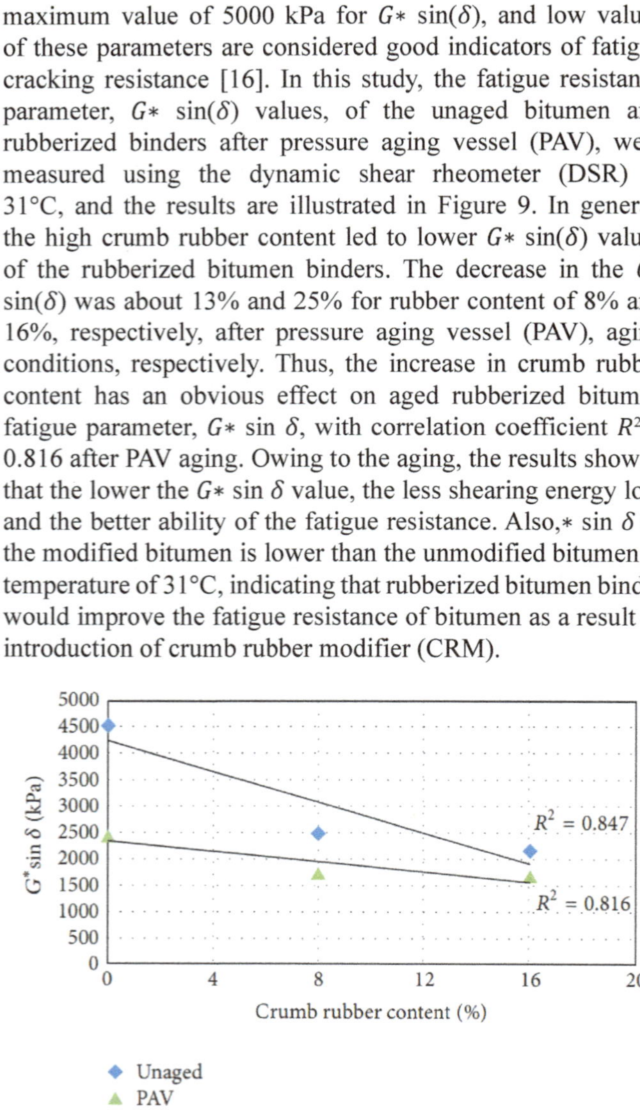

Figure 9: G^* sin δ results versus crumb rubber content.

CONCLUSION

After the introduction to the problem, a review of the literature, and analysis and discussion of the findings, this section intends to summarize the overall conclusions achieved through this research. Therefore, the significant findings of the current study are as follows.

1. The results showed that the use of rubberised bitumen binder reduced the aging effect on physical and rheological properties of the bitumen binder as illustrated through lower aging index of viscosity, lower aging index of $G*/$ sin δ at 76° C, and an increase in Tan δ with increased crumb rubber modifier content, indicating that the antioxidant released in asphalt from the tire rubber particles is a significant factor in reducing the aging effect as correctly observed by the data results.
2. In addition, the results showed that the softening point increment (ΔS) and penetration aging ratio (PRA) of the rubberized bitumen binder decreased significantly due to crumb rubber modification.
3. The aging index of viscosity reflected better resistance of crumb rubber modified bitumen to aging as it showed lower value after short-term aging, RTFOT, and long-term aging (PAV), respectively. Thus, the use of viscosity index eliminated the variability caused by difference in viscosity of the original bitumen and modified bitumen which provided a more reliable picture of the hardening (aging) rate.
4. Furthermore, the higher crumb rubber content, the lower $G*$ sin δ at 31° C after PAV aging, which led to higher resistance to fatigue cracking.
5. Crumb rubber modifier increased the stiffness of the bitumen binder improving its resistance to rutting.

REFERENCES

1. J. C. Petersen, "Chemical composition of asphalt as related to asphalt durability—state of the art,"Transportation Research Record, vol. 999, pp. 13–30, 1984.
2. D. R. Brown, D. Jared, C. Jones, and D. Watson, "Georgia›s experience with crumb rubber in hot-mix asphalt," Journal of Transportation Research Record, vol. 1583, pp. 45–51, 1997. ·
3. G. W. Maupin Jr., "Hot mix asphalt rubber applications in Virginia," Journal of Transportation Research Record, vol. 1530, pp. 18–24, 1996.
4. E. Charania, J. O. Cano, and R. H. Schnormeier, "Twenty- year study of asphalt rubber pavement in Phoenix, Arizona," Transportation Research Board, vol. 1307, pp. 29–38, 1991.
5. M. Stroup-Gardiner, B. Chadbourn, and D. E. Newcomb, "Babbitt, Minnesota: case study of pretreated crumb rubber modified asphalt concrete," Transportation Research Record, vol. 1530, pp. 34–42, 1996.
6. A. A. Mohamed, O. Husaini, M. O. Hamzah, and H. Ismail, "Rheological properties of crumb rubber modified bitumen containing antioxidant," The Arabian Journal For Science and Engineering, vol. 34, article 1B, 2009.

7. M. Abdelaziz and M. R. Karim, "Rheological evaluation of ageing properties of rubber crumb modified bitumen," Journal of the Eastern Asia Society for Transportation Studies, vol. 5, pp. 820–833, 2003.

8. S. Liu, W. Cao, J. Fang, and S. Shang, "Variance analysis and performance evaluation of different crumb rubber modified (CRM) asphalt," Construction and Building Materials, vol. 23, no. 7, pp. 2701–2708, 2009.

9. S. J. Lee, C. K. Akisetty, and S. N. Amirkhanian, "The effect of crumb rubber modifier (CRM) on the performance properties of rubberized binders in HMA pavements," Construction and Building Materials, vol. 22, no. 7, pp. 1368–1376, 2008.

10. K. D. Jeong, S. J. Lee, S. N. Amirkhanian, and K. W. Kim, "Interaction effects of crumb rubber modified asphalt binders," Construction and Building Materials, vol. 24, no. 5, pp. 824–831, 2010.

11. L. Raad and S. Saboundjian, "Fatigue behavior of rubber-modified pavements," Transportation Research Record, vol. 1639, pp. 73–82, 1998.

12. H. R. Soleymani, H. Zhai, and H. Bahia, "Role of modified binders in rheology and damage resistance behavior of asphalt mixtures," Transportation Research Record, vol. 1875, pp. 70–79, 2004.

13. S. Aflaki and M. Memarzadeh, "Using two-way ANOVA and hypothesis test in evaluating crumb rubber modification (CRM) agitation effects on rheological properties of bitumen," Construction and Building Materials, vol. 25, no. 4, pp. 2094–2106, 2011.

14. H. U. Bahia and D. A. Anderson, "Glass transition behavior and physical hardening of asphalt binders,"Journal of the Association of Asphalt Paving Technologists, vol. 62, pp. 93–129, 1993.

15. X. Lu and U. Isacsson, "Effect of ageing on bitumen chemistry and rheology," Construction and Building Materials, vol. 16, no. 1, pp. 15–22, 2002.

16. The Asphalt Institute, Performance Graded Asphalt Binder Specification and Testing, Superpave Series (SP-1), Lexington, Lexington, Ky, USA, 2003.

17. P. Cong, S. Chen, J. Yu, and S. Wu, "Effects of aging on the properties of modified asphalt binder with flame retardants," Construction and Building Materials, vol. 24, no. 12, pp. 2554–2558, 2010.

18. M. N. Siddiqui and M. F. Ali, "Studies on the aging behavior of the Arabian asphalts," Fuel, vol. 78, no. 9, pp. 1005–1015, 1999.

19. A. K. Apeagyei, "Laboratory evaluation of antioxidants for asphalt binders," Construction and Building Materials, vol. 25, no. 1, pp. 47–53, 2011.

CHAPTER 15

CORROSION EFFECTS ON THE STRENGTH PROPERTIES OF STEEL FIBRE REINFORCED CONCRETE CONTAINING SLAG AND CORROSION INHIBITOR

Sivakumar Anandan, Sounthararajan Vallarasu Manoharan, and Thirumurugan Sengottian

INTRODUCTION

The process of corrosion in reinforced concrete elements requires more focus in the present construction practice due to the reduction in service life performance. Steel inclusion in concrete elements cannot be avoided as a result of corrosion process and necessitates more careful technique for designing the concrete elements. Right selection of concrete constituents and careful mixture proportion can provide a highly integrated concrete, which can substantially abate the process of corrosion initiation. More promising methods for protecting steel bars in concrete were found to exist. Several research studies showed the advantages of using anticorrosion agents, sacrificial anode by cathodic protection, and epoxy coating on the steel bars in civil engineering application. However, the cost of concrete construction is an important factor for executing infrastructure project and, hence, additional protection techniques can increase the cost. Hence, it is better to find a cost effective technique with the proper selection of concrete which can protect the rebar against corrosion [1, 2]. Addition of mineral admixtures with inclusion of steel fibres causes significant improvement on the micostructural properties and can lead to refined pore structure. Moreover, the process of corrosion occurs due to poor cover concrete materials. This serves as an entry point of corrosion initiation and, hence, the process of corrosion being accelerated in a short period. However, the inclusion of steel is a major application of fibre reinforced concrete and finds potential merits for providing high quality construction without rebar and, hence, any costly treatment method cannot be suitable for surface treatment of steel fibres [3, 4]. The process of corrosion in fibre reinforced concrete is phenomenal due to homogenized steel fibre distribution and, hence, more careful attention has to be provided for designing concrete constituents. In this direction, the cement replacements such as fly ash, ground granulated blast furnace slag, and silica fume were typically used in concrete construction to save cost [5–8]. The addition of such supplementary materials is found to be one choice for reducing the corrosion potential of steel fibre incorporated concrete mixtures. Many studies conducted in this direction have focused on rebar corrosion of reinforced concrete element in which different industrial waste materials had been incorporated to improve the pore properties of concrete [9, 10]. This beneficial addition of binder materials showed an enhanced performance of corrosion process. In another study, the process of corrosion initiation was observed during its exposure to aggressive chloride environment and continues over longer period leading to aggressive corrosion process [11–13]. Studies also indicated that the high volume fly ash with steel fibre incorporated rebar concrete had shown a favorable reduction in the corrosion potential. The thickness of corrosion was identified in the corrosion process for different fly ash substituted concrete [14, 15]. The rate of

Citation: Sivakumar Anandan, Sounthararajan Vallarasu Manoharan, and Thirumurugan Sengottian, "Corrosion Effects on the Strength Properties of Steel Fibre Reinforced Concrete Containing Slag and Corrosion Inhibitor," International Journal of Corrosion, vol. 2014, Article ID 595040, 7 pages, 2014. doi:10.1155/2014/595040.

change of strain was measured using strain gauge principle in which the fly ash based fibre reinforced concrete showed lesser corrosion potential and type of mixes. Studies conducted in the past had shown that much focus is required to study the effect of corrosion in steel fibre incorporated concrete. Also, the accelerated corrosion process occurring in stressed concrete elements needs further investigation to understand the structural efficiency. The incorporation of binder materials such as slag in cement also needs to be carefully investigated.

Research Significance

The present research study evaluates the effects of ground granulated blast furnace slag and corrosion inhibitor on the process of corrosion in steel fibre reinforced concrete. The significance of filler binder materials (slag) for better improvement on the strength properties (such as compressive and flexural) of steel fibre concrete is studied. Enhanced interfacial characteristics and bonding stress between cement and the fibres are also investigated systematically.

EXPERIMENTAL INVESTIGATIONS

The details of concrete making materials used in the present study are as follows.

Cement

Ordinary Portland cement of grade 53 which had a specific gravity of 3.14, fineness modulus 2.56, consistency limits 32%, initial setting time 165 min, and final setting time of cement 255 min was used.

Aggregates Used

A crushed granite waste such as manufactured sand (M-sand) conforming to zone III as per IS 383-1970 [16] was used as fine aggregate filler. The specific gravity value of M-sand was 2.45 and the loose bulk density values of M-sand were 1600 kg/m^3. Crushed granite was used as coarse aggregate passing through 20 mm and 12 mm down having specific gravity of 2.64 and conforms to IS 383-1970.

GGBS

The physical properties of ground granulated blast furnace slag are as follows: specific gravity 3.43 and fineness modulus 3.35. The chemical properties are as follows: carbon (C) 0.24%, manganese (Mn) 0.58%, sulphur (S) 0.05%, phosphorous (P) 0.06%, free silica 6.10%, and iron (Fe) 92.97%. Slag was replaced for cement at 20%, 40%, and 60% by weight of binder.

Chemical Admixtures

In order to improve the workability properties of concrete and to avoid the harshness of the concrete at low water content, the addition of chemical admixtures was found to be essential. In the present study, a polycarboxylate ether based superplasticizer was used at an optimum dosage of 1.5% (by weight of cement). The mix water used for preparing concrete specimens was free from chlorides and sulphates. The study was also conducted with addition of anticorrosion inhibitors at 0.3%, 0.6%, and 0.9% by weight of steel fibres.

Glued Steel Fibres

Glued steel fibres as shown in Figure 1 were used in the present study and the various properties of the material are as follows: length 60 mm, diameter 0.75 mm, aspect ratio 80, and failure strain 3 to 5%. The steel fibres were consisting of hooked ends to provide adequate end anchorages and bond strength in the concrete matrix. The glued steel fibres were bundled together using water soluble glue and, upon addition in concrete, the glue dissolved and dispersed homogenously in the concrete.

Figure 1: Glued steel fibres used in the study.

Concrete Mixture Proportions and Curing

A high strength concrete grade of M40 was designed as per IS 10262-2009 [17] and the detailed mixture proportions of concrete are given in Table 1. The conventional concrete mix contained 100% manufactured sand (M-sand). The slag replacement level of 0%, 20%, 40%, and 60% of cement was replaced. The desired workability of all concrete mixes was maintained using superplasticizer at a maximum dosage restricted up to 1.5% by weight of the binder. A horizontal pan mixer of 40 litres capacity was used for mixing the constituent material. Fresh concrete was casted in steel mould of size 100 mm × 100 mm and beam mould of size of 100 × 100 × 500 mm for testing compressive and flexural strength properties, respectively. The specimens were remoulded after a day and cured in water tank at 35° ± 3° C for different curing days (28, 56, 90, and 180 days).

Table 1: Various concrete mixture proportions used in the study

Mix ID	w/b ratio	Steel fibres (V_f) %	Slag %	Anticorrosion inhibitors	SP %	Cement	Slag	Manufactured sand	Coarse aggregate	Water
						kg/m³				
GGC1	0.3	0	0	—	1.5	449	0	712	1165	135
GSF5	0.3	1.5	0	—	1.5	449	0	712	1165	135
GSF6	0.3	1.5	20	—	1.5	359	90	712	1165	135
GSF7	0.3	1.5	40	—	1.5	269	180	712	1165	135
GSF8	0.3	1.5	60	—	1.5	179	270	712	1165	135
AC1	0.3	0	0	0.3	1.5	449	0	712	1165	135
AC2	0.3	0	0	0.6	1.5	449	0	712	1165	135
AC3	0.3	0	0	0.9	1.5	449	0	712	1165	135

Note: V_f: volume fraction, w/b: water to binder ratio, and SP: superplasticizers.

Experimental Testing Procedure

Accelerated Corrosion Process

Alternate wetting and drying cycles are done in order to accelerate the deterioration process. Similarly, the initial loading of concrete reflects the real-time situation wherein the concrete sustains dead load and results in opening up of stress induced microcracks. This actually provides the real quantification on the process of accelerated corrosion mechanism. Hence, the initial stress on the concrete can open up the microcracks and promulgates the water and chloride ions to penetrate the concrete specimens which can initiate the corrosion process of steel fibres. Initially, the specimens were cured for required 28 days and later stressed initially up to 40% and 60% of the ultimate load. Later, the specimens were subjected to alternate wetting and drying cycle for 28 days. This process is continued up to 6 months and subsequent testing was carried on the concrete specimens. After stressing, the specimens were immersed in chloride-free water (curing tank) for 1 day and later taken out for drying in hot air oven at 100°C for 1 day. This process is continued for 28 days till testing.

Similar methodology was also adopted for the other set of concrete specimens wherein the specimens were cured in sodium chloride solution prepared at 3% concentration and the snapshot of cubes and beams immersed in the solution are shown in Figures 2(a) and 2(b), respectively; similar cycles were also adopted to evaluate the strength degradation at different curing period till 6-month duration. Compressive and flexural strength loss for different time periods was recorded and systematically analyzed. Also, the thickness of steel fibres after corrosion was measured after 6 months for different concrete mixes subjected to accelerated corrosive environment.

(a) (b)

Figure 2: (a) Cube specimens cured in NaCl (3%) solution. (b) Beam specimens cured in NaCl (3%) solution.

Mechanical Testing

Compressive test was conducted in universal compressive testing machine of capacity 2000 kN at a loading rate of 2.5 kN/sec (as shown in Figure 3) and flexural testing was carried out in a displacement control machine at 0.5 mm/minute (as shown in Figure 4). Flexural testing was conducted as per standard specifications in a third point loading test setup. The loss in diameter of steel fibres due to corrosion was measured using a screw gauge. Initially, the rusted steel fibres were surface cleaned with an emery paper and then measured at different points along the length of fibre.

Figure 3: Initial stress applied to concrete specimens.

Figure 4: Snapshot of flexural test setup.

EXPERIMENTAL TEST RESULTS AND DISCUSSIONS

Effect of Corrosion due to Alternate Wetting and Drying Cycle

Test results of various steel fibre concrete specimens subjected to initial stress and cured in water and salt solution are provided in Figure 5. Initial stress applied on the specimens provides accelerated corrosion process and helps to assess the deterioration process in plain, slag, and corrosion inhibitor substituted concrete specimens. The experimental test results on the compressive properties of all slag substituted concrete mixes investigated showed a reasonable strength attainment above 40 N/mm² (28 days) which was higher than reference concrete. However, there was a significant increasing trend observed in the case of 40% slag concrete containing 1.5% V_f (volume fraction) of steel fibre reinforced concrete mixes. A high strength was recorded up to 49.80 N/mm² at 28 days for GSF7 concrete mix containing 40% slag with 1.5% (V_f) of steel fibres, which was 13.96% higher than that of control concrete (GGC1) as shown in Figure 5. The increase in diameter due to corrosion causes the subsequent increase on the compressive strength and the failure pattern as shown in Figure 6. It is understood that the strength increase was reported in steel fibre substituted concrete mixes due to better bonding of corroded fibres in the matrix. Also, the discrete fibre orientation and the corresponding increase in volume of corroded steel fibres had caused matrix interlocking. This eventually led to the increase in the strength of concrete and was reported at longer exposure up to 6 months of wetting and drying cycles. A similar trend was observed in the case of 20% slag with steel fibres at 1.5% V_f (volume fraction). Also, the addition of corrosion inhibitor provided a corresponding increase in strength as compared to that of slag substituted concrete mixes. It was clearly noticed that the addition of slag and corrosion inhibitor showed favorable reduction in the corrosion in 40% of stressed concrete specimens (as shown in Figure 7). It can be inferred from the test results that the improvement on the matrix properties of concrete systems occurred due to closer granular packing with the binder particles. This eventually contributed for a well-developed matrix system with low porosity and high densification of the mortar phase leading to improved compressive properties. With the addition of 20% slag, the compressive strength increase was observed and the optimum addition of 0.6% of anticorrosion inhibitor was also found to be showing improved performance (as shown in Figure 8). It is noted that the stressed concrete can be detrimental and, hence, addition of slag is favorable and the addition of admixture was found to reduce the corrosion process. In normal water, cured concrete specimens subjected to alternate wetting and drying had not showed considerable reduction in strength. Similarly, at higher stress (60% of ultimate stress), the incorporation of high volume slag replacement up to 60% is needed and had shown promising improvements. It is understood that the corrosion of steel fibres had caused a great reduction in plain concrete compared to slang and corrosion inhibitor substituted concretes. However, the corrosion effects are accelerated in chloride environment due to faster deterioration.

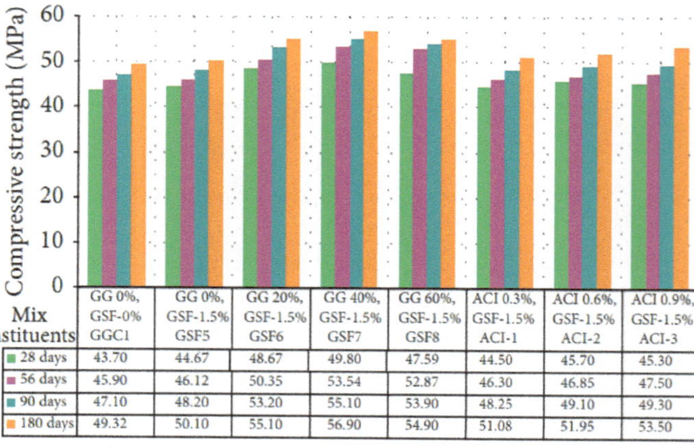

Mix constituents	GG 0%, GSF-0% GGC1	GG 0%, GSF-1.5% GSF5	GG 20%, GSF-1.5% GSF6	GG 40%, GSF-1.5% GSF7	GG 60%, GSF-1.5% GSF8	ACI 0.3%, GSF-1.5% ACI-1	ACI 0.6%, GSF-1.5% ACI-2	ACI 0.9%, GSF-1.5% ACI-3
28 days	43.70	44.67	48.67	49.80	47.59	44.50	45.70	45.30
56 days	45.90	46.12	50.35	53.54	52.87	46.30	46.85	47.50
90 days	47.10	48.20	53.20	55.10	53.90	48.25	49.10	49.30
180 days	49.32	50.10	55.10	56.90	54.90	51.08	51.95	53.50

Figure 5: Compressive strength of concrete for alternate wetting and drying of concrete specimens in chloride-free water (stressed at 40% ultimate load).

Figure 6: Failure pattern of corroded steel fibre concrete specimen.

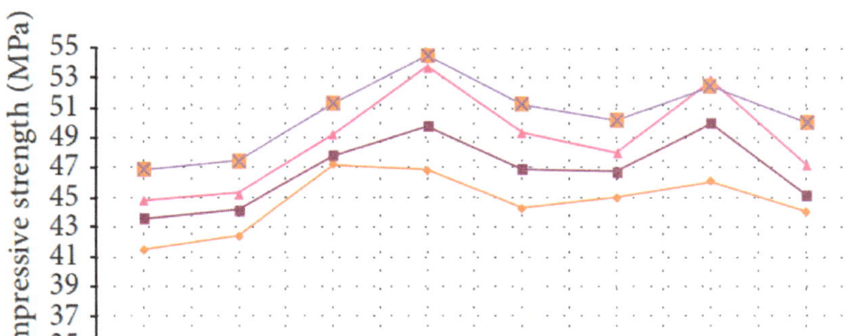

Mix constituents	GG 0% GGC1	GG 0%, GSF-1.5% GSF5	GG 20%, GSF-1.5% GSF6	GG 40%, GSF-1.5% GSF7	GG 60%, GSF-1.5% GSF8	ACI 0.3%, GSF-1.5% ACI-1	ACI 0.6%, GSF-1.5% ACI-2	ACI 0.9%, GSF-1.5% ACI-3
28 days	41.52	42.44	47.20	46.84	44.30	45.01	46.08	44.10
56 days	43.61	44.20	47.83	49.80	46.90	46.74	50.01	45.21
90 days	44.87	45.34	49.30	53.78	49.40	48.03	52.90	47.30
180 days	46.90	47.50	51.35	54.49	51.23	50.20	52.50	50.10

Figure 7: Compressive strength of concrete cured in NaCl (3%) solution for various mixes (40% stressed).

Mix constituents	GG 0% GGC1	GG 0%, GSF-1.5% GSF5	GG 20%, GSF-1.5% GSF6	GG 40%, GSF-1.5% GSF7	GG 60%, GSF-1.5% GSF8	ACI 0.3%, GSF-1.5% ACI-1	ACI 0.6%, GSF-1.5% ACI-2	ACI 0.9%, GSF-1.5% ACI-3
28 days	38.25	40.20	43.80	44.82	42.85	43.65	41.40	42.51
56 days	40.27	40.55	44.84	48.19	47.85	47.08	45.95	43.90
90 days	42.15	42.40	46.80	50.85	48.62	49.30	46.45	45.73
180 days	43.23	43.98	49.70	52.90	50.15	51.45	47.90	48.31

Figure 8: Compressive strength of concrete cured in NaCl (3%) solution for various concrete mixes (60% stressed).

Flexural Strength

The flexural strength test results for various steel fibre concretes are presented in Figure 9. During the initial loading up to 40% stress, the flexural strength was found to be unaffected in 40% slag substituted concrete. Similarly, the favorable improvement was noticed in anticorrosion inhibitors (ACI) concrete mixes. The wetting and drying of concrete specimens in normal water cured specimens showed a reduction in reference concrete and, with the addition of slag, it was better improved. Test results denoted clearly that the corrosion occurring in concrete results in effective bond developed between matrix and steel fibre interface. This actually provides enough pullout resistance and thereby increased the flexural load capacity. As noticed, the strength improvement was observed in flexural strength; on the other hand, the strength reduction was noticed in compressive strength. It is understood that the accelerated corrosion causes intrusion of sodium chloride ions which had not caused enough deceleration on the strength gain. The effect of concrete deterioration is shown in Figure 10 and the corroded beam specimens tested in flexure are shown in Figure 11. The experimental trends shown in Figure 12 show flexural results of various steel fibre concretes subjected to initial stress and cured in salt solution. It was noted that the strength loss was not observed even though the severity of corrosion occurring in salt water was found to be higher. Surface erosion of plain concrete occurred in plain concrete compared to steel fibre concrete possibly due to leaching of calcium ions leaving roughened surface. From the results it was also noted that corrosion effects in steel fibre concrete were not reported in slag concrete and corrosion inhibitor added concrete. This subsequently showed higher strength gain compared to reference concrete without slag and corrosion inhibitor.

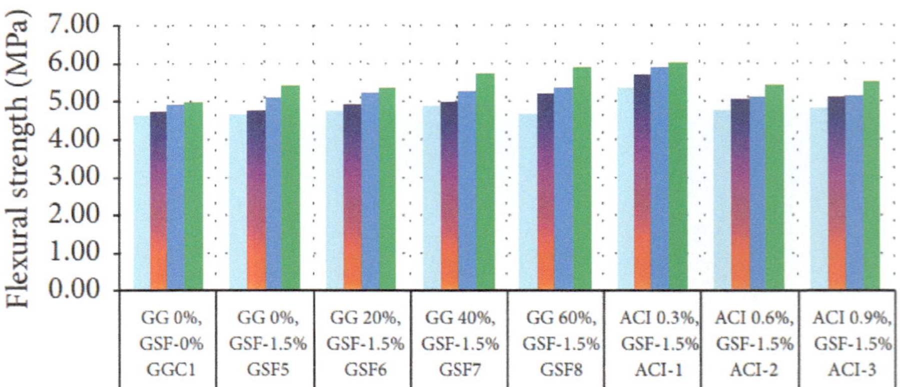

	GG 0%, GSF-0% GGC1	GG 0%, GSF-1.5% GSF5	GG 20%, GSF-1.5% GSF6	GG 40%, GSF-1.5% GSF7	GG 60%, GSF-1.5% GSF8	ACI 0.3%, GSF-1.5% ACI-1	ACI 0.6%, GSF-1.5% ACI-2	ACI 0.9%, GSF-1.5% ACI-3
28 days	4.63	4.68	4.75	4.88	4.70	5.34	4.76	4.85
56 days	4.74	4.75	4.91	4.97	5.20	5.70	5.04	5.09
90 days	4.93	5.10	5.21	5.25	5.34	5.90	5.12	5.15
180 days	5.01	5.42	5.34	5.72	5.89	6.01	5.43	5.50

Figure 9: Flexural strength of concrete specimens for alternate wetting and drying cycles for various concrete mixes (subjected to 40% initial stress).

Figure 10: Plain concrete specimens exposed to sodium chloride solution (3%).

Figure 11: Flexural testing of corroded steel fibre concrete specimens.

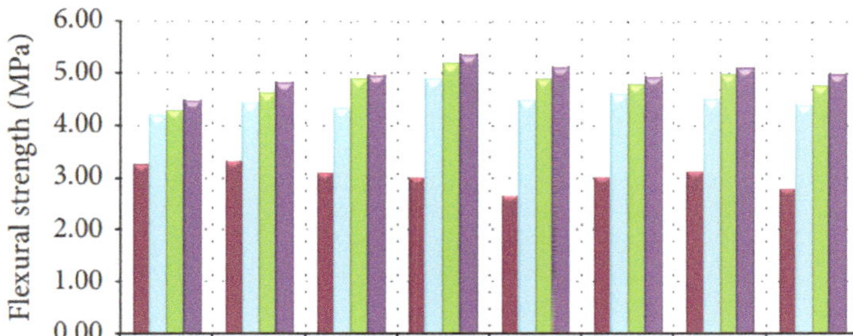

	GG 0%, GSF-0% GGC1	GG 0%, GSF-1.5% GSF5	GG 20%, GSF-1.5% GSF6	GG 40%, GSF-1.5% GSF7	GG 60%, GSF-1.5% GSF3	ACI 0.3%, GSF-1.5% ACI-1	ACI 0.6%, GSF-1.5% ACI-2	ACI 0.9%, GSF-1.5% ACI-3
28 days	3.24	3.33	3.08	3.00	2.65	2.99	3.10	2.80
56 days	4.21	4.45	4.34	4.90	4.50	4.60	4.50	4.39
90 days	4.30	4.62	4.89	5.20	4.90	4.79	4.98	4.75
180 days	4.45	4.81	4.95	5.34	5.10	4.93	5.10	5.01

Figure 12: Flexural strength of concrete specimens cured in sodium chloride solution for various concrete mixes (subjected to 40% initial stress).

The effects of corrosion in different concrete mixes were analyzed after failure and the degree of corrosion occurring in concrete specimens were shown in Figure 13. The reduction in diameter of steel fibres showed the degree of corrosion and was represented in Figure 14. It can be easily seen in plain concrete mixes that the degree of corrosion in steel fibres was higher up to 52.00% (GSF5). With the incorporation of slag up to 60%, the reduction of corrosion in steel fibres was found to be 25.33% (GSF8). Also, the corrosion potential was reduced significantly with the addition of corrosion inhibitor in plain steel fibre concrete and showed a reduction up to 6.66% (ACI-3). The reduction in diameter of various steel fibres in different concrete mixes showed the degree of corrosion occurring in concrete (as seen in Figure 15). This can possibly give the qualitative and quantitative assessment on the degree of corrosion occurring in steel fibre concrete specimens.

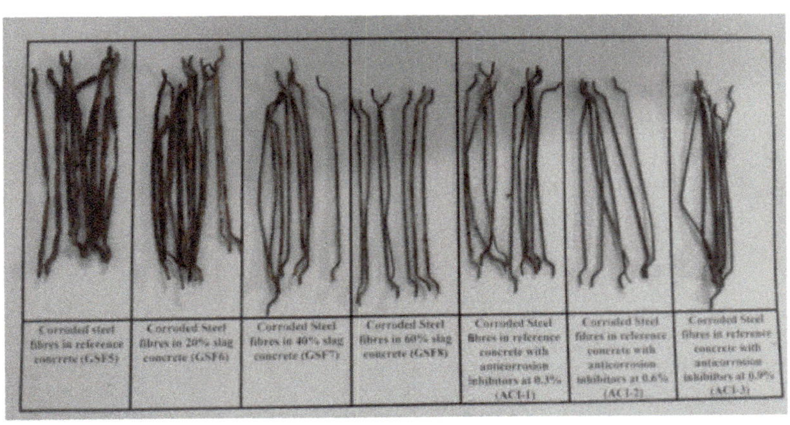

Figure 13: Snapshot of corroded steel fibres in different concrete mixes containing slag and corrosion inhibitor.

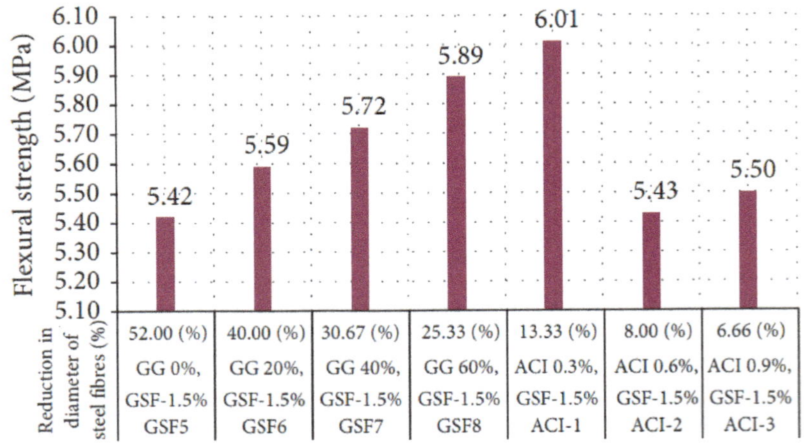

Figure 14: Variation of flexural strength with respect to % of reduction in thickness of steel fibres.

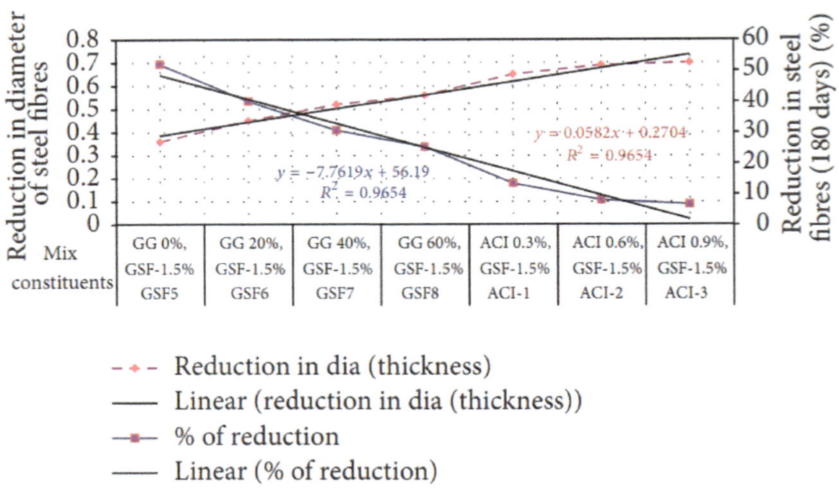

Figure 15: Variation of reduction in steel fibre diameter and the percentage reduction in steel fibre diameter for various concrete mixes.

CONCLUSIONS

Based on the experimental investigation, the following conclusions are drawn within the limitations of the test results.

1. Compressive properties of steel fibre concrete specimens were found to be affected when exposed to accelerated corrosion process due to rapid deterioration.
2. Alternate wetting and drying cycles of all plain steel fibre concrete mixes in normal water curing and salt curing showed rapid corrosion process after 28 days of curing period.
3. The influence of initial stress on the steel fibre concrete specimens reported faster initiation of corrosion process in steel fibres present in the concrete system. This provides a reliable estimate for predicting the corrosion potential in different concrete systems.
4. Compressive properties were found to be favourably improved in slag incorporated steel fibre concretes compared to plain steel fibre concretes.
5. Compared to concrete specimens subjected to normal water curing, the specimens immersed in salt solution showed faster deterioration in terms of concrete surface deterioration as well as corrosion of steel fibres.
6. Strength increase was noticed in all steel fibre concrete specimens incorporating slag and corrosion inhibitors. This could be possibly due to improved micostructural properties of slag concrete and the corrosion inhibitors possibly reduced the intensity of steel fibre corrosion.
7. Compressive strength was found to be higher (54.49 MPa) in 40% slag substituted concretes as well as with the addition of corrosion inhibitor at 0.6% showing 52.50 MPa.
8. Flexural strength results also showed improvement with the slag substitution (60%) as well as corrosion inhibitor (0.9%) added steel fibre concrete mixes. A maximum flexural strength of 5.89 Mpa was noticed in slag substituted concrete mixes subjected to severe wetting and drying cycles accompanied with initial stress up to 40%.
9. The intensity of corrosion was measured with the reduction of steel fibre diameter and can predict the degree of corrosion measured in terms of percentage of corrosion potential. A maximum reduction in corrosion percentage up to 25.33% and 6.67% was noticed in slag and corrosion inhibitor substituted concrete mixes, respectively.

10. Corrosion potential was found to be reduced in the case of slag substituted steel fibre concrete mixes up to a maximum of 25.33%.

REFERENCES

1. M. Otieno, H. Beushausen, and M. Alexander, "Effect of chemical composition of slag on chloride penetration resistance of concrete," Cement and Concrete Composites, vol. 46, pp. 56–64, 2014.

2. S. Teng, T. Y. D. Lim, and B. Sabet Divsholi, "Durability and mechanical properties of high strength concrete incorporating ultra fine ground granulated blast-furnace slag," Construction and Building Materials, vol. 40, pp. 875–881, 2013.

3. M. O›Connell, C. McNally, and M. G. Richardson, "Performance of concrete incorporating GGBS in aggressive wastewater environments," Construction and Building Materials, vol. 27, no. 1, pp. 368–374, 2012.

4. P. Pipilikaki, M. Katsioti, and J. L. Gallias, "Performance of limestone cement mortars in a high sulfates environment," Construction and Building Materials, vol. 23, no. 2, pp. 1042–1049, 2009.

5. C. D. Atiş and O. Karahan, "Properties of steel fiber reinforced fly ash concrete," Construction and Building Materials, vol. 23, no. 1, pp. 392–399, 2009.

6. S. Pavía and E. Condren, "Study of the durability of OPC versus GGBS concrete on exposure to silage effluent," Journal of Materials in Civil Engineering, vol. 20, no. 4, pp. 313–320, 2008.

7. A. Oner and S. Akyuz, "An experimental study on optimum usage of GGBS for the compressive strength of concrete," Cement and Concrete Composites, vol. 29, no. 6, pp. 505–514, 2007.

8. Building Research Establishment, BRE Special Digest 1: Concrete in Aggressive Ground. Part 1: Assessing the Aggressive Chemical Environment, BRE, 2nd edition, 2003.

9. ACI Committee, "Protection of metals in concrete against corrosion," ACI 222R-01, American concrete Institute, Farmington Hills, Mich, USA, 2001.

10. S. U. Al-Dulaijan, M. Maslehuddin, M. M. Al-Zahrani, A. M. Sharif, M. Shameem, and M. Ibrahim, "Sulfate resistance of plain and blended cements exposed to varying concentrations of sodium sulfate,"Cement and Concrete Composites, vol. 25, no. 4-5, pp. 429–437, 2003.

11. G. J. Osborne, "Durability of Portland blast-furnace slag cement concrete," Cement and Concrete Composites, vol. 21, no. 1, pp. 11–21, 1999.

12. R. S. Gollop and H. F. W. Taylor, "Microstructural and microanalytical studies of sulfate attack. V. Comparison of different slag blends," Cement and Concrete Research, vol. 26, no. 7, pp. 1029–1044, 1996.

13. V. Pavlík and S. Unčík, "The rate of corrosion of hardened cement pastes and mortars with additive of silica fume in acids," Cement and Concrete Research, vol. 27, no. 11, pp. 1731–1745, 1997.

14. N. de Belie, H. J. Verselder, B. de Blaere, D. van Nieuwenburg, and R. Verschoore, "Influence of the cement type on the resistance of concrete to feed acids," Cement and Concrete Research, vol. 26, no. 11, pp. 1717–1725, 1996.

15. E. K. Attiogbe and S. H. Rizkalla, "Response of concrete to sulfuric acid attack," ACI Materials Journal, vol. 85, no. 6, pp. 481–488, 1988.

16. BIS (Bureau of Indian Standards), IS 383: Specification for Coarse and Finer Aggregates from Natural Sources for Concrete, Bureau of Indian Standards, New Delhi, India, 1970.

17. BIS (Bureau of Indian Standards), IS 10262-2009: Concrete Mix Proportioning—Guidelines, New Delhi, India, 2009.

CHAPTER 16

STUDY ON STRENGTH AND MICROSTRUCTURE OF CEMENT PASTES CONTAINING LIMESTONE POWDER UNDER FLOWING ACID SOLUTION CONDITION

Shuhua Liu, Lihua Li, Zhigang Wang, Jiafeng Wang, and Meijuan Rao

INTRODUCTION

Concrete and mortar are always affected by physical and chemical attack under environmental water conditions and therefore their microstructure will be deteriorated, strength of the structure will decrease, and finally the architectures will be destroyed [1, 2]. There are mainly five kinds of environmental water attack, among which the acid attack is the most common. Acidic materials are such widely spread as in the atmosphere, in the soil contiguous with buildings, in the industrial wastes, and in the marine environments. Cement paste as an alkali compound is easily to be corroded under acid attack.

Modern cements often incorporate several mineral admixtures, one of which is limestone powder. The use of Portland cement containing limestone powder is a common practice in Europe. European standard EN 197 identifies two types of Portland limestone cements (PLC): type II/A-L containing 6%–20% and Type II /B-L containing 21%–35% [3–5]. The use of limestone powder can improve properties of concrete, decrease the costs, and reduce CO_2 and NO_x emissions during cement manufacture [6–8]. But there are still problems concerning the application of concrete mixed with limestone powder. The main composition of limestone powder is calcium carbonate which is easily to be suffered from acid attack. Although they have been implemented about this kind of concrete, most of the researches were limited to the static erosion environment [9–12]. While in practical projects, the concrete structures such as piers and foundations are usually located where the seepage of groundwater is positive or the erosion of the flowing water is intense. How the flowing acid attack resistance of cement paste changes after incorporating limestone powder still remains uncertain.

In this paper, cement pastes with different mix proportions and different contents of limestone powder were prepared. These pastes were then soaked in the flowing acetic acid solution with pH value of 4 and sulfuric acid solution with pH value of 2, respectively. The changes of strength and microstructure of pastes after different flowing acid attack periods were investigated by using strength test, X-ray diffraction (XRD), and scanning electron microscopy (SEM) techniques, which will indicate the effect of limestone powder on flowing acid attack resistance mechanism of cement paste.

EXPERIMENTAL

Raw Materials

The mixtures were prepared with ordinary Portland cement PO 42.5 (the Chinese standard GB175-2007), limestone powder, fly ash, and silica fume. Limestone powder, produced from carboniferous limestone with a very high purity (95% of $CaCO_3$ content), was added as filler. The particle size

Citation: Shuhua Liu, Lihua Li, Zhigang Wang, Jiafeng Wang, and Meijuan Rao, "Study on Strength and Microstructure of Cement Pastes Containing Limestone Powder under Flowing Acid Solution Condition," ISRN Ceramics, vol. 2012, Article ID 719636, 6 pages, 2012. doi:10.5402/2012/719636.

distribution of Portland cement, limestone powder, and fly ash measured by laser diffraction is shown in Figure 1. Obviously, the particle size of limestone powder is much smaller than those of Portland cement and fly ash. The dominant particle size of limestone powder is below 5 μm.

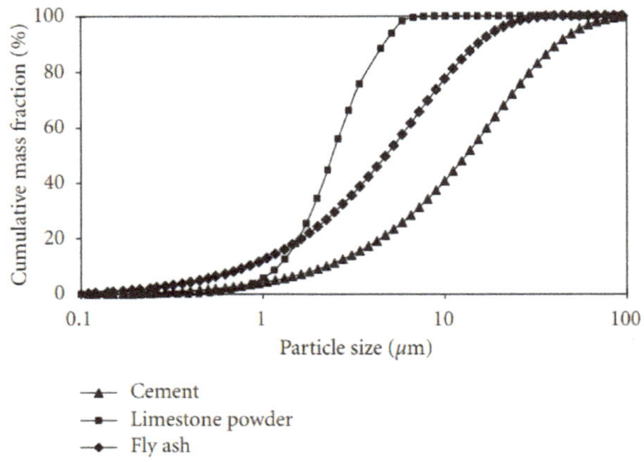

Figure 1: Particle size distributions of cement, limestone powder and fly ash.

Mix Proportions

In order to study the effect of limestone powder on flowing acid attack resistance of cement paste, two different water-binder ratio serials including the low value of 0.3 and the high value of 0.5 were set. By changing the content of limestone powder and mixing with fly ash or silica fume, 10 different pastes were prepared. The specific mix proportions are shown in Table 1.

Table 1: Mix proportion of the cement pastes

Sample	Water-binding materials	Water	Cement	Limestone powder	Flying ash	Silica fume
L-1	0.3	0.3	1	0	0	0
L-2	0.3	0.3	0.7	0.3	0	0
L-3	0.3	0.3	0.5	0.5	0	0
L-4	0.3	0.3	0.5	0.25	0.25	0
L-5	0.3	0.3	0.5	0.4	0	0.1
H-1	0.5	0.5	1	0	0	0
H-2	0.5	0.5	0.7	0.3	0	0
H-3	0.5	0.5	0.5	0.5	0	0
H-4	0.5	0.5	0.5	0.25	0.25	0
H-5	0.5	0.5	0.5	0.4	0	0.1

Paste specimens with size of $20 \times 20 \times 20 mm^3$ were cast according to Table 1. After molding 24 hours, the pastes were removed from the mold and then placed for standard curing till 28 d when their strength was tested as reference. Thereafter, they were soaked in two different solutions: the flowing acetic acid solution with pH value of 4 and the flowing sulfuric acid solution with pH value of 2. The strength change, XRD, and SEM analysis were studied, respectively. The actual devices simulating flowing solution environment are shown in Figure 2.

(a) acetic acid attack

(b) sulfuric acid attack

Figure 2: Devices of simulating flowing (a) acetic acid solution and (b) sulfuric acid solution.

XRD

XRD measurements were implemented on a Philips X'Pert diffractometer equipped with a graphite monochromator using Cu Kα radiation and operating at 40 kV and 20 mA. Step scanning was performed with a scan speed of 2°/min and sampling interval of 0.02°/2θ. XRD was used to identify the hydrates in the cement pastes containing limestone powder.

RESULTS AND DISCUSSIONS

Strength

The compressive strength of the pastes soaked in flowing acid solution for different periods is shown from Figure 3 to Figure 6. It can be concluded from the four diagrams that compressive strength of the pastes would firstly decrease and then increase slightly and finally decrease again. The strength would decrease correspondingly with the increase of water-binder ratio and the content of limestone powder. Strength of the pastes incorporating fly ash and silica fume declines more slowly than that of pastes only containing limestone powder.

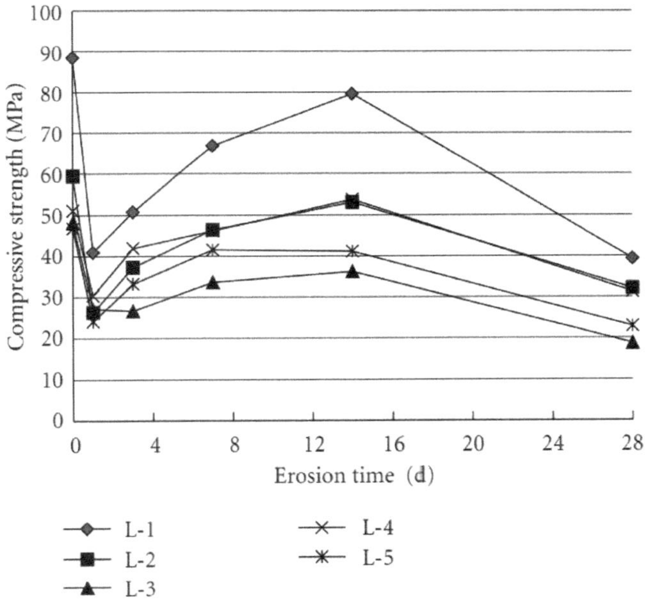

Figure 3: Compressive strength of pastes with W/B = 0.3 soaked in flowing acetic acid solution over different time.

Figures 3 and 4 show the early strength of pastes decreases in flowing acetic acid solution because calcium hydroxide and calcium carbonate reacted with the acid and the product calcium acetate will leach out. Then unhydrated cement particles continue their hydration which leads pastes' strength to recover slightly. At last, with the further acetic acid attack into the interior, calcium hydroxide will be consumed and calcium carbonate will be dissolved. Thereafter other hydration products decomposed and as a result the strength descended again.

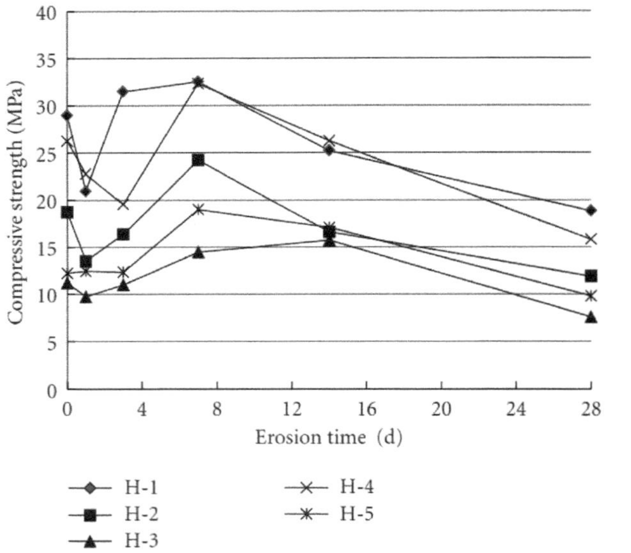

Figure 4: Compressive strength of pastes with W/B = 0.5 soaked in flowing acetic acid solution over different time.

Figures 5 and 6 show the early strength of pastes decreases in the flowing sulfuric acid solution, possibly because calcium hydroxide and calcium carbonate in the pastes reacted with the acid and the products leach out. Then unhydrated cement particles' proceeding hydration and the swelling effect of the generated gypsum lead the pastes strength to recover slightly. At last, with the further sulfuric acid attack into the interior, simultaneously the generated gypsum destroys the pastes' structure and consequently the strength descends again. In comparison with the flowing acetic acid solution, strength of the pastes soaked in the flowing sulfuric acid solution declines more slowly possibly due to the depression effect of the flowing liquid onto the gypsum's crystallization which is commonly the main reason to cause expansion of the paste under sulfuric acid attack. As is indicated in Figure 2, the feculent sulfuric acid solution, unlike the clear appearance presented in the acetic acid solution, suggests gypsum was dissolved out. The pictures provide a convincing demonstration of the explanation.

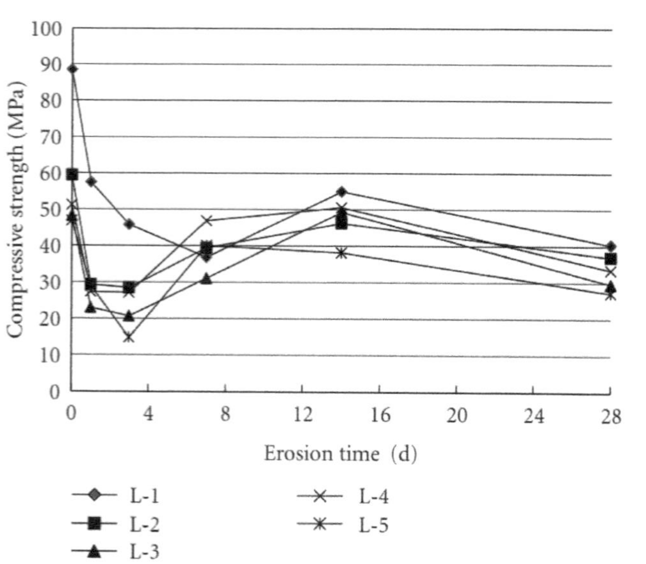

Figure 5: Compressive strength of pastes with W/B = 0.5 soaked in flowing sulfuric acid solution over different time.

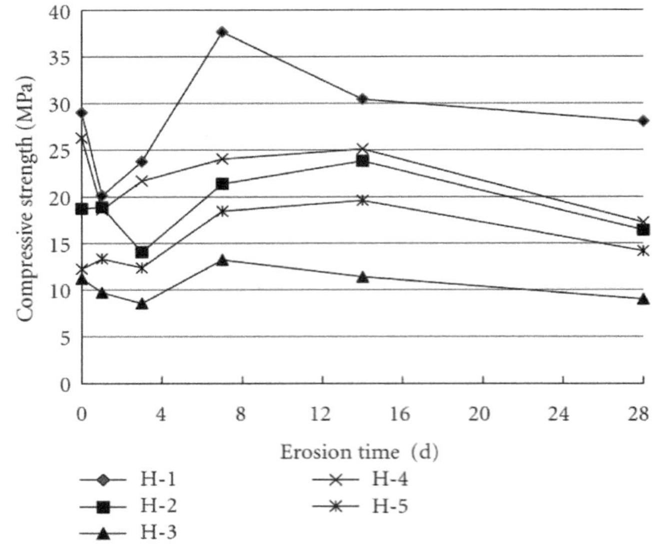

Figure 6: Compressive strength of pastes with W/B = 0.5 soaked in flowing sulfuric acid solution over different time.

As for the influence of water-binder ratio, pastes with high W/B value are more easily to be damaged because of their loose microstructure and high porosity. As for the effect of limestone powder, the higher the content, the more paste strength decreases because calcium carbonate, dominant composition of the limestone powder, could react with the acid and the reaction products will leach out. When mixed with an additional cementing material, fly ash, or silica fume, the descending tendency of strength is not that intense because the content of limestone powder decreases. What is more, both fly ash and silica fume could react with calcium hydroxide and the hydration products are beneficial to enhance the strength. Subsequently, the content of calcium hydroxide decreases and so is its degree of reaction with the sulfuric acid.

Microstructure

Figure 7 shows the results of XRD analysis of hydrates of cement pastes with W/B value of 0.5 soaked in the flowing acetic acid solution at 28 d. There are calcium carbonate peak caused by the mixing of limestone powder and calcium hydroxide peak caused by the hydration. The higher the content of limestone powder, the higher the calcium carbonate peak. Calcium carbonate peak is much higher than calcium hydroxide peak. Calcium hydroxide peak is very low because they are consumed in their reactions with the acetic acid.

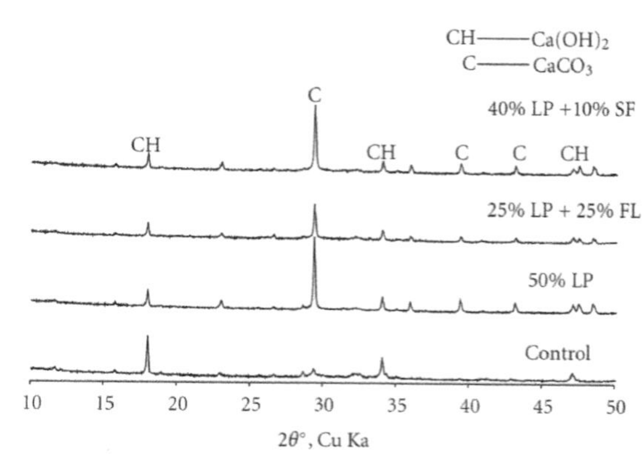

Figure 7: XRD patterns of hydrates of cement pastes with W/B = 0.5 soaked in flowing acetic acid solution at 28 d.

Figure 8 shows the results of XRD analysis of hydrates of cement pastes with W/B value of 0.5 soaked in the flowing sulfuric acid solution at 28 d. There are calcium carbonate peak, calcium hydroxide peak, gypsum peak and ettringite peak. Calcium carbonate peak is the highest, and the higher the content of limestone powder, the higher the calcium carbonate peak. Calcium hydroxide peak is very low because they are consumed in their reactions with the sulfuric acid which leads to the generation of gypsum. The gypsum could then react with C-A-H and ettringite is generated. The expansion of pastes caused by gypsum and ettringite should account for the destruction of microstructure and the loss of strength. The figure also indicates that the content of gypsum is relatively low because of the depression effect of the flowing solution onto the gypsum's crystallization. This is consistent with the explanation in the results of strength tests.

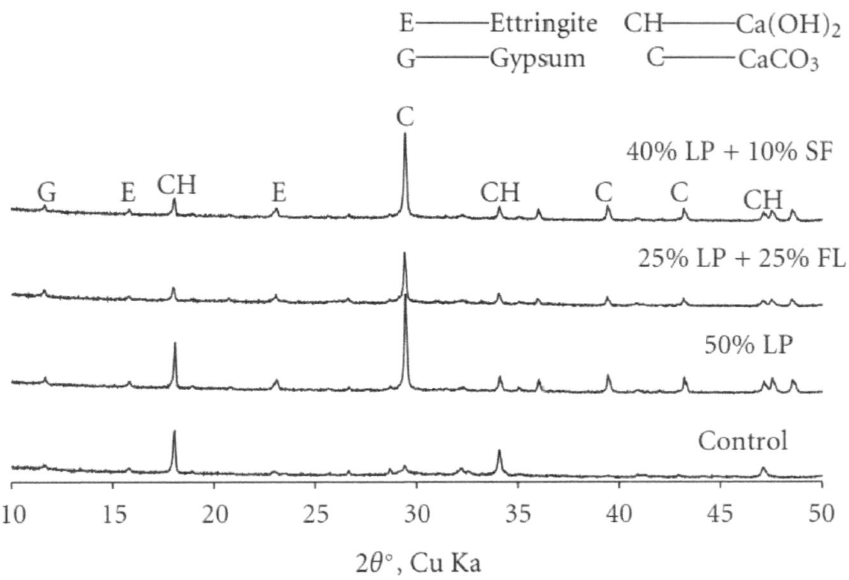

Figure 8: XRD patterns of hydrates of cement pastes with W/B = 0.5 soaked in flowing sulfuric acid solution at 28 d.

Fragments of specimens broken off and washed with acetone were examined by SEM. Figure 9 shows the SEM pictures of sample H-3 soaked in the flowing acetic acid solution at 28 d. C-S-H gel, ettringite and calcium carbonate crystals can be found. But calcium hydroxide crystals are not found because they were consumed in their reactions with the acid. This is consistent with the results of XRD analysis.

Figure 9: SEM pictures of sample H-3 soaked in flowing acetic acid solution at 28 d.

Figure 10 shows the SEM pictures of sample H-3 soaked in the flowing sulfuric acid solution at 28 d. C-S-H gel, calcium hydroxide and ettringite crystal can be found in it. But the gypsum is not that much because of the distorting effects of the flowing solution onto the gypsum's crystallization which is consistent with the results of XRD analysis.

(a) (b)

Figure 10: SEM pictures of sample H-3 soaked in flowing sulfuric acid solution at 28 d.

CONCLUSIONS

1. The compressive strength of the cement pastes would firstly decrease and then increase slightly and finally decreases again under flowing acid attack. The strength will decrease correspondingly with the increase of the water binder ratio and content of the limestone powder. Strength of pastes mixed with fly ash of silica fume decreases slowly than those of pastes only containing limestone powder.
2. Cement pastes with high value of W/B are more easily to be damaged because of their loose structure and high porosity. Calcium carbonate in limestone powder could react with the acid and the higher content of it, the faster the strength decreases. When mixed with fly ash or silica fume, they could both react with calcium hydroxide and C-S-H is generated which not only is good to enhance the pastes' strength but also weakens the degree of reactions between the hydration products and the acid.
3. Microstructural analysis manifests on condition of the flowing acetic acid attack, destruction of the microstructure, and loss of the strength are caused by the reactions between calcium hydroxide and the acid which make the deterioration proceed from the exterior to the interior. While under the flowing sulfuric acid attack, crystallizations of the gypsum and ettringite lead to the expansion of the pastes. The flowing solution environment inhibits the crystallization of gypsum and therefore delays the attack process.

ACKNOWLEDGMENT

This project was supported by the Fundamental Research Funds for the Central Universities (111039).

REFERENCES

1. S. Liu and P. Yan, "Experimental study on sodium sulfate attack on mortar containing limestone powder and fly ash [C]," in Proceedings of the International Conference on Durability of Concrete Structures, pp. 542–548, Hangzhou, China, November 2008.
2. S. Liu and P. Yan, "Effect of limestone powder on microstructure of concrete," Journal of Wuhan University of Technology, vol. 25, no. 2, pp. 328–331, 2010.
3. European Committee for Standardization, "Cement: composition, specifications and conformity criteria, Part 1: Common cements [S]," EN 197-1, EN/TC51/WG 6 rev., 2000.
4. M. Schmidt, "Cement with interground materials—capabilities and environmental relief, part 1,"Zement-Kalk-Gips, vol. 45, no. 4, pp. 87–92, 1992.
5. B. Schiller and H. G. Ellerbrock, "The grinding and properties of cements with several main constituents," Zement-Kalk-Gips, vol. 45, no. 9, pp. 223–231, 1992.
6. J. Baron and C. Douvre, "Technical and economical aspects of the use of limestone filler additions in cement," World Cement, vol. 18, no. 3, pp. 100–104, 1987.
7. S. Sprung and E. Siebel, "Assessment of the suitability of limestone for producing Portland limestone cement (PKZ)," Zement-Kalk-Gips, vol. 44, no. 1, pp. 1–11, 1991.
8. V. Bonavetti, H. Donza, G. Menéndez, O. Cabrera, and E. F. Irassar, "Limestone filler cement in low w/c concrete: a rational use of energy," Cement and Concrete Research, vol. 33, no. 6, pp. 865–871, 2003.
9. A. Bertron, J. Duchesne, and G. Escadeillas, "Attack of cement pastes exposed to organic acids in manure," Cement and Concrete Composites, vol. 27, no. 9-10, pp. 898–909, 2005.
10. B. Lothenbach, G. Le Saout, E. Gallucci, and K. Scrivener, "Influence of limestone on the hydration of Portland cements," Cement and Concrete Research, vol. 38, no. 6, pp. 848–860, 2008.
11. J. Dweck, P. M. Buchler, A. C. V. Coelho, and F. K. Cartledge, "Hydration of a Portland cement blended with calcium carbonate," Thermochimica Acta, vol. 346, no. 1-2, pp. 105–113, 2000.
12. E. F. Irassar, V. L. Bonavetti, and M. González, "Microstructural study of sulfate attack on ordinary and limestone Portland cements at ambient temperature," Cement and Concrete Research, vol. 33, no. 1, pp. 31–41, 2003.

CHAPTER 17

MECHANICAL ANALYSIS OF STRESS DISTRIBUTION IN A CARBON FIBER-REINFORCED POLYMER ROD BONDING ANCHOR

Peng Feng, Pan Zhang, Xinmiao Meng and Lieping Ye

Keywords: Carbon fiber-reinforced polymer (CFRP) rod, stress distribution, bonding anchor, anchorage efficiency

INTRODUCTION

Carbon fiber-reinforced polymer (CFRP) has a high strength, is light weight, with excellent fatigue performance and non-corrosive properties and is, thus, widely used in the structures of buildings and bridges in various forms. As an effective form of CFRP, pultruded CFRP rods or cables can better utilize the unique advantages mentioned above. A cable composed of CFRP rods, as shown in Figure 1a, is made up of unidirectional carbon fibers and the epoxy resin, manufactured by the pultrusion process. The section of a CFRP rod by scanning electron microscope (SEM) in Figure 1b shows that the unidirectional fibers are embedded in the polymer resin.

The specific strength of the CFRP cable is fairly high compared to that of a steel cable, while the density of CFRP is one-fourth that of steel. This makes it more suitable for large-span structures, such as large-span roof systems [1,2,3] and cable-stayed bridges [4,5]. The corrosion of steel cables in the long run can also be avoided by using CFRP cables.

However, the anchorage system is the key factor of whether CFRP cables can reach tensile strength in service. Compared to steel, the anisotropy and brittleness of CFRP make it more difficult to anchor. Currently, anchorage systems can be roughly divided into the clamp anchor, bonding anchor and barrel-wedge anchor [6]. The CFRP rod bonding anchorage system is made up of a metal barrel, CFRP rods and bonding medium, usually epoxy resin or cement grout. The surface of the CFRP rod can be threaded or sandblasted. The anchoring force is via chemical bonding, mechanical interlocking and friction at the interface between the bonding medium and the CFRP rod.

Several CFRP rod anchorage systems were studied. Parametric analysis of the barrel-wedge-type anchor has been reported. A four-wedge type anchor used to grip a single CFRP rod was proposed [7]. The effect of the friction coefficient, the barrel-wedge angle difference and the soft metal sleeve on the stress distribution and stiffness of the anchor was studied [8]. Similar research has shown that the friction coefficient, preload value and sleeve material has a significant effect on the anchorage performance [9,10,11]. Mayah also mentioned that the use of arc barrel-wedge surfaces with varying inclined angles in a longitudinal direction might reduce concentrated stresses at the loading end [12]. For the bonding anchor, the bond-slip relationship influences the stress distribution along the rod. The bond-slip relationship between the CFRP rod and the cement grout was investigated [13,14], and the results revealed that the surface condition of the rod, the expansion of the cement grout, the anchorage length and the elastic modulus of the barrel can influence the anchorage performance to different degrees.

There is some research on multi-rod anchors, but not as much as single-rod ones, because it is more difficult to grip a

Citation: Feng, P.; Zhang, P.; Meng, X.; Ye, L. Mechanical Analysis of Stress Distribution in a Carbon Fiber-Reinforced Polymer Rod Bonding Anchor. Polymers 2014, 6, 1129-1143. doi:10.3390/polym6041129.

collection of rods as opposed to only one. To anchor multiple rods, the bonding anchor is preferred over the barrel-wedge-type anchor. The CFCC (Carbon Fiber Composite Cable) strand in Japan already has a mature post-tensioning anchorage system using highly expansive cementing material used in the Penobscot Narrows Cable Stayed Bridge [15,16]. However, a reduction in strength may occur due to bending and torsion in the manufacturing process of the CFRP strands, which does not apply to parallel rods. A new way to anchor CFRP parallel rods is by changing the elastic modulus of the adhesive along the barrel [17]. Ceramic particles in different diameters have been mixed into the resin to change the elastic modulus of the adhesive. The modulus was low to high from the loading end to the free end, and the concentrated shear stress was significantly reduced by this method.

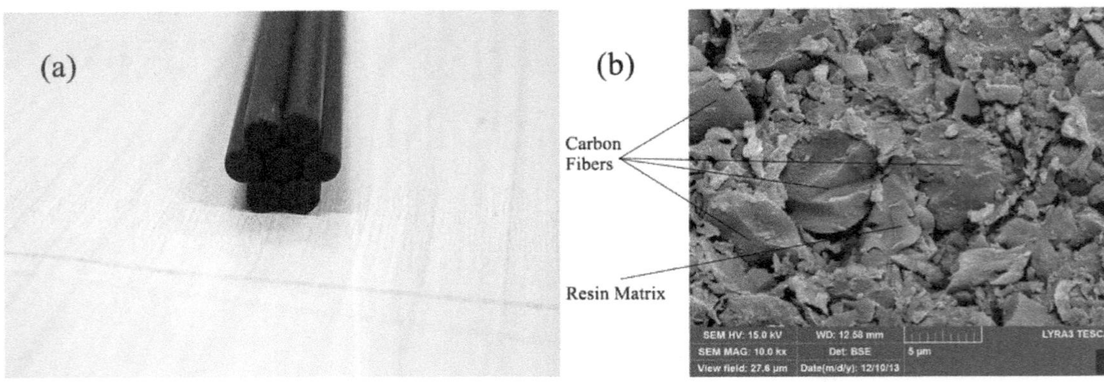

Figure 1. (**a**) A cable composed of carbon fiber-reinforced polymer (CFRP) pultruded rods; (**b**) scanning electron microscope (SEM) image of a rod section.

Based on the existing research, the bonding stress distribution of CFRP single-rod and multi-rod bonding anchors is analyzed by a theoretical model and a finite element program in this paper.

MECHANICAL ANALYSIS OF SHEAR STRESS DISTRIBUTION

The ultimate loading capacity of a CFRP rod anchor generally depends on the shear stress distribution along the rod. The more uniform the stress distribution, the higher the anchorage efficiency.

A theory was proposed to calculate the ultimate capacity of a straight-pipe bonding anchor [18]. The shear stress distribution along the rod was simplified to three parts. The bearing capacity of each part can be calculated through the integration of shear stress, and thus, the whole capacity of the anchor can be calculated. The error was less than 15% *versus* experimental data. Based on the theory above, a simplified model was suggested, such that each part of the shear stress distribution was considered to be a straight line. The ultimate capacity can be gained in the same way [19].

An elastic theoretical model is proposed in this paper to calculate shear stress referring to [18,19]. It generalizes the regularities of the shear stress distribution at the CFRP-adhesive interface. This theory can calculate the shear stress distribution along a single rod or multiple rods in a straight-pipe bonding anchor within an elastic stage. The fundamental assumptions are listed as follows:

1. All the materials are elastic: the stress state of the anchorage system remains elastic.
2. Deformation of the metal barrel is not considered, because the stiffness of the barrel is much higher than that of the adhesive. Therefore, the internal surface of the barrel is considered fixed.
3. Interface slippage is ignored, so that the displacement compatibility can be founded at the interface between the barrel and adhesive (hereinafter, called the first interface) and the interface between the adhesive and CFRP rods (hereinafter, called the second interface).

Thus, the theory is suitable for preliminary shear stress analysis in an elastic state.

Single Rod

The cross-section of a single-rod anchor is shown in Figure 2a, and the theoretical model of a single rod is shown in Figure 2b. The length of the anchor is set as l, while the diameter of the CFRP rod and the radial thickness of the adhesive are d and t, respectively.

Tensile load (T_0) is applied at the loading end of the CFRP rod. We assume that the axial displacement of the rod has a uniform distribution in one section that is set as $u(x)$. The axial force distribution is set as $T(x)$. The interfacial shear stress at the first interface and the second interface is $\tau(x)$ and $\tau_b(x)$, while x represents the axial coordinate. The value is zero at the free end and one at the loading end of the anchor.

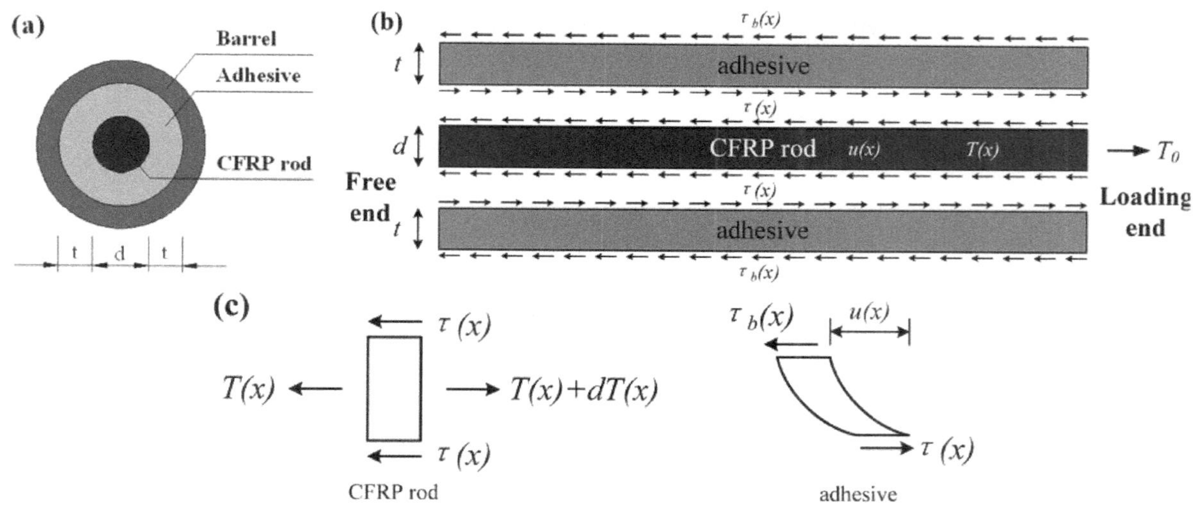

Figure 2. (**a**) Cross-section of a single-rod anchor; (**b**) theoretical model of a single rod; (**c**) stress state of a single-rod anchor.

Figure 2c shows the stress state of the CFRP rod and adhesive within infinitesimal length dx in the axial direction. For the CFRP rod, the equilibrium and geometric equation is easily obtained:

$$\tau(x) = \frac{1}{\pi d}\frac{dT(x)}{dx} \tag{1}$$

$$T(x) = E_c A\frac{du(x)}{dx} \tag{2}$$

E_c is the longitudinal elastic modulus of the CFRP rod, and A is the section area. For the adhesive, shear stress at the first and second interface can be considered equilibrant forces to obtain Equation (3):

$$\pi d\tau(x)dx = \pi(d + 2t)\tau_b(x)dx \tag{3}$$

Assuming that the shear strain has a linear distribution in the radial direction, the average shear strain of the adhesive can be gained with Equation (3):

$$\bar{\gamma}(x) = \frac{1}{2G}[\tau(x) + \tau_b(x)] = \frac{d+t}{d+2t}\frac{\tau(x)}{G} \tag{4}$$

Here, G represents the shear modulus of the adhesive. The relative displacement between the first and second interface is$u(x)$, and therefore, we can get the geometric equation of adhesive with Equation (4):

$$\frac{u(x)}{t} = \bar{\gamma}(x) = \frac{d+t}{d+2t}\frac{\tau(x)}{G} \tag{5}$$

With Equations (1), (2) and (5) and setting $u(x)$ as the variable, we get a differential equation that is satisfied at any infinitesimal point:

$$u''(x) - k^2u(x) = 0 \tag{6}$$

In Equation (6), $k^2 = \frac{d+2t}{d+t}\frac{4G}{E_c dt}$. The general solution of Equation (6) is $u(x) = C_1e^x + C_2e^{-x}$. With boundary conditions at the free end of $u(0) = 0$ and the loading end $T(l) = T_0$, the value of C_1 and C_2 can be calculated as below:

$$C_1 = \frac{d+t}{d+2t}\frac{T_0kt}{\pi Gd(e^{kl}+e^{-kl}-2)};\ C_2 = -\frac{d+t}{d+2t}\frac{T_0kt}{\pi Gd(e^{kl}+e^{-kl}-2)} \tag{7}$$

Therefore, the shear stress distribution can be gained:

$$\tau(x) = \frac{d+2t}{d+t}\frac{G}{t}(C_1e^x + C_2e^{-x}) \tag{8}$$

Multiple Rods

The stress distribution in multiple rods can also be derived from theory. A common arrangement is multiple rods symmetrically anchored with one rod in the center and other rods arranged in a ring (Figure 3a). The diameter of the rods is d. The radial thickness of the adhesive between the center rod and the side rods is t_1, while it is t_2 between side rods and the barrel. The other parameters are the same as the single-rod anchor system described in Section 2.1.

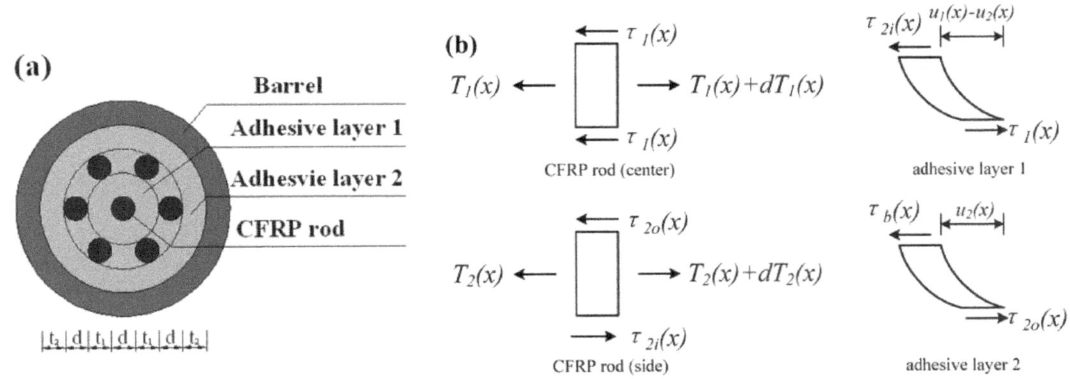

Figure 3. (**a**) Cross-section of a multi-rod anchor; (**b**) stress state of a multi-rod anchor.

In contrast to the single rod, the shear stress distribution on the surface of the side rods is not uniform along the perimeter. To simplify the theory, the adhesive is considered to be two layers of concentric rings, and the shear stress on the side rods is divided into the inner stress distribution, $\tau_{2i}(x)$, and the outer stress distribution, $\tau_{2o}(x)$. Stress along the center rod is set as $\tau_1(x)$. Axial displacements of the center rod and side rods are expressed as $u_1(x)$ and $u_2(x)$, while the axial forces are $T_1(x)$ and $T_2(x)$. The stress state of the CFRP rod and adhesive within an infinitesimal length is shown in Figure 3b.

According to the equilibrium and interaction relationships from Figure 3b, the inner shear stress distribution on the side rods is in the opposite direction to the others. This indicates that the stress state of multi-rod anchors is more complicated than the single-rod case. The equilibrium and geometric equations for the center rod and side rods can also be described.

$$\tau_1(x) = \frac{1}{\pi d}\frac{dT_1(x)}{dx};\, T_1(x) = E_c A\frac{du_1(x)}{dx};$$
$$\frac{\tau_{2o}(x) - \tau_{2i}(x)}{2} = \frac{1}{\pi d}\frac{dT_2(x)}{dx};\, T_2(x) = E_c A\frac{du_2(x)}{dx} \qquad (9)$$

Through simplification, we get:

$$E_c A\frac{d^2u_1(x)}{dx} = \pi d\tau_1(x) \qquad (10)$$

$$E_c A\frac{d^2u_2(x)}{dx} = \frac{\pi d}{2}[\tau_{2o}(x) - \tau_{2i}(x)] \qquad (11)$$

The shear force for each layer of adhesive is also considered to be approximately equal:

$$\pi d\tau_1(x)dx = \pi(d + 2t_1)\tau_{2i}(x)dx \qquad (12)$$
$$\pi(3d + 2t_1)\tau_{2o}(x)dx = \pi(3d + 2t_1 + 2t_2)\tau_b(x)dx \qquad (13)$$

It can be seen from Figure 3b that the relative shear displacement of adhesive Layer 1 is $u_1(x) - u_2(x)$ and $u_2(x)$ for adhesive Layer 2. We can derive the geometric equation of the adhesive with Equations (12) and (13) using the method described above:

$$G\frac{u_1(x) - u_2(x)}{t_1} = \frac{\tau_1(x) + \tau_{2i}(x)}{2} = \frac{d + t_1}{d + 2t_1}\tau_1(x) \qquad (14)$$

$$G\frac{u_2(x)}{t_2} = \frac{\tau_{20}(x) + \tau_b(x)}{2} = \frac{3d + 2t_1 + t_2}{3d + 2t_1 + 2t_2}\tau_{2o}(x) \qquad (15)$$

With Equations (10)–(15), the displacements, $u_1(x)$ and $u_2(x)$, remain as variables to be solved, and Equations (16) and (17) can be obtained:

$$\frac{d^2u_1(x)}{dx^2} = C_1u_1(x) - C_1u_2(x) \qquad (16)$$

$$\frac{d^2u_2(x)}{dx^2} = C_2u_1(x) + C_3u_2(x) \qquad (17)$$

In the equations above, the coefficients are listed respectively:

$$C_1 = \frac{4G}{E_c dt_1}\frac{d + 2t_1}{d + t_1}, C_2 = -\frac{2G}{E_c dt_1}\frac{d}{d + t_1}, C_3 = \frac{2G}{E_c d}[\frac{3d + 2t_1 + 2t_2}{(3d + 2t_1 + t_2)t_2} + \frac{d}{(d + t_1)t_1}] \qquad (18)$$

With Equations (16) and (17), the displacement can be solved. By eliminating $u_2(x)$, we get:

$$u_1^{(4)}(x) - (C_1 + C_3)u_1^{(2)}(x) + C_1(C_2 + C_3)u_1(x) = 0 \qquad (19)$$

The boundary conditions at the free end and the loading end are $u_1(0) = 0$, $u_2(0) = 0$, $T_1(0) = T_0$ and $T_2(0) = T_0$, which can be converted into conditions that only contain u_1, as listed in Equation (20):

$$u_1(0) = 0;\, u_1'(l) = \frac{T_0}{EA};\, u_1''(0) = 0;\, u_1'''(l) = 0 \qquad (20)$$

With Equations (19) and (20), $u_1(x)$ can be calculated to find the shear stress distribution.

FINITE ELEMENT ANALYSIS OF SHEAR STRESS DISTRIBUTION

The finite element method was used to study the stress distribution in the CFRP rod bonding anchor and to test the accuracy of the theoretical model. The anchor was modeled by the Solid45 element in ANSYS software, and nodes on the interface were merged, because slippage is not considered.

Single Rod Anchor

The CFRP rod, adhesive grout and steel tube were modeled, respectively (Figure 4a). Figure 4b shows the finite element model after meshing with a total element number of 192,000.

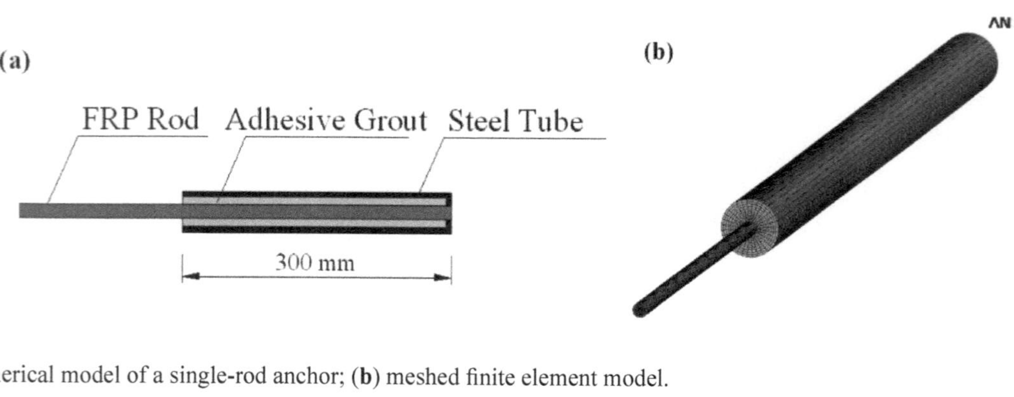

Figure 4. (**a**) Numerical model of a single-rod anchor; (**b**) meshed finite element model.

The material properties of CFRP and steel are presented in Table 1. The length of the anchor is 300 mm, and the diameter of the CFRP rod is 5 mm. The thickness of the steel barrel is 2.5 mm. Three kinds of parameter combinations are selected, as shown in Table 2. The elastic modulus and thickness of the adhesives are 5 GPa and 2 mm, 5 GPa and 5 mm and 2 GPa and 5 mm, respectively.

Table 1. Material properties (GPa)

Material	E_x	E_y, E_z	v_{xy}, v_{xz}	v_{yz}	G_{xy}, G_{xz}	G_{yz}
CFRP rod	181	10.05	0.28	0.3	7.17	7.17
Steel	200	/	0.27	/	78.74	/

Table 2. Numerical cases for the single-rod anchor

No.	1	2	3
Modulus of adhesive (GPa)	5	5	2
Thickness of adhesive (mm)	2	5	5

In the FE (Finite Element) model, the outer surface of the steel tube is restrained in all degrees. A tensile stress of 1500 MPa is then applied to the loading end of the CFRP rod. To compare the numerical and theoretical datasets, each case is also analyzed by theory in Section 2.1. Figure 5a,b displays the displacement and shear stress distribution of the CFRP rod in axial directions by the theory and finite element method. The solid line represents the finite element results, while the dotted line represents the theoretical results. The displacement and stress distribution results generally agree well, while there is some deviation in the shear stress value at the loading end in Figure 5b. This may be caused by stress discontinuity or approximation error in the application of the theory.

From Figure 5a, we notice that the stiffness of the single-rod anchor increases as the modulus of the adhesive increases or the thickness of the adhesive decreases. Figure 5b also shows that the peak value of shear stress increases as the modulus of the adhesive increases or the thickness of the adhesive decreases.

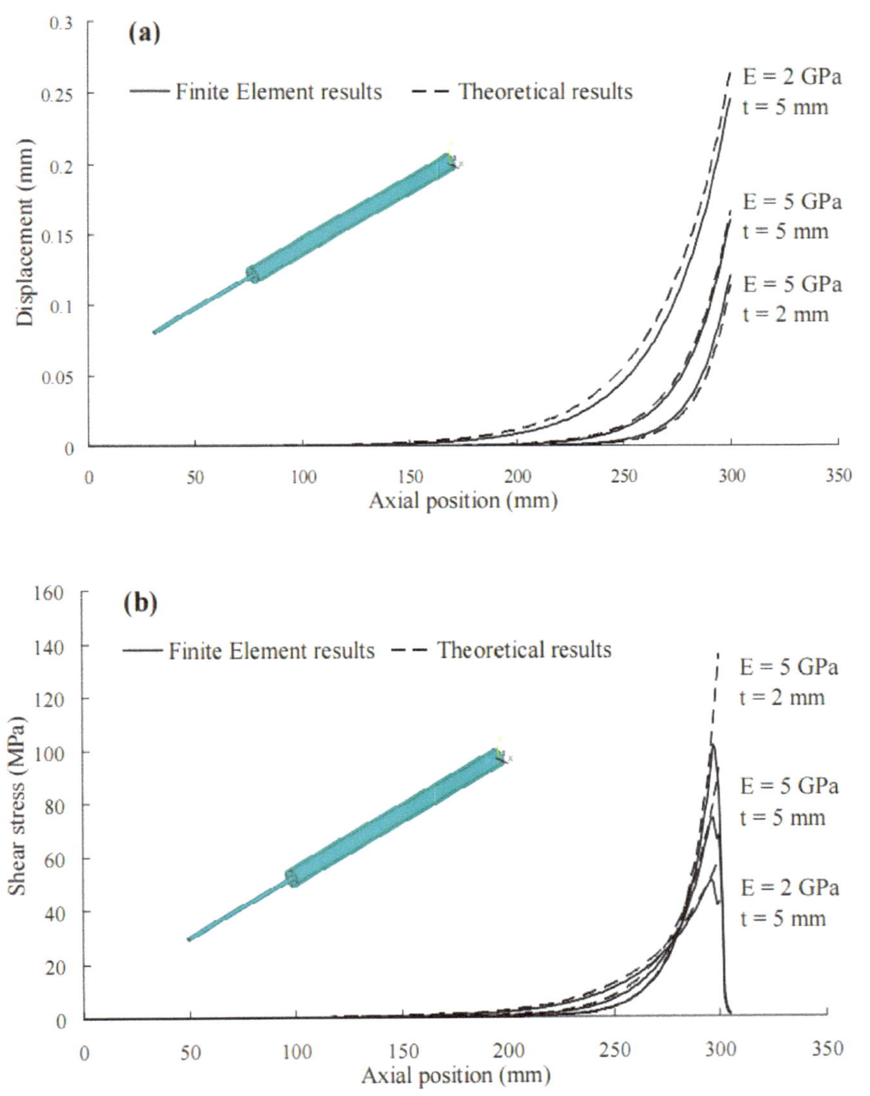

Figure 5. (**a**) Displacement distribution on the rod; (**b**) shear stress distribution on the rod.

Multi-Rod Anchor

Figure 6a models a seven-rod bonding anchor with a side view. The rods have a plum-shaped distribution in the adhesive within a certain spacing. There are a total of 245,520 elements after meshing. The material properties of the CFRP and steel are the same as above. The elastic modulus of the adhesive is 5 GPa. The length of the anchor is also 300 mm, and the diameter of the CFRP rod is 5 mm. The outside diameter of the anchor is 45 mm with a 5 mm-thick barrel.

Two kinds of parameter combinations are selected (Table 3). The dimensions of adhesive Layers 1 and 2 are 5 mm + 5 mm and 8 mm + 2 mm, respectively. The outside diameter of the anchor remains constant.

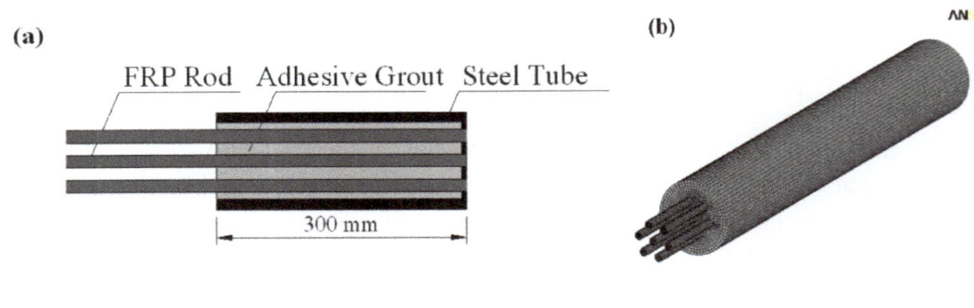

Figure 6. (**a**) Numerical model of a seven-rod anchor; (**b**) meshed finite element model.

Table 3. Numerical cases for the seven-rod anchor

No.	Thickness of adhesive Layer 1 (mm)	Thickness of adhesive Layer 2 (mm)
1	5	5
2	8	2

Here, the boundary conditions are the same as the single-rod anchor, and the tensile stress is still set as 1500 MPa on each rod. We used FE to obtain the axial displacement, the shear stress on the surface of the center rod, as well as the shear stress on the inner and outer surface of the side rods. Mathmetica 6.0 solved differential Equation (19). For each case, the displacement distribution gained by theory and the finite element method is shown in Figure 7a,b, while the shear stress distribution is shown in Figure 8a,b.

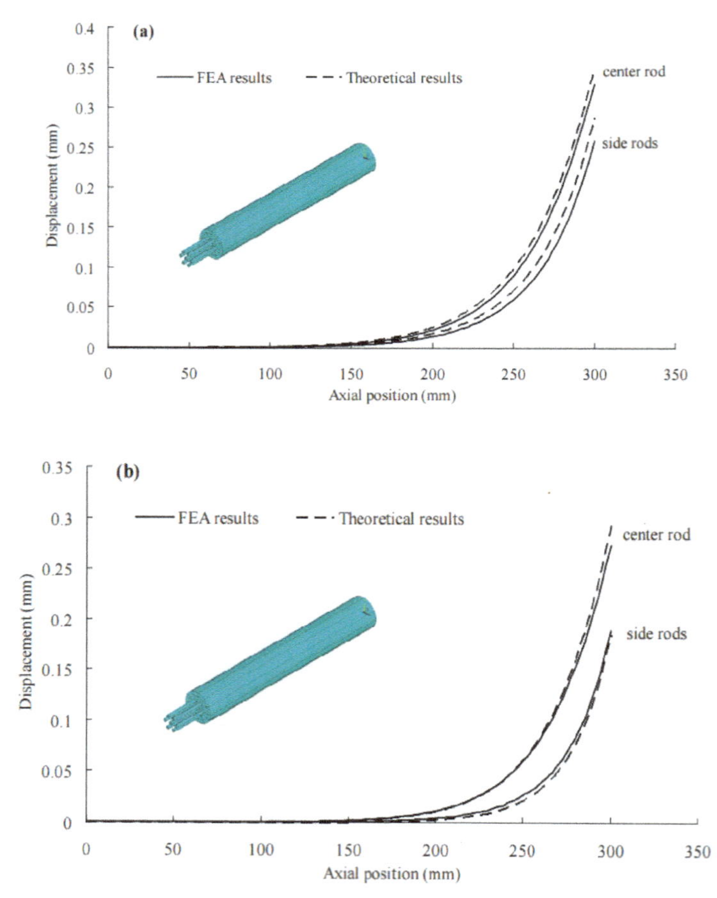

Figure 7. (**a**) Displacement distribution of Case 1; (**b**) displacement distribution of Case 2.

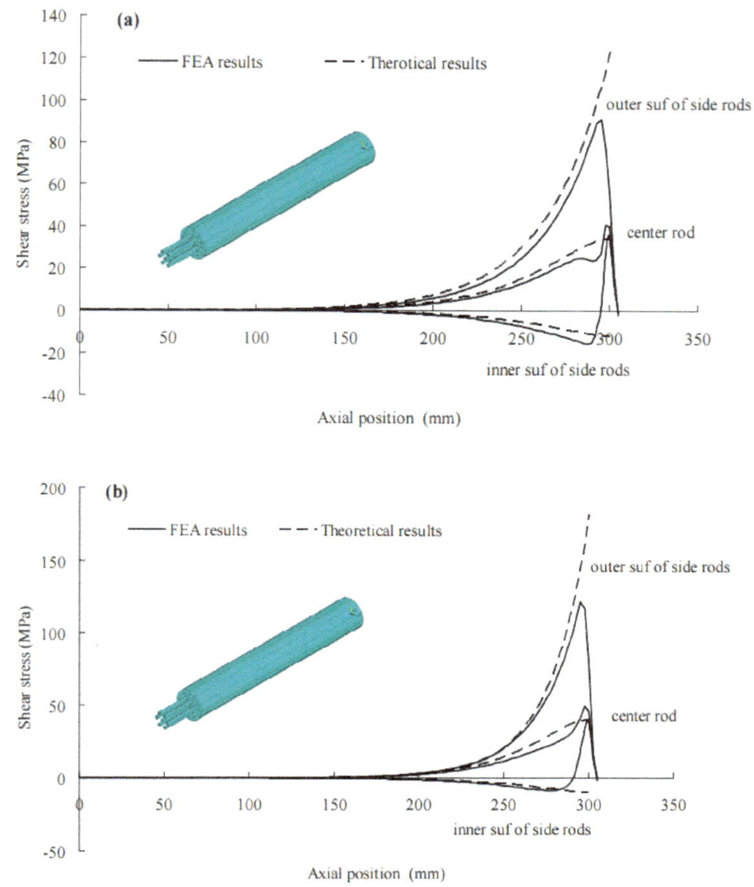

Figure 8. (**a**) Shear stress distribution of Case 1; (**b**) shear stress distribution of Case 2.

As can be seen, the rod in the center has a shear stress peak with a relatively low value at the loading end. The six rods on the side have a complex shear stress distribution. The shear stress in different positions on the same section of side rods may have opposite directions that are negative inside the rod and positive outside the rod. The peak value is also much bigger than the center rod. As the space between the center rod and side rods increases, the stiffness of the anchor and the shear stress peak increase.

The theoretical and finite element results are generally the same, except for some local positions. Some deviation is seen at the peak point of shear stress close to the end cross-section of the anchorage, and the shear stress on the inner surface of the side rod changes its direction in the FE results. This may be caused by stress discontinuity or approximation error in the application of the theory, which has been explained in Section 3.1. In reality, flexural deformation exists in the side rods, because of an asymmetric stress state, and the stress concentration at the loading end makes the stress state more complicated in this area.

Zhang *et al.* [18] displayed the experimental bond stress distribution of an FRP tendon bond anchorage. The stress peak appears at the loading end when the load is relatively low. Analysis in [19] shows that the shear stress on the side rods is higher than on the center rod. Similar experimental investigation were conducted on the multiple rod bonding anchor in [20]. The tensile stiffness and the anchorage efficiency is improved when the spacing of the rods gets smaller. Therefore, the experimental results in these references are in good agreement with the theoretical and FE results above.

PARAMETRIC ANALYSIS OF IN-CONE BONDING ANCHOR

A straight-pipe bonding anchor for multiple CFRP rods has some disadvantages (Figure 6). Slippage at the first interface is not allowed, otherwise the rods and adhesive will be pulled out. This leads to a longer bond length and a lower anchorage efficiency [19]. However, the inner cone bonding anchor is more widely used than straight-pipe ones, because radial pressure can be provided at the interface to grip the CFRP rods by an interference fit when slippage occurs at the first interface. Fang *et al.*, have studied the inner cone bonding anchors. The loading capacity of the double-rod inner cone anchor filled with RPC (Reactive Powder Concrete) is influenced by the inner inclined angle of the anchor and the rod spacing [21]. The authors also tested the performance of a bonding anchor with nine 12.6 mm-diameter CFRP rods inside [22]. The capacity reduction factor and the critical bond length were discussed through the experimental results.

Design and Modeling of the Inner Cone Anchor

Additional finite element analysis was performed to investigate the mechanical properties of the inner cone multi-rod bonding anchor. The seven-rod bonding anchor is designed by an anchorage manufacturer in China (Figure 9a) with a 300 mm bond length and a 76 mm outside diameter. The 5 mm-diameter CFRP rods are installed symmetrically in the inner cone barrel by fixtures, and the inner inclined angle is 3%.

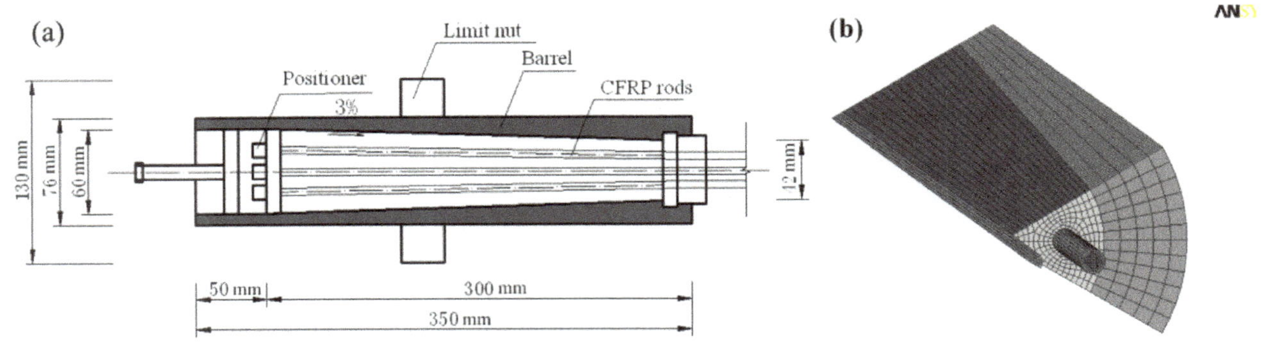

Figure 9. (**a**) Primary design of the inner cone bonding anchor; (**b**) 1/6 finite element model.

The ANSYS Solid185 element is applied to model the inner cone barrel, adhesive and CFRP rods, and the 1/6 model is shown in Figure 9b with 39,200 elements. The contact element is used at the first interface to permit slippage, while nodes at the second interface are coupled, and thus, the slippage between adhesive and rods is ignored. A tensile load is then applied to seven CFRP rods until the stress of the rods reaches 2000 MPa.

The coefficient of friction (COF) at the first interface and the inner inclined angle were varied to study their influence on the stiffness, radial stress value and shear stress distribution along the rod. The material properties of the rod are the same as in Table 1. The elastic modulus of the adhesive is 5 GPa, and the yield stress of the steel barrel is 345 MPa. Seven numerical cases are listed in Table 4 with various parameters. Because the change of the inclined angle in the longitudinal direction can reduce the concentrated stress [12], a straight-parabolic inner cone anchor is also added to the inner decline angle group for investigation. The anchor has a straight length of 60 mm from the loading end at the inner surface of the barrel with a 0° angle. The angle then increases along the remaining 240 mm to the free end, where the inner diameter changes parabolically from 42 mm to 60 mm.

Table 4. Numerical cases of the inner cone bonding anchor

Case No.	1	2	3	4	5	6	7
COF at the first interface	0	0.05	0.1	0.2	0.05	0.05	0.05
Inner inclined angle	3%	3%	3%	3%	2%	4%	Straight-parabolic

Numerical Results of Inner Cone Anchor

The convergence of the FEM (Finite Element Method) is studied through mesh-refinement of the FE model. The meshing of the straight-parabolic inner cone anchor model is refined with the element number from 39,200 to 148,140. After mesh-refinement, the tensile load value is 50.8 kN when the displacement reaches 15 mm, while it is 50.5 kN in the original model. Under the same stress state of 2000 MPa, the maximum radial compressive stress at the first interface is 111.5 MPa compared to the original value 104.3 MPa. The relative error of the load and stress value is 0.6% and 7%, which is acceptable. Therefore, the convergence of the FEM results is reliable.

The loading curves of the bonding anchor with different first-interface coefficients of friction are illustrated in Figure 10a. The stiffness of the anchor increases when the COF increases from zero to 0.2. As the COF increases, resistance will be stronger while slippage occurs at the first interface. Notice that the load level plateaus and begins to fluctuate when it nears 30 kN. This is because the steel barrel has yielded (equivalent stress reaches 345 MPa in FEA (Finite Element Analysis)), and a higher load capacity cannot be provided. The tensile stress of the rods is 1,300 MPa, which is much lower than the tensile strength. Therefore, the COF at the first interface should not be too low in the anchorage design to avoid the yielding of the barrel and compressive failure of the adhesive. This would lead to a low load capacity.

Figure 10b shows the radial compressive stress distribution at the first interface with different COF values under a certain stress level. The x-coordinate is zero at the free end. The compressive stress value reaches a maximum if there is no friction, with some stress concentration near the loading end, which may be caused by the warping of the adhesive. The stress decreases as the COF increases, because the tensile force is balanced by the horizontal component of the radial compressive and frictional forces. Under a certain tensile load, the frictional force will rise if the COF value increases. This will cause a reduction of compressive force or stress. Therefore, the COF at the first interface should also not be too high, or the radial compression may not be adequate to grip the CFRP rods.

Loading curves with different inner inclined angles are displayed in Figure 11a. As the angle becomes larger from 2% to 4%, the stiffness of the anchor also rises. The straight-parabolic anchor has a stiffness approximately equal to the 3% case. Corresponding to Figure 11a, the radial stress distribution at the first interface under a certain tensile load is illustrated in Figure 11b. The stress concentration can be observed at the loading end, except for the straight-parabolic case. The stress is higher at the loading end than the free end, and the distribution becomes more uneven when the inclined angle increases. Therefore, increasing the inner inclined angle can improve the stiffness of the anchor, but it also makes the radial stress distribute near the loading end and reduces the effective bond length, which is not expected in the design. Thus, a suitable inclined angle should always be selected.

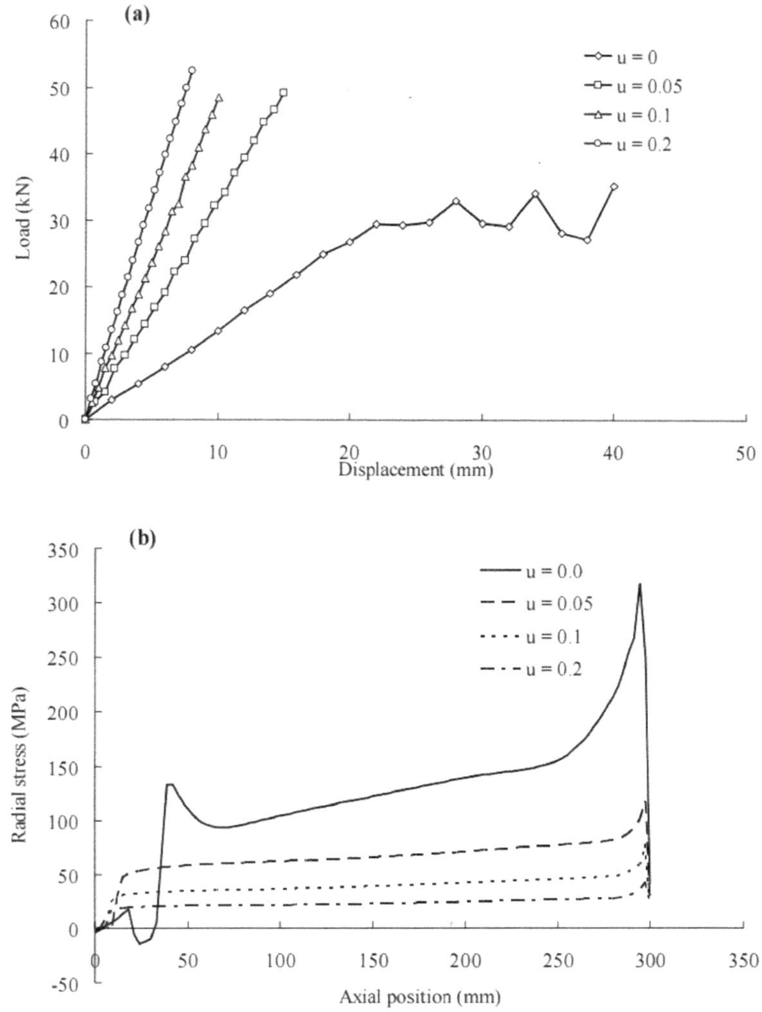

Figure 10. (**a**) The loading curve of the inner cone anchor with a varying coefficient of friction (COF); (**b**) the radial stress distribution of the inner cone anchor with a varying COF.

Meanwhile, the radial stress of the straight-parabolic anchor has a larger value at the free end *versus* the loading end. The stress concentration is eliminated due to the 60 mm straight component. A horizontal section of the curve can be observed, because part of the steel barrel has yielded at this point. However, the tensile load can still increase, and premature failure can be avoided, because the yielding of the barrel first occurs at the free end.

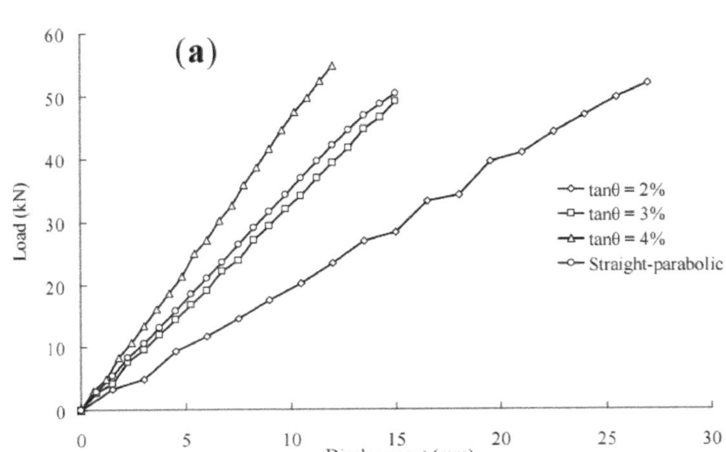

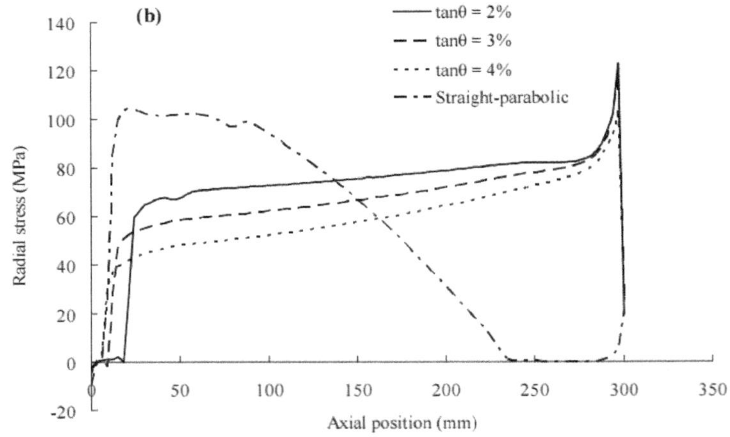

Figure 11. (**a**) The loading curve of inner cone anchor with a varying inclined angle; (**b**) the radial stress distribution of inner cone anchor with a varying inclined angle.

Figure 12 shows the shear stress distribution of Case 2 and Case 7 in Table 4 at the second interface between the CFRP rods and the adhesive for comparison. The shear stress on the surface of the center rod and the inner and outer surface of the side rods is illustrated, respectively. The difference in the stress distribution between the center rod and the side rods has been discussed in Section 3.2. It is not uniform and has a direction opposite that of the inner surface of the side rods. From Figure 12, we can conclude that the shear stress peak value moves towards the free end in the straight-parabolic anchor *versus* the 3% inclined angle anchor. Thus, the anchorage efficiency can be improved, and the local premature failure of the rods at the loading end can also be prevented.

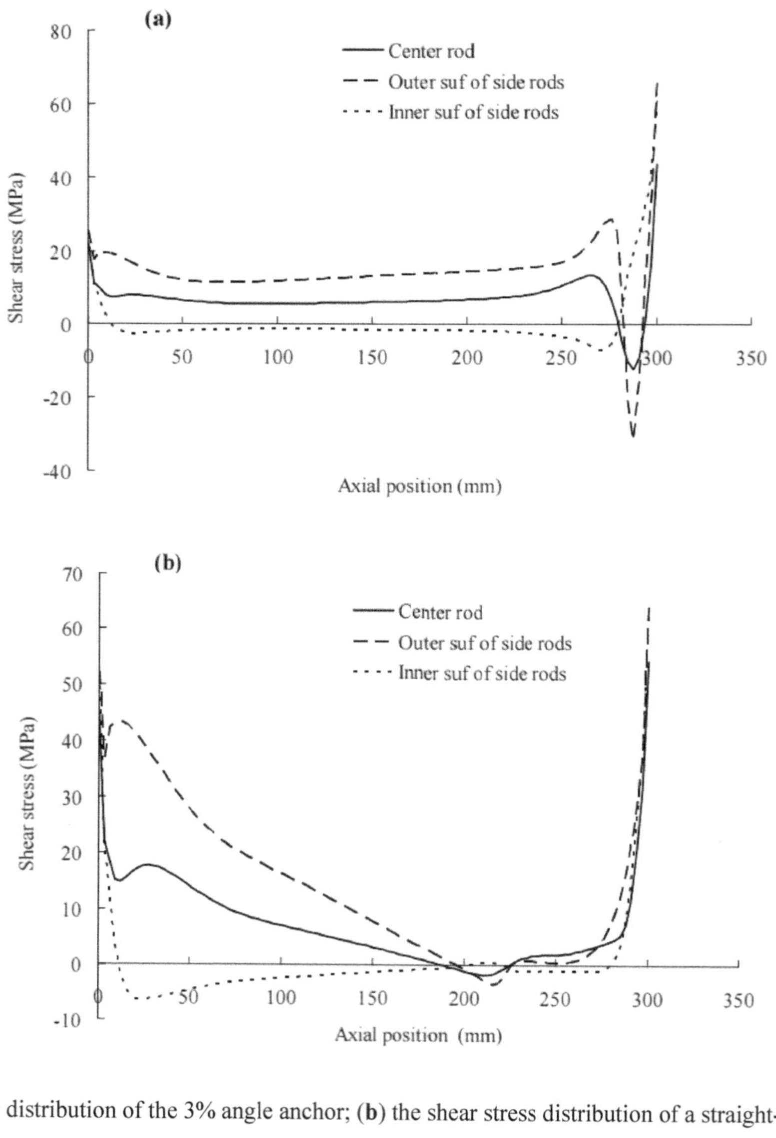

Figure 12. (**a**) The shear stress distribution of the 3% angle anchor; (**b**) the shear stress distribution of a straight-parabolic angle anchor.

Campell *et al.* [8] did some experiments on the wedge-type anchor, and the results show that the stiffness of the anchorage and radial stress at the barrel-wedge interface increases as the coefficient of friction decreases through greasing. The experimental results of [20] also indicate that the stiffness and loading capacity of the bonding anchor increases as the inclined angle increases from 0° to 3°, which are in good agreement with the FE results above. The radial stress concentration at the loading end of the barrel-wedge anchorage can also be seen in the numerical analysis of [9,11].

CONCLUSIONS

Based on the research of this paper, the following conclusions can be made:

1. With fundamental assumptions of elasticity and non-slip, a theoretical model for single and multiple rod bonding anchors is proposed by calculating the shear stress distribution on the rods. The theoretical equations are derived from the equilibrium and geometric conditions of the CFRP rods and adhesive. The theoretical model is suitable for preliminary shear stress analysis in the elastic state, and it can provide support for further study on decreasing the shear stress at the loading end cross-section of the anchorage.
2. Finite element results show good agreement with the theoretical results, thus confirming the accuracy of the theory. For the single rod bonding anchor, as the modulus of the adhesive increases or the thickness of the adhesive decreases, the stiffness of the single rod and the peak value of the shear stress will increase. For multi-rod anchors, the stiffness and shear stress peak value will also increase as the space between the center rod and side rods increases. Anyway, the side rods have a complex stress state under tensile force, which may lead to premature failure. Therefore, the modulus and thickness of the adhesive, as well as the space between the rods need to be considered in the anchorage design. The modulus of the adhesive should be high enough to ensure the stiffness of the anchor, but not too high to cause shear or sliding failure. A thin adhesive layer can help to improve the anchorage performance on the condition that the bond strength is guaranteed. Small spacing between the rods is preferred in a multi-rod anchor, because the longitudinal radial stress is more uniform, and the influence on the side rods caused by asymmetry becomes smaller. The size of the anchorage can also be smaller.
3. In the parametric analysis of the in-cone bonding anchor, as the coefficient of friction at the barrel-adhesive interface increases, the stiffness will increase and the radial compression will decrease, such that it will be more difficult for the steel barrel to yield under a certain load. The stiffness of the anchor also increases when the inner inclined angle increases, with the side effect that the radial pressure is more uneven. In the anchorage design, the inner inclined angle should be appropriate under the overall consideration of the stiffness and radial pressure.
4. Straight-parabolic in-cone anchoring is shown to be an efficient way to reduce the concentrated shear stress and to improve the anchorage efficiency by moving the stress peak from the loading end to the free end. It can be a feasible type of anchorage for CFRP cable made up of parallel CFRP rods in engineering.

ACKNOWLEDGMENTS

This work is supported by the China 863 Project (No. 2012AA03A204) and China 973 Project (2012CB026200), and the first author acknowledges the support of Beijing Higher Education Young Elite Teacher Project (YETP0078).

AUTHOR CONTRIBUTIONS

The work presented here was carried out in collaboration between all authors. Peng Feng, Pan Zhang and Lieping Ye defined the research theme. Peng Feng and Pan Zhang designed the research methods, carried out the finite element analysis and theoretical analysis, analyzed the data and wrote the paper. Xinmiao Meng co-worked on the data collection and their interpretation and presentation. All authors have contributed to, seen and approved the manuscript.

REFERENCES

1. Feng, P.; Ye, L.P.; Teng, J.G. Large-span woven web structure made of fiber-reinforced polymer. *J. Compos. Constr.* **2007**, *11*, 110–119.
2. Qi, Y.J.; Feng, P.; Ye, L.P. Fundamental mechanical model and analysis of single layer FRP woven web structures. *Eng. Mech.* **2012**, *29*, 180–188. (In Chinese).
3. Qi, Y.J.; Feng, P.; Ye, L.P.; Yang, J.Q. Model test of construction and formation of large-span FRP woven web structures. *J. Build. Struct.* **2012**, *33*, 117–126. (In Chinese).
4. Qi, Y.J.; Feng, P.; Ye, L.P. Comparison of Theoretical Solutions and Ultimate Span-Length between FRP Cables and Steel Cables. *J. Civ. Archit. Environ. Eng.* **2011**, *33*, 52–59. (In Chinese).
5. Wang, X.; Wu, Z.S. Integrated high-performance thousand-metre scale cable-stayed bridge with hybrid FRP cables.*Compos. Part B Eng.* **2010**, *41*, 166–175.
6. Nanni, A.; Bakis, C.E.; O'Neil, E.F.; Dixon, T.O. Performance of FRP tendon-anchor systems for prestressed concrete structures. *PCI J.* **1996**, *41*, 34–43.
7. Sayed-Ahmed, E.Y.; Shrive, N.G. A new steel anchorage system for post-tensioning applications using carbon fiber reinforced plastic tendons. *Can. J. Civ. Eng.* **1998**, *25*, 113–127.
8. Campbell, T.I.; Shrive, N.G.; Soudki, K.A.; Al-Mayah, A.; Keatley, J.P.; Reda, M.M. Design and evaluation of a wedge-type anchor for fibre reinforced polymer tendons. *Can. J. Civ. Eng.* **2000**, *27*, 985–992.
9. Al-Mayah, A.; Soudki, K.A.; Plumtree, A. Experimental and analytical investigation of a stainless steel anchorage for CFRP prestressing tendons. *PCI J.* **2001**, *46*, 88–99.
10. Al-Mayah, A.; Soudki, K.; Plumtree, A. Effect of sandblasting on interfacial contact behavior of carbon-fiber-reinforced polymer-metal couples. *J. Compos. Constr.* **2005**, *9*, 289–295.

11. Al-Mayah, A.; Soudki, K.; Plumtree, A. Effect of sleeve material on interfacial contact behavior of CFRP-metal couples.*J. Mater. Civ. Eng.* **2006**, *18*, 825–830.

12. Al-Mayah, A.; Soudki, K.; Plumtree, A. Novel anchor system for CFRP rod: Finite-element and mathematical models.*J. Compos. Const.* **2007**, *11*, 469–476.

13. Benmokrane, B.; Zhang, B.; Chennouf, A. Tensile properties and pullout behavior of AFRP and CFRP rods for grouted anchor applications. *Constr. Build. Mater.* **2000**, *14*, 157–170.

14. Zhang, B.; Benmokrane, B. Pullout bond properties of fiber-reinforced polymer tendons to grout. *J. Mater. Civ. Eng.***2002**, *14*, 399–408.

15. Enomoto, T.; Ushijima, K. Use of CFCC tendons and reinforcements in concrete structures for durability. In Proceedings of Asia-Pacific Conference on FRP in Structures (APFIS) 2012, Hokkaido, Japan, 2–4 February 2012.

16. Rohleder, W.J.; Tang, B.; Doe, T.A.; Grace, N.F.; Burgess, C.J. Carbon Fiber-Reinforced Polymer Strand Application on Cable-Stayed Bridge, Penobscot Narrows, Maine. *Transp. Res. Rec. J. Transp. Res. Board* **2008**, *2050*, 169–176.

17. Meier, U. Carbon Fiber Reinforced Polymer Cables: Why? Why Not? What If? *Arab. J. Sci. Eng.* **2009**, *37*, 399–411.

18. Zhang, B.; Benmokrane, B.; Chennouf, A. Prediction of tensile capacity of bond anchorages for FRP tendons. *J. Compos. Constr.* **2000**, *4*, 39–47.

19. Mei, K.H. Analysis of mechanical behavior of CFRP cable bonding anchors. *Bridge Constr.* **2007**, *3*, 80–83. (In Chinese).

20. Zhu, G.P.; Ye, H.W.; Qiang, S.Z. Experimental investigation and mechanical behavior analysis of multiple CFRP tendons anchorage system. *Eng. Mech.* **2011**, *28*, 165–170. (In Chinese).

21. Fang, Z.; Jiang, T.Y.; Liang, D. The Anchorage Behavior of CFRP Tendons in RPC. *J. Hunan Univ.* **2007**, *34*, 1–5. (In Chinese).

22. Fang, Z.; Zhang, K.; Tu, B. Experimental investigation of a bond-type anchorage system for multiple FRP tendons.*Eng. Struct.* **2013**, *57*, 364–373.

CHAPTER 18

EPOXY ENHANCED BY RECYCLED MILLED CARBON FIBRES IN ADHESIVELY-BONDED CFRP FOR STRUCTURAL STRENGTHENING

Chao Wu, Peng Feng, Yu Bai and Ye Lu

Keywords: Polymer-matrix composites, mechanical properties, scanning electron microscopy (SEM), recycled milled carbon fibres, electrical resistivity

INTRODUCTION

Steel structures are susceptible to corrosion and fatigue damages and there is an urgent need to repair and retrofit such deteriorated steel structures. The use of adhesively bonded carbon fibre reinforced polymer (CFRP) composites in strengthening of steel structures may be an advantageous solution, because of the excellent mechanical properties of CFRP such as high strength to weight ratios (with an E-modulus of 240 GPa and nominal tensile strength of 3800 MPa) and excellent resistance to corrosion and fatigue. In addition, interruption of structural service may be minimized during the strengthening process in this way. The concept of using CFRP in structural rehabilitation was initiated from aerospace engineering in the 1970s [1]. Relevant practices of such a concept in civil engineering are increasing rapidly [2]. Recent studies indicated that combining FRP and steel in different configurations, such as adhesively bonding [3,4] and confining [5,6] may form a reliable load-carrying system and can be utilized in steel structure strengthening in a safe and economic manner.

In such strengthened systems, CFRP composites are externally bonded to steel members, therefore the structural adhesive used is of a great importance [7,8,9]. Epoxy is a structural adhesive frequently used in strengthening of steel structures and it is reported that failure often occurs through steel and adhesive interface debonding, or adhesive layer failure (cohesive failure) as a result of relatively low strength of epoxy, or delamination of CFRP composites [4,10]. There is a need to improve the mechanical performance of adhesive epoxy through enhancement of certain mechanical properties using appropriate additives without significantly decreasing other ones.

Efforts have been made in the modification of epoxy to achieve better mechanical performance by adding rigid particles such as glass beads, and silica while the results have turned out to be unsuccessful in terms of the improvement of physical bonding [11]. It was found that the toughness of the epoxy was obviously increased by implementing a carboxyl-terminated copolymer of butadiene and acrylonitrile (CTBN). However a decrease of glass transition temperature was observed for such a modified epoxy and therefore not suitable for applications at higher temperature [12]. Recently studies have focused on the introduction of carbon nanotubes (CNT) to modify the epoxy. The integration between CNT and resin matrix has the potential to ameliorate the weaknesses of conventional laminated composites including delamination, lack of impact damage resistance, and low transverse mechanical properties [13]. Compared to original epoxy, improvements in flexural strength, glass transition temperature, and decomposition temperature have been reported [14]. However the demand of high volume and high rate production of CNTs, as well as the associated costs, hinder the applications of CNT-modified

Citation: Wu, C.; Feng, P.; Bai, Y.; Lu, Y. Epoxy Enhanced by Recycled Milled Carbon Fibres in Adhesively-Bonded CFRP for Structural Strengthening. Polymers 2014, 6, 76-92. doi:10.3390/polym6010076.

epoxy, in large scale particularly in civil infrastructure [15]. Yet, it is a challenge to achieve a high percentage of CNT (especially for single-walled CNT) at the desired level because of dispersion problems [13].

Recycled milled carbon fibres (MCF), on the other hand, present a promising and more practical way to enhance structural epoxy adhesives. MCF refers to a carbon fibre with a length of 1mm or less and much shorter than chopped carbon fibres, which can be recycled from used carbon materials. The traditional solutions of disposal of such carbon fibre waste are to use land-fill or waste incineration [16]. Such solutions represent a waste of natural resources and, considering the fact that carbon fibre does not decompose naturally, environmental issues may well be very critical. Several recycling technologies have been developed to reclaim carbon fibres from CFRP materials used in aerospace and military products for other structural applications [17]. Although the exact recycling process to produce milled carbon fibres was patented [18], it is known that a pyrolysis process is involved [19,20] to burn off the polymer matrix and free the carbon fibres through elevated temperatures.

Although investigations on enhancement of epoxy using chopped carbon fibres indicated an improvement of its toughness up to a fibre volume fraction of 36% [21], studies on the mechanical performance of modified epoxy using MCFs are limited. It was reported that the direct-current (DC) conductivity was improved after the addition of MCFs [22] while the abrasive wear rate of the resulting composite was reduced [23]. Such a MCF enhanced epoxy was made using a centrifugation technique. MCFs were firstly mixed with epoxy resin and hardener and fully stirred with a glass rod and then the sample was placed in a mould and rotated to form a gradient of MCFs [22]. The results from the literature suggest reduced electrical resistance and promising mechanical performance of the MCF enhanced epoxy. However, adhesively-bonded systems using MCF enhanced epoxy have not been reported yet.

In this paper, MCFs were used to enhance an epoxy and the resulting epoxy was used to prepare a steel/CFRP adhesively-bonded system in a form of double strap joints. Different percentages of weight ratio of MCFs were investigated in the epoxy specimens and joint specimens. Both specimens were tested in tension to examine the mechanical properties such as stress-strain response, E-modulus and ultimate load. Failure modes were identified and discussed with respect to the effects introduced by the addition of MCFs. In addition, scanning electron microscopy (SEM) was used to understand the changes of mechanical performance after introducing MCFs to the epoxy and steel/CFRP adhesively-bonded joints. Finally, the electrical resistivity of the MCF modified epoxy was tested as a preliminary study for its potential application in health monitoring of a FRP/steel bonding system.

EXPERIMENTAL INVESTIGATIONS

Materials

Araldite 420 epoxy was used for MCF addition and was used as the adhesive to prepare the steel/CFRP double strap joints. Araldite420 epoxy (produced by Huntsman Advanced Materials Ltd., Melbourne, Australia) consists of two components 420A (Bisphenol A/epochlorohydrin resin, liquid based epoxy resin) and 420B [Diethylene glycol, di(3-aminopropyl)ether based epoxy resin as hardener]. Previous experiments indicated a nominal tensile strength of 32 MPa, tensile modulus of 1.9 GPa and tensile ultimate strain of 2.4% for the epoxy [24].

The milled carbon fibre (MF100) used to modify the epoxy was recycled and supplied by ELG Carbon Fibre Ltd. (West Midlands, UK). It is with an average particle diameter of 7.5 μm and length of 100 μm. The fibre density was reported as 1800 kg/m^3 and the tensile strength and tensile modulus were 3150 and 200 GPa respectively [25].

Both carbon fibre sheet (CF130), and carbon fibre laminate (MBrace 210/3300) were used to prepare steel/CFRP double strap joints. CFRP sheet and laminate were supplied by BASF (Melbourne, Australia). For CFRP sheet (MBrace CF130), the nominal elastic modulus is 240 GPa and its nominal tensile strength is 3800 MPa [26]. For CFRP laminate (MBrace Laminate 210/3300), the nominal elastic modulus is 210 GPa and its nominal tensile strength is 3300 MPa [27]. The steel elastic modulus was measured as 200 GPa and the yield stress and ultimate stress were measured as 359 and 430 MPa respectively [28].

Mechanical Experiments

Three scenarios were examined in mechanical experiments as summarized in Table 1. Scenario EP investigated the mechanical properties of the epoxy after MCF modification. Dog-bone specimens were prepared with different weight ratios of milled carbon fibres to epoxy (see Table 1). It was suggested that Araldite 420A and 420B should be mixed with a ratio by weight of 100:40 to form the epoxy matrix before application [29]. To achieve sufficient time for mixing and vacuum in order to minimise the amount of bubble in the epoxy, Araldite 420A was first mixed with the MCFs by hand mixing and then placed in the vacuum pump for two days before the addition of hardener 420B. After that, the 420B hardener was mixed by hand mixing into the mixture. This had to be a gentle process otherwise bubble would have been produced and made the vacuum pumping process useless. Once this process was completed, the MCF enhanced epoxy was poured into the mould to form the dog-bone epoxy specimens as shown in Figure 1a, according to the geometry specified in ASTM D638 [30]. The pouring process was as slowly as possible to minimise the formation of air bubble. The MCF enhanced epoxy was left in the mould to cure at room temperature for 10 days according to the manufacturer's instructions [29]. The specimens without MCF were deep green and became dark black after the MCF addition. Totally, twelve specimens were prepared in this scenario, and therefore three identical specimens (named as EP*x*-*y*, where *x* represents the weight ratio and *y* is from 1 to 3 as the specimen number) were repeated for each MCF weight ratio. For the EP specimen, the tensile test was performed using an Instron machine (Instron, Melbourne, Australia) with a load capacity of 50 kN. The strain gages with resistance of 120.0 ± 0.3% and gage factor of 2.095 ± 0.5% at room temperature were installed at the centre of the specimens to measure the load-strain developments (see Figure 2a). The specimens were loaded in tension at a displacement control rate of 1 mm/min.

Table 1. Mechanical experiments and key parameters

Scenarios	Materials	Specimens	Weight ratio of milled carbon fibre	Carbon fibre sheet or laminate layers
EP (epoxy)	milled carbon fibre, epoxy	dog-bone coupons	0%, 1.5%, 3% and 5%	NA
DJS (double strap joints with CFRP sheets)	milled carbon fibre, epoxy, CFRP sheet, steel plate	double strap joints	0%, 1.5%, 3% and 5%	1 and 3
DJL (double strap joints with CFRP laminates)	milled carbon fibre, epoxy, CFRP laminate, steel plate	double strap joints	0% and 5%	1

Scenario DJS investigated the mechanical performance of steel strengthened by CFRP sheets using origin and enhanced epoxies. Double strap joints were prepared in this scenario with different weight ratios of milled carbon fibres. In addition, two fibre sheet layouts—one layer and three layers of CFRP sheets—were used in the joint configuration as shown in Figure 1b. The different fibre sheet layouts were to examine the change of failure modes and the load carrying capacity corresponding to each failure mode, with the existence of milled carbon fibre modification.

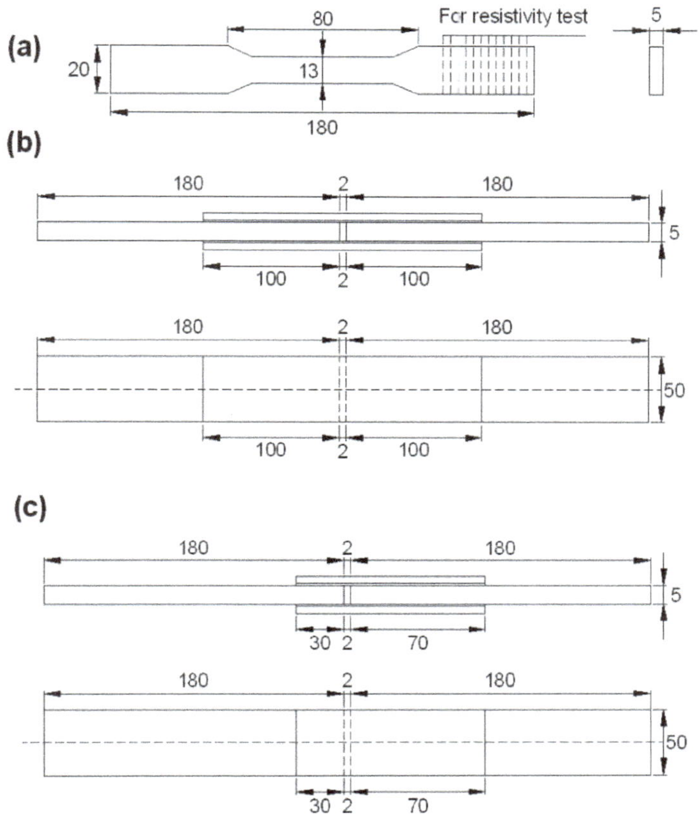

Figure 1. Dimensions and configurations for (**a**) EP specimens; (**b**) DJS specimens; and (**c**) DJL specimens (unit in mm, not in scale).

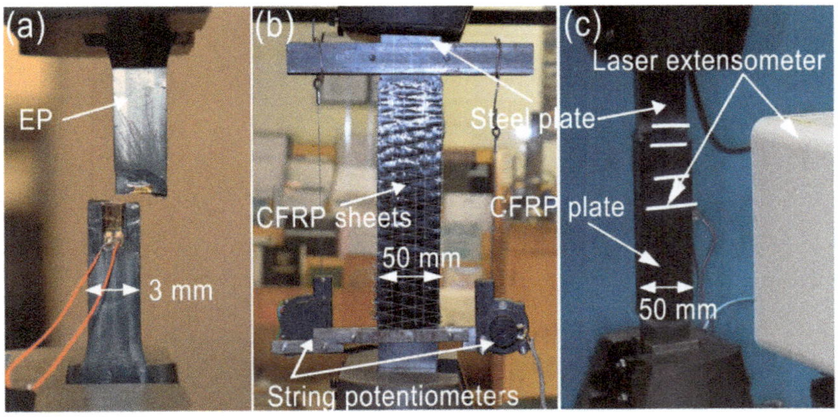

Figure 2. Experimental setup and instrumentation for (**a**) EP specimens; (**b**) DJS specimens; and (**c**) DJL specimens.

The joint specimens were prepared using modified epoxy adhesives (with different weight ratios of milled carbon fibre, see Table 1), steel plates and CFRP sheets. Two steel plates with a dimension of 5 mm × 50 mm × 180 mm were used to fabricate the double strap joint, after surface preparation (cleaning and sand-blasting). Then the prepared epoxy was applied on each surface of the steel plate, and one layer of CFRP sheet was placed and then again the same epoxy was applied on its surface or between CFRP layers to form the polymer matrices of the CFRP composite. The resulting specimens are illustrated in Figure 1b where a bond length of 100 mm was used for all the specimens. This bond length was greater than the effective bond length of 40 mm for the one-layer CFRP joints and of 50 mm for the three-layer CFRP joints with identical geometry prepared using original epoxy [28]. The specimens were designated as DJS*x*-*y*-*z*, where *x* represents the weight ratio of MCF, *y* is the number of CFRP sheets (1 or 3) and *z* is the specimen number (1 or 2) in each specified condition.

Scenario DJL was also a double strap joint of steel plates bonded with CFRP laminate. The steel plates had the same dimension as those in scenario DJS, whereas the bond length of CFRP laminate was 30 mm, as shown in Figure 1c. The bond length was chosen shorter than its effective bond length (60 to 70 mm) as reported in [31], to ensure failure occurs because of the adhesive bonding. CFRP on the other side was set at 70 mm so that the failure would occur on the side of the 30 mm bond length. Eight double strap joints were prepared, four of which were with original epoxy, and the other four were with modified epoxy (5% weight ratio of MCF). The preparation process was the same as that in scenario DJS and the specimens were labelled as DJL*x*-*y*, where *x* represents the weight ratio of MCF (0% or 5%), *y* is the specimen number (1–4).

A Shimadzu Universal Hydraulic Testing Machine (Shimadzu, Melbourne, Australia) was used to test the DJS specimens in tension to failure (see Figure 2b). This testing was carried out at a constant rate of 2 mm/min. Two string potentiometers, LX-PA2 manufactured by Unimeasure, with accuracy of 1.25×10^{-3} mm, were used to measure the elongation of a joint region of 205 mm (see Figure 2b), which was then averaged by the readings from the two string potentiometers. The same setup and loading speed were adopted for DJL joints, and a laser extensometer was used to measure the relative slip through the 30 mm bond length (see Figure 2c).

Scanning Electron Microscopy

SEM was used to observe the microstructure of the epoxy after MCF addition with different weight ratios. The samples used in SEM were cut from the damaged EP specimens after testing and the observation target was the failure surface.

The samples were examined at the Monash Centre for Electron Microscopy, using an accelerating voltage of 10 KV. Different magnifications ranging from 100 to 2500 were adopted for each specimen with different weight ratios (0%, 1.5%, 3% and 5%), in order to identify any differences among them.

Electrical Resistivity

The electrical resistance (R) of original epoxy and epoxy with 5% weight ratio of MCF was tested using an Agilent 34461A 6½ digit multimeter (Agilent Technologies, Melbourne, Australia). The samples for electrical resistance test were sliced from dog-bone coupons prepared in scenario EP (see Figure 1a) using a diamond saw. Four samples of original epoxy and eight samples of epoxy with 5% weight ratio of MCF were prepared. All samples were in a rectangular shape with a thickness less than 1.5 mm. The detailed dimensions are listed in Table 3. The electrical resistance was measured in the thickness direction. In order to ensure the testing consistency, all samples of the same MCF weight ratio (0% or 5%) were cut from the same dog-bone coupon. Grade 1200 sand paper was used to polish the two sides of each sample. Then all samples were washed in a water bath (constant temperature control of 25 °C) using a horn sonicator (model VCX 500, from Sonics & Materials Inc., Melbourne, Australia) under 30% energy output for 30 min. The purpose was to clean the surface debris from saw cutting and sand paper polishing. The samples were left to dry for 12 h at room temperature. Finally, the two sides of each sample were painted as electrodes using silver conductive paint from RS Components Pty. Ltd. (Melbourne, Australia). After the measurement of electrical resistance (*R*), the electrical resistivity (ρ) can be calculated by:

$$\rho = Rab/t \quad (1)$$

where *a*, *b* and *t* are the length, width and thickness of the rectangular sample, respectively.

RESULTS AND DISCUSSION

Results from EP Specimens

All the EP specimens failed in a similar way to the breakage at the middle region of the dog-bone shape as shown inFigure 2a. The stress and strain curves of typical EP specimens with different weight ratios of milled carbon fibre are summarized in Figure 3, where the stress values were calculated using the measured loads divided by the sectional area at the failure position. It can be observed that the slope of the curves at the elastic region (linear part), *i.e.*, the E-modulus, is improved with the increase of the MCF weight ratio, as well as the ultimate stress at failure (*i.e.*, the tensile strength).

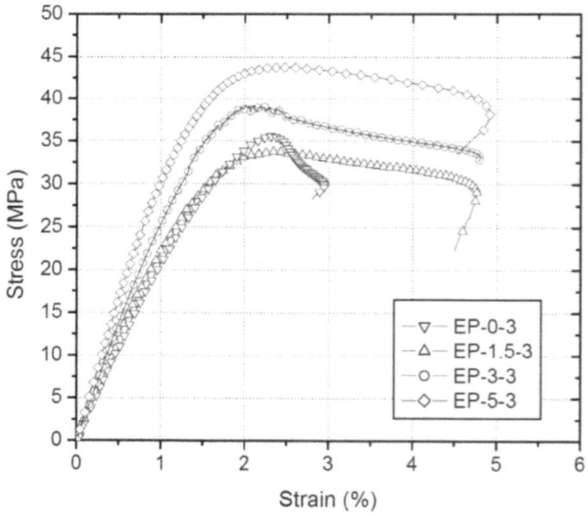

Figure 3. Stress-strain curves of typical original epoxy specimen (EP0-3) and enhanced epoxy specimens with different weight ratios of MCFs (EP1.5-3, EP3-3 and EP5-3).

The detailed values of E-modulus and tensile strength of all the specimens (average values and standard derivations) are summarized in Table 2. The E-modulus of the original epoxy showed an average E-modulus of 2.06 GPa and very similar to the value (1.9 GPa) reported by the manufacturer. The improvement of E-modulus was not obvious (only 6.8%) at the weight ratio of 1.5% of milled carbon fibres while such an improvement became significant (31.1%) when the weight ratio increased to 3%, and became 50.5% when only 5% by weight of MCFs were used to modify the epoxy. More than 5% by weight of MCFs made the specimen preparation very difficult because of the loss of mobility of the epoxy resin. Surfactant may be necessary to prepare the epoxy specimens modified using MCFs with a weight ratio more than 5%.

Table 2. Elastic modulus, strength and ultimate stain of modified epoxy with different weight ratios of milled carbon fibre (values are averages of three identical specimens)

Specimen	Weight ratio	E-modulus (MPa)	Strength (MPa)
EP0-*y*	0%	2.06 ± 0.06 (100%)	34.8 ± 1.1 (100%)
EP1.5-*y*	1.5%	2.20 ± 0.10 (106.8%)	32.8 ± 1.4 (94.3%)
EP3-*y*	3%	2.70 ± 0.17 (131.1%)	37.3 ± 1.7 (107.2%)
EP5-*y*	5%	3.10 ± 0.37 (150.5%)	40.1 ± 3.3 (115.2%)

The tensile strength of enhanced epoxy specimens demonstrated similar variation to that of E-Modulus. No improvement was found for the specimens with a weight ratio of 1.5% MCFs but rather a slight decrease (see Table 2). It was believed that the slight decrease of strength for specimens EP1.5-*y* in comparison to EP0-*y* in Table 2 was because of data scattering rather than any physical mechanism. Considerable tensile strength enhancement (15.2%) was identified for the specimens EP5-*y*with a weight ratio of 5% of MCFs.

The strain at the ultimate load of specimens EP0-*y* was found to be 2.37 ± 0.03% and this value is consistent with that (2.4%) reported previously [24]. Similar but slight smaller strain values at the ultimate loads were found for the other specimens (EP1.5-*y*, EP3-*y* and EP5-*y*) in a range from 2.15% to 2.37% with larger derivations. After the ultimate loads were reached, the strains kept increasing without significant loss of applied stress for the epoxy after MCF addition and this behaviour demonstrated a "yielding" stage in their stress-strain curves, implying that a ductile performance was introduced to the epoxy after MCF addition. As a result, the strains at the final breakage failure were largely improved to be in a range from 3.5% to 4.7% for the MCF enhanced epoxy specimens, in comparison to that (3.0%) for the original epoxy, as also evidenced in Figure 3.

Results from DJS Specimens

Failed specimens of steel/CFRP double strap joints with one layer of carbon fibre sheet are shown in Figure 4a, where the same failure mode was identified. These joints failed through the delamination of the carbon fibre sheet at the centre of the joints, regardless of the different MCF weight ratios used in the epoxy adhesive. Only part of the carbon fibres was pulled out from the full bond length of 100 mm and the steel surface in the joint region was still covered by most of the carbon fibre sheet. No adhesive interface debonding or cohesive failure was identified and this result suggested that the bonding provided by the epoxy between the carbon fibre sheet and the steel surface was almost intact. The failure modes of steel/CFRP double strap joints with three layers of carbon fibre sheets were shown in Figure 4b. Again all the specimens with three layers of carbon fibre sheets failed mostly through delamination of the layers of carbon fibre sheets, independent of the MCF weight ratios. It was found that the steel surface in the joint region was largely covered by carbon fibre sheet while only a small surface area was exposed for the joints DJ5-3 (see Figure 4b). Such a failure mode may imply that the joint load-carrying capacity is dominated by the interlaminar shear strength of the carbon fibre sheets, rather than the epoxy strength of the adhesive layer.

Typical load displacement curves of DJS joints are shown in Figure 5. All the joint specimens with one or three layers of carbon fibre sheets showed a linear load-displacement development until the final failure. Although the epoxy demonstrated a ductile failure process after MCF addition, such behaviour was not observed for the joints with different MCF weight ratios. This may be attributed to failure mode of delamination, rather than cohesive failure.

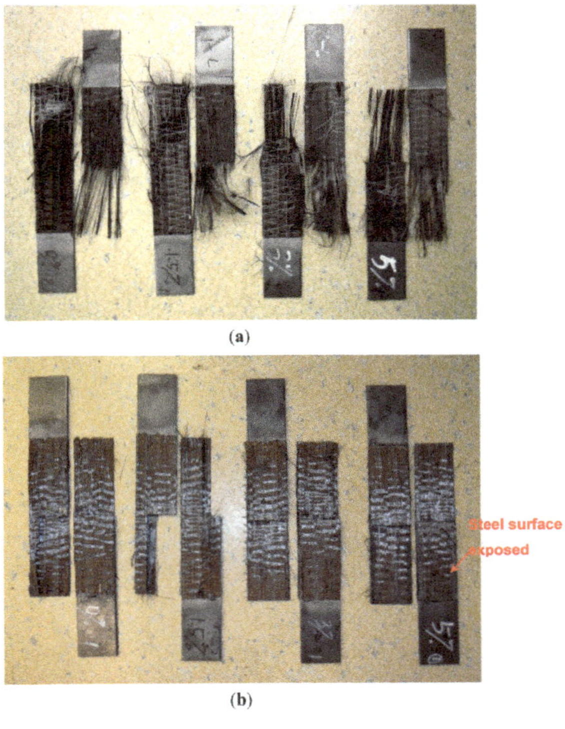

Figure 4. Failure modes of DJS specimens with (**a**) one layer of carbon fibre sheet (DJS0-1, DJS1.5-1, DJS3-1, and DJS5-1); and (**b**) three layers of carbon fibre sheet (DJS0-3, DJS1.5-3, DJS3-3, and DJS5-3).

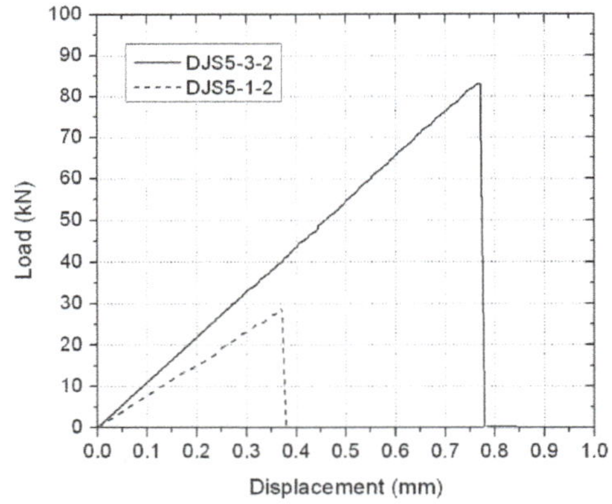

Figure 5. Load-displacement curves for typical DJS specimens.

The change of joint stiffness with the weight ratio of MCF is shown in Figure 6a and that of joint ultimate load is shown in Figure 6b. No obvious improvement of joint stiffness or joint ultimate load was observed after the MCF addition to the specimens with one and three layers of carbon fibre sheets, again as a result of the delamination failure mode of the carbon fibre sheet. The slight variation of results in Figure 6a,b is due to data scattering rather than any solid physical mechanism.

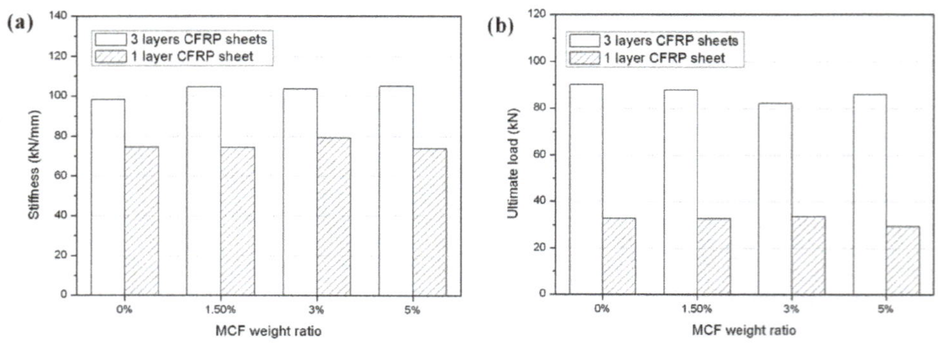

Figure 6. Comparison of (**a**) stiffness; and (**b**) ultimate load of DJS specimens with different MCF weight ratios.

Results from DJL Specimens

A shorter bond length of 30 mm in DJL specimens changed the failure mode to adhesive-CFRP laminate interface failure, as shown in Figure 7. It can be seen that, the same interface failure happened for joints with either original epoxy or modified epoxy with 5% weight ratio of MCF. This failure mode explains the similar ultimate failure load and stiffness of all DJL joints. For example, the average measured ultimate load of DJL with 0% MCF was 100.6 kN, compared to 102.0 kN for DJL with 5% MCF. The stiffness of the DJL specimens was calculated from the load-displacement curves of DJL joints, and the average stiffness of DJL with 0% MCF was 854.9 kN/mm, compared to 904.6 kN/mm for DJL with 5% MCF.

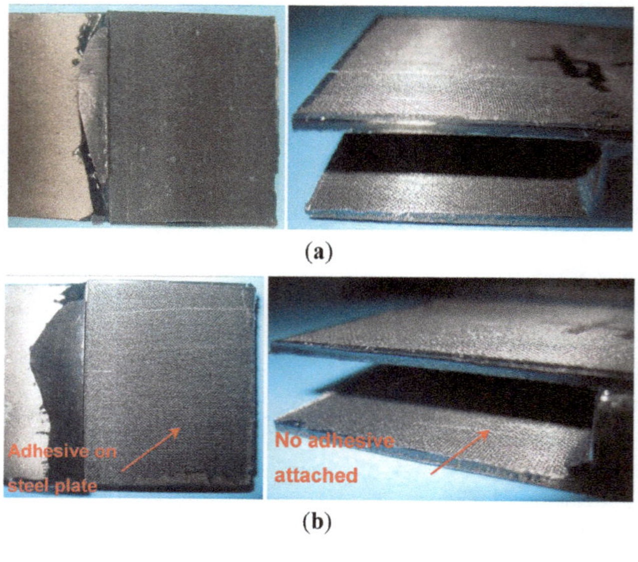

Figure 7. Failure modes of DJL specimens with (**a**) original epoxy adhesive (DJL0-4); and (**b**) epoxy with 5% MCF weight ratio (DJL5-4).

Similar double strap joints tests were conducted by Korayem *et al.* [31], where the epoxy (Araldite 2011) was modified using a 3% weight ratio of multi-walled carbon nanotubes (MWCNT). It was also concluded that no obvious improvement in ultimate load and stiffness was observed due to the fact that little failure occurred in the adhesive layer (CFRP delamination and adhesive/adherent interface failure dominated). Therefore, careful design of double strap joints to achieve adhesive failure (cohesive failure) is necessary to reveal the potential enhanced bond performance of the CFRP/steel system using modified epoxy.

SEM Results

Figure 8a shows the failure location of the sample from the original epoxy without MCF addition at a magnification of 250. The rough surface was a result of fracture in tension, and a number of small pieces of epoxy were evidenced also inFigure 8a, as a result of relatively brittle failure of the specimen.

The microscopic images are shown in Figure 8b for the sample from the epoxy with an MCF weight ratio of 1.5 (EP1.5) and in Figure 8c for EP3. It was found in Figure 8b that a number of carbon fibres were partially embedded into the damage surfaces and more fibres were evidenced in Figure 8c corresponding to a denser distribution, because of a larger MCF weight ratio in specimens EP3. The carbon fibres were randomly dispersed in the matrix during specimen preparation and therefore different orientations to the damage surface were noticed. In addition to these fibres, a number of holes with very similar diameters to the fibres were also identified from Figure 8b,c—they were produced because the corresponding fibres were pulled out from the matrix and left (still embedded) in the other damage surface. The bond failure between milled carbon fibres and the epoxy matrix contributed to the improvement of E-modulus and strength of the modified epoxy specimens. Furthermore, this additional failure mechanism and associated energy dissipation also explains a more ductile failure process for the MCF enhanced epoxy than the original one, overall resulting in a smoother damaged surface inFigure 8b,c than the fracture surface shown in Figure 8a.

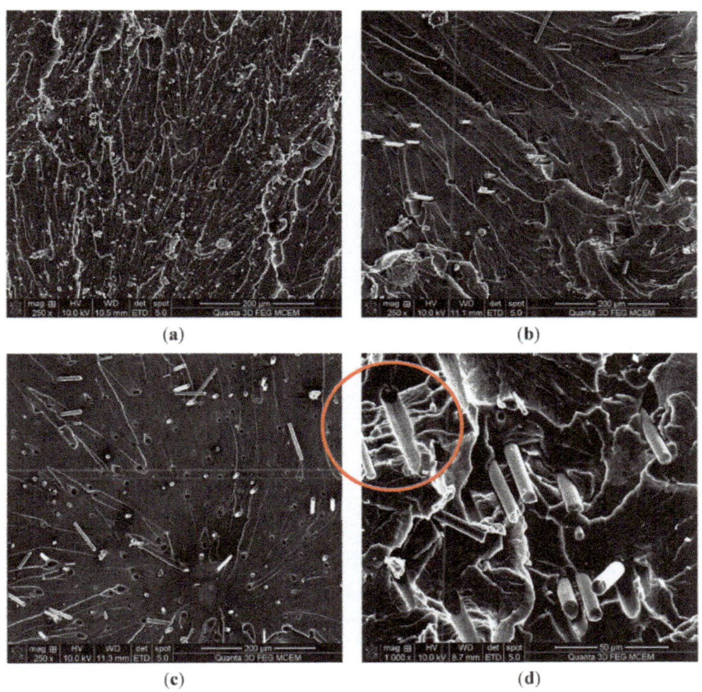

Figure 8. Typical scanning electron microscopy (SEM) images of epoxy specimens with different weight ratios of MCFs (**a**) 0%; (**b**) 1.5%; (**c**) 3%; and (**d**) 5%.

Figure 8d shows the microscopic image of the damaged surface for the epoxy with 5% MCF addition (EP5) at a higher magnification of 1000, where the top left corner of the image is of further interest. From there a fibre nearly parallel to the damaged surface was found pulled out and the bond surface between the fibre and matrix was revealed. A few cracking traces were initiated from the bond surface and along the fibre direction—they evidenced the stress transfer from the fibre to the surrounding matrix through the bonding in between. Both matrix and fibre surfaces after de-bonding remained smooth, implying that the failure might have occurred at the interface. In this figure, the diameter of carbon fibres was measured as 7.22 μm and close to the value (7.5 μm) reported by the manufacturer. It should be noted that, fibre pull out was observed for all EP specimens. This is most probably because of the high tensile strength of MCFs (3150 MPa) and the short bond length of MCFs (100 μm as the fibre length).

Electrical Resistivity Results

The dimensions and measured electrical resistivity of all samples are listed in Table 3. The label "E" refers to original epoxy and "EM" means modified epoxy with 5% weight ratio of MCF. It should be noted that, the resistance of original epoxy samples exceeded the measurement upper limit of the multimeter (1 × 108 Ohm), therefore, this value was used to calculate the lower limit resistivity of the original epoxy with Equation (1).

Table 3. Dimensions and measured electrical resistivity of epoxy samples with 0% or 5% recycled milled carbon fibre (MCF) weight ratio

	Samples	Length *a* (mm)	Width *b* (mm)	Thickness *t* (mm)	Resistivity ρ (Ohm.m)
Epoxy with 0% MCF weight ratio	E1	20.0	5.77	0.69	$>1.67 \times 10^7$
	E2	19.96	5.81	0.69	$>1.68 \times 10^7$
	E3	19.99	5.82	0.93	$>1.25 \times 10^7$
	E4	19.99	5.82	1.23	$>9.46 \times 10^6$
Epoxy with 5% MCF weight ratio	EM1	19.94	5.95	1.46	$>8.13 \times 10^6$
	EM2	19.94	5.94	1.33	$>8.91 \times 10^6$
	EM3	20.0	5.96	1.52	$>7.84 \times 10^6$
	EM4	19.96	5.95	1.17	8.44×10^1
	EM5	20.04	5.97	1.35	$>8.86 \times 10^6$
	EM6	19.95	5.90	1.05	2.23×10^2
	EM7	19.96	5.88	1.28	4.00×10^2
	EM8	19.96	5.93	1.23	1.10×10^2

It can be seen in Table 3 that, four EM samples exhibited much lower resistivity compared to original epoxy, which ranged from 84 to 400 Ohm.m. On the other hand, the resistance of the other four EM samples also exceeded the measurement capacity of the multimeter, indicating the same order of resistivity as that of original epoxy. The resistivity variation of EM samples might be due to the different MCF conductive network within the EM samples. The preparation process of the dog-bone coupons in scenario EP also contributed to a non-uniform and imperfect random distribution of MCFs within the epoxy. When the local distribution of MCF is good enough to form a network for electrons to pass, the resistivity will decrease accordingly as evidenced in the EM4 and EM6 to EM8 specimens.

Interestingly, in another similar study [32], a resistivity of the same order of magnitude as that in this paper was obtained for epoxy (diglycidyl ether of bisphenol A) modified with carbon black. For example, when 2.0% weight ratio of carbon black was added, the measured resistivity was 1300 Ohm.m. The resistivity was further reduced to 257 Ohm.m with the use of 2.0% weight ratio of carbon black in combination with copper chloride. The resistivity reduction of the EM samples in this paper showed a great potential for health monitoring of adhesive bonding systems, e.g., FRP-strengthened civil structures using the MCF modified epoxy, where the change in electrical resistivity can be correlated with the degradation of the system subjected to service loads. It also should be noted that, the increase in electrical conductivity of the MCF-modified epoxy may incur a problem of galvanic corrosion if it is used for CFRP strengthening of steel structures. Therefore, this MCF-modified epoxy should be used in areas where galvanic corrosion is not a problem (e.g., CFRP strengthening concrete structures) or where galvanic corrosion can be effectively avoided (by adding a layer of glass fibre before bonding CFRP on steel structures, *etc.*).

CONCLUSIONS

In this study, recycled MCF was used to enhance an epoxy adhesive in order to improve its mechanical and electrical performance. Original and enhanced epoxy specimens were tested in tension and the mechanism of the improved mechanical properties was revealed by SEM. Double strap steel/CFRP joints adhesively bonded using the enhanced epoxy were further examined. From this work, the following conclusions can be drawn:

1. Addition of MCFs to the epoxy adhesive considerably enhanced its E-modulus and tensile strength when the MCF weight ratio is greater than 1.5%. The improvement of tensile strength was 7.2% for a weight ratio of 3% and 15.2% for a weight ratio of 5%, and that of E-modulus was 30.1% for a MCF weight ratio of 3% and 50.5% for a weight ratio of 5%. It was also found that the failure process of the MCF enhanced epoxy specimens became more ductile, resulting in a much larger ultimate strain at the final breakage in tension.
2. SEM indicated that the MCFs were randomly dispersed in the matrix with various orientations to the damaged surface. Those short fibres were pulled out from the fracture surfaces during the tensile failure of the enhanced epoxy specimens. This bond failure between MCFs and the epoxy matrix, in addition to the tensile fracture of the original epoxy, contributed to the improvement of E-modulus and tensile strength of the modified epoxy. The associated energy dissipation during the de-bonding of MCFs and the epoxy matrix explains a more ductile failure process.
3. Although a considerable improvement of stiffness and strength was observed for the MCF enhanced epoxy, steel/CFRP double strap joints (with both CFRP sheet and laminate) adhesively-bonded using the epoxy did not show corresponding increases in joint stiffness and ultimate load. This is because the failure occurred either at the carbon fibre sheet through delamination (DJS specimens) or at the interface between the carbon fibre laminate and the adhesive (DJL specimens), rather than in the adhesive layer where the mechanical properties of the epoxy are more dominant. This result suggests that cohesive failure is essential for an adhesively bonded system to fully deploy the mechanical enhancement of the adhesive in use.
4. Results indicated that the resistivity of MCF modified epoxy was in the range from 80 to 400 Ohm.m,

corresponding to a reduction of nearly four orders of magnitude in comparison to that of the original epoxy. However, the variation of the resistivity of the MCF modified epoxy was considerable, attributable to the sensitivity of resistance measurement to the local MCF distribution and the resulting conductivity network formed by MCFs. Due to the increase in electrical conductivity of this MCF-modified epoxy, special care should be taken when galvanic corrosion is a major concern.

ACKNOWLEDGMENTS

The authors wish to acknowledge Long Goh and Mark Taylor for their assistance in conducting the experiments at the Monash University Civil Engineering Laboratory. Financial support was received from the Australian Research Council through the Discovery Early Career Researcher Award scheme (DE120101913) and Monash University. The second author acknowledge China 973 project 2012CB026200. The authors also acknowledge use of the facilities and the assistance of Xi-Ya Fang at the Monash Centre for Electron Microscopy and the work from Kevin Lazuardy (Monash University, Australia) and Kezhen Hu (Central South University, China) in their final year projects.

REFERENCES

1. Myhre, S.H.; Beck, C.E. Repair concepts for advanced composite structures. *J. Aircr.* **1979**, *16*, 720–728.
2. Hollaway, L.C. A review of the present and future utilisation of FRP composites in the civil infrastructure with reference to their important in-service properties. *Constr. Build. Mater.* **2010**, *24*, 2419–2445.
3. Shaat, A.; Schnerch, D.; Fam, A.; Rizkalla, S. Retrofit of Steel Structures Using Fibre-Reinforced Polymers: State-of-the-Art. In Proceedings of the 83rd Transportation Research Board Annual Meeting, Washington, DC, USA, 11–15 January 2004.
4. Zhao, X.L.; Zhang, L. State of the art review on FRP strengthened steel structures. *Eng. Struct.* **2007**, *29*, 1808–1823.
5. Feng, P.; Zhang, Y.H.; Bai, Y.; Ye, L.P. Combination of bamboo filling and FRP wrapping to strengthen steel members in compression. *J. Compos. Constr.* **2013**, *17*, 347–356.
6. Feng, P.; Zhang, Y.H.; Bai, Y.; Ye, L.P. Strengthening of steel members in compression by mortar-filled FRP tubes. *Thin Walled Struct.* **2013**, *64*, 1–12.
7. Wu, C.; Zhao, X.L.; Duan, W.H.; Al-Mahaidi, R. Bond characteristics between ultra high modulus CFRP laminates and steel. *Thin Walled Struct.* **2012**, *51*, 147–157.
8. Wu, C.; Zhao, X.L.; Chiu, W.K.; Al-Mahaidi, R.; Duan, W.H. Effect of fatigue loading on the bond behaviour between UHM CFRP plates and steel plates. *Compos. Part B Eng.* **2013**, *50*, 344–353.
9. Yang, J.Q.; Smith, S.T.; Feng, P. Effect of FRP-to-metal bonded joint configuration on interfacial stresses: Finite element investigation. *Thin Walled Struct.* **2013**, *62*, 215–228.
10. Buyukozturk, O.; Gunes, O.; Karaca, E. Progress on understanding debonding problems in reinforced concrete and steel members strengthened using FRP composites. *Constr. Build. Mater.* **2004**, *18*, 9–19.
11. Chikhi, N.; Fellahi, S.; Baker, M. Modification of epoxy resins for improvement of adhesion: A critical review. *J. Adhes. Sci. Technol.* **2003**, *17*, 1655–1668.
12. Mizutani, K. Transparency and toughness characterization of epoxy resins modified with liquid chloroprene rubber. *J. Mater. Sci.* **1993**, *28*, 2178–2182.
13. Ashrafi, B.; Guan, J.; Mirjalili, V.; Zhang, Y.F.; Chun, L.; Hubert, P.; Simard, B.; Kingston, C.T.; Bourne, O.; Johnston, A. Enhancement of mechanical performance of epoxy/carbon fibre laminate composites using single-walled carbon nanotubes. *Compos. Sci. Technol.* **2011**, *71*, 1569–1578.
14. Zhou, Y.; Pervin, F.; Lewis, L.; Jeelani, S. Fabrication and characterization of carbon/epoxy composites mixed with multi-walled carbon nanotubes. *Mater. Sci. Eng. A* **2008**, *475*, 157–165.
15. Thostenson, E.T.; Li, C.; Chou, T.W. Nanocomposites in context. *Compos. Sci. Technol.* **2005**, *65*, 491–516.
16. Composite Recycling and Disposal—An Environmental R&D Issue. Available online: http://www.boeingsuppliers.com/environmental/TechNotes/TechNotes2003-11.pdf (accessed on 11 November 2013).
17. Feraboli, P.; Kawakami, H.; Wade, B.; Gasco, F.; DeOto, L.; Masini, A. Recyclability and reutilization of carbon fibre fabric/epoxy composites. *J. Compos. Mater.* **2012**, *46*, 1459–1473.
18. Nishimura, Y.; Ejiri, H. Milled Carbon Fibre and Process for Producing the Same. EP0644280 B1, 23 December 1998.
19. Davidson, J. Carbon Fibre Composites: Advancements in Reclamation Processes and Recycled Material Forms. In Proceedings of the Composites Innovation 2007 Conference, Barcelona, Spain, 4–5 October 2007.
20. Allen, B.E. Characterization of Reclaimed Carbon Fibres and Their Integration into New Thermoset Polymer Matrices via Existing Composite Fabrication Techniques. Master's Thesis, North Carolina State University, Raleigh, NC, USA, 2 July 2008.
21. Jacob, G.C.; Starbuck, J.M.; Fellers, J.F.; Simunovic, S.; Boeman, R.G. Crashworthiness of various random chopped carbon fibre reinforced epoxy composite materials and their strain rate dependence. *J. Appl. Polym. Sci.* **2006**, *101*, 1477–1486.
22. Chand, N.; Nigrawal, A. Investigations on D.C. conductivity behaviour of milled carbon fibre reinforced epoxy graded composites. *Bull. Mater. Sci.* **2008**, *31*, 665–668.
23. Chand, N.; Naik, A.M. Development and high stress abrasive wear behavior of milled carbon fibre-reinforced epoxy gradient composites. *Polym. Compos.* **2008**, *29*, 736–744.

24. Fawzia, S. Bond Characteristics between Steel and Carbon Fibre Reinforced Polymer (CFRP) Composites. Ph.D. Thesis, Monash University, Clayton, Australia, September 2007.

25. ELG Carbon Fibre Ltd. *Material Date Sheet of Milled Carbon Fibre*; ELG Carbon Fibre Ltd.: West Midlands, UK, 2012.

26. BASF The Chemical Company. *Product Data of MBRACE® CF130*; BASF The Chemical Company: Shakopee, MN, USA, 2010.

27. BASF The Chemical Company. *Product Data of MBRACE® Laminate 210/3300*; BASF The Chemical Company: Shakopee, MN, USA, 2010.

28. Nguyen, T.C.; Bai, Y.; Zhao, X.L.; Al-Mahaidi, R. Mechanical characterization of steel/CFRP double strap joints at elevated temperatures. *Compos. Struct.* **2011**, *93*, 1604–1612.

29. Huntsman. Structural Adhesives Aerospace Adhesives Araldite 420A/B. Available online: http://www.silmid.com/getattachment/0fe65000-9138-4ebb-8e64-4829b5a52e13/Araldite-420Tech-Data.aspx (accessed on 27 December 2013).

30. American Society for Testing and Materials. *Standard Test Method for Tensile Properties of Plastics*; ASTM D638-10; American Society for Testing and Materials: West Conshohocken, PA, USA, 2010.

31. Korayem, A.H.; Li, C.Y.; Zhang, Q.H.; Zhao, X.L.; Duan, W.H. Investigation on the Behavior of CFRP-to-Steel Joint Using Carbon Nanotubes Modified Epoxy Adhesive. In Proceedings of the Fourth Asia-Pacific Conference on FRP in Structures, Melbourne, Australia, 11–13 December 2013.

32. Zhang, D.; Ye, L.; Wang, D.; Tang, Y.; Mustapha, S.; Chen, Y. Assessment of transverse impact damage in GF/EP laminates of conductive nanoparticles using electrical resistivity tomography. *Compos. Part A: Appl. Sci. Manuf.* **2012**,*43*, 1587–1598.

INDEX

R

S

T

U

V

W

X